From Waste to Value

From Waste to Value investigates how streams of organic waste and residues can be transformed into valuable products, to foster a transition towards a sustainable and circular bioeconomy. The studies are carried out within a cross-disciplinary framework, drawing on a diverse set of theoretical approaches and defining different valorisation pathways.

Organic waste streams from households and industry are becoming a valuable resource in today's economies. Substances that have long represented a cost to companies and a burden for society are now becoming an asset. Waste products, such as leftover food, forest residues and animal carcasses, can be turned into valuable products such as biomaterials, biochemicals and biopharmaceuticals. Exploiting these waste resources is challenging, however. It requires that companies develop new technologies and that public authorities introduce new regulation and governance models.

This book helps policy-makers govern and regulate bio-based industries, and helps industry actors to identify and exploit new opportunities in the circular bioeconomy. Moreover, it provides important insights for all students and scholars concerned with renewable energy, sustainable development and climate change.

Antje Klitkou has a PhD from Humboldt University, Berlin (1993). She has worked at the Nordic Institute for Studies in Innovation, Research and Education (NIFU), Oslo, Norway, since 2002, as Senior Researcher until 2014 and as Research Professor since 2014. She has research interests in research and innovation policy and the transition to a sustainable bioeconomy, energy and transport. She has been Project Coordinator for several international research projects, such as "SusValueWaste: Sustainable path creation for innovative value chains for organic waste products" (2015–19) on the transition to the bioeconomy, funded by the Research Council of Norway; "Technology opportunities in Nordic energy system transitions – TOP-NEST" (2011–15), funded by Nordic Energy Research; and "InnoDemo: Role of demonstration projects in innovation: Transition to sustainable energy and transport" (2013–15), funded by the Research Council of Norway. She has published several journal articles on bioeconomy topics.

Arne Martin Fevolden is a Senior Researcher at NIFU. He holds a Master's degree (cand. polit) in Economic Sociology and a PhD (Dr. polit.) in Innovation Studies from the University of Oslo. Much of his earlier research was concerned with the development of the ICT (information and communications industry) and defence industries. His more recent research has looked at the controversies surrounding biofuels and at potential valorisation pathways for urban waste streams.

Marco Capasso is a Senior Researcher at NIFU. He has worked in the scientific fields of econometrics, economic geography and economics of innovation, and his research has been published in a wide variety of scientific journals, including *Industrial and Corporate Change*, *International Statistical Review*, *Regional Studies* and *Small Business Economics*. He is currently participating in the projects "Sustainable path creation for innovative value chains for organic waste products", funded by the Research Council of Norway, and "Where does the green economy grow? The geography of Nordic sustainability transitions", funded by the Nordic Green Growth Research and Innovation Programme.

Routledge Studies in Waste Management and Policy

From Waste to Value
Valorisation Pathways for Organic Waste Streams in Circular Bioeconomies
Edited by Antje Klitkou, Arne Martin Fevolden and Marco Capasso

Waste Prevention Policy and Behaviour
New Approaches to Reducing Waste Generation and Its Environmental Impacts
Ana Paula Bortoleto

Decision-making and Radioactive Waste Disposal
Andrew Newman and Gerry Nagtzaam

Nuclear Waste Politics
An Incrementalist Perspective
Matthew Cotton

The Politics of Radioactive Waste Management
Public Involvement and Policy-Making in the European Union
Gianluca Ferraro

For more information about this series, please visit: www.routledge.com/Routledge-Studies-in-Waste-Management-and-Policy/book-series/RSWMP.

From Waste to Value

Valorisation Pathways for Organic Waste Streams in Circular Bioeconomies

Edited by Antje Klitkou, Arne Martin Fevolden and Marco Capasso

First published 2019 by Routledge

2 Park Square, Milton Park, Abingdon, Oxon, OX14 4RN
605 Third Avenue, New York, NY 10017

Routledge is an imprint of the Taylor & Francis Group, an informa business

First issued in paperback 2020

British Library Cataloguing-in-Publication Data
A catalogue record for this book is available from the British Library

Library of Congress Cataloging-in-Publication Data
Names: Klitkou, Antje, editor. | Fevolden, Arne, editor. | Capasso, Marco, editor.
Title: From waste to value : valorisation pathways for organic waste streams in bioeconomies / edited by Antje Klitkou, Arne Martin Fevolden, and Marco Capasso.
Description: Abingdon, Oxon : New York, NY : Routledge, 2019. | Series: Routledge studies in waste management and policy | Includes bibliographical references and index.
Identifiers: LCCN 2018057710| ISBN 9781138624979 (hbk) | ISBN 9780429460289 (ebk)
Subjects: LCSH: Organic wastes–Recycling.
Classification: LCC TD804 .F76 2019 | DDC 628.1/68–dc23
LC record available at https://lccn.loc.gov/2018057710

ISBN: 978-1-138-62497-9 (hbk)
ISBN: 978-0-367-73077-2 (pbk)

Typeset in Bembo
by Wearset Ltd, Boldon, Tyne and Wear

Contents

Figures

Tables

Contributors

Simon Bolwig is a Senior Researcher and leader of the research group Climate Change and Sustainable Development, Department of Management Engineering, Technical University of Denmark. He contributed to the Fifth IPCC Assessment Report, Mitigation of Climate Change (2014), where he was lead author on Chapter 4, Sustainable Development and Equity, and contributing author on Chapter 11, Agriculture, Forestry and Other Land Use. He has published several journal articles on bioeconomy topics.

Andreas Brekke is a Senior Research Scientist at Ostfold Research. His main line of work is in the field of life cycle assessment (LCA) and he has contributed to books such as *Life Cycle Assessment (LCA) of the Building Sector: Strengths and Weaknesses* (2014) and *Resourcing in Business Logistics* (2006). He is author and co-author of a wide range of articles on issues connected to LCA, waste and biomaterials.

Markus M. Bugge works as a Senior Researcher at NIFU (Nordic Institute for Studies in Innovation, Research and Education), and is Associate Professor at the Department of Technology Systems at the University of Oslo. He holds a PhD from the Department of Social and Economic Geography at Uppsala University, and has extensive experience from his research on the systemic and evolutionary underpinnings of knowledge production, innovation and innovation policies.

Marco Capasso is a Senior Researcher at NIFU (Nordic Institute for Studies in Innovation, Research and Education), Oslo (Norway). He has worked in the scientific fields of econometrics, economic geography and economics of innovation, and his research has been published in a wide variety of scientific journals, including *Industrial and Corporate Change*, *International Statistical Review*, *Regional Studies* and *Small Business Economics*. He is currently participating in the projects "Sustainable path creation for innovative value chains for organic waste products", funded by the Research Council of Norway, and "Where does the green economy grow? The geography of Nordic sustainability transitions", funded by the Nordic Green Growth Research and Innovation Programme.

Linn Meidell Dybdahl works at the Nordic Institute for Studies in Innovation, Research and Education (NIFU). Her research interest is innovation towards a more sustainable society, both on an organisational and systemic level. She is particularly interested in sustainable business model innovation and sustainability transitions. In addition, Dybdahl is engaged in the field of responsible research and innovation, promoting inclusive, reflexive and responsive research designs, to co-produce knowledge that can solve the grand challenges of our times. Her educational background is interdisciplinary, including training in innovation, entrepreneurship, communication, project management and psychology.

Arne Martin Fevolden is a Senior Researcher at the Nordic Institute for Studies in Innovation, Research and Education (NIFU). He holds a Master's degree (cand. polit) in Economic Sociology and a PhD (Dr. polit.) in Innovation Studies from the University of Oslo. Much of his earlier research was concerned with the development of the ICT (information and communications industry) and defence industries. His more recent research has looked at the controversies surrounding biofuels and at potential valorisation pathways for urban waste streams.

Jay Sterling Gregg is a Senior Researcher at the Technical University of Denmark within the Climate Change and Sustainable Development Programme. He has researched and published on a wide range of topics within global climate change and sustainable development, including land use, bioenergy, biogeochemical cycles, forest management and waste-to-energy. He was responsible for developing a model for the regional and global potential for agricultural residues, including the ecological effects of different residue harvest rates, which was implemented into the Global Change (integrated) Assessment Model (GCAM).

Teis Hansen is a Senior Lecturer at the Department of Human Geography, Lund University, and Senior Researcher at the Nordic Institute for Studies in Innovation, Research and Education (NIFU) in Oslo. He has core research expertise in innovation in the bioeconomy and the role of geography in sustainability transition processes. He currently leads a €2.2 million project on regional green growth in the Nordic countries.

Maaike Karlijn Happel is a PhD Researcher at Wageningen University in the Netherlands. She has a background in Anthropology and Agricultural Sciences. She has been a Research Assistant at the Climate Change and Sustainable Development Group of the Danish Technical University, where she researched sustainable value chains within the Nordic brewery industry as part of the SusValueWaste project. She is currently working on the socio-ecological modelling of circular agriculture and sustainable soil practices in the Netherlands via the SINCERE project in collaboration with Leiden University. Through agent-based modelling and social network analysis, she aims to map farmers' drivers, risk attitudes and social learning processes regarding the adoption of organic residues as a soil conditioner. She is also the coordinator for science outreach platform Science & Cocktails Amsterdam.

Eric James Iversen (PhD) is a Research Professor at the Nordic Institute for Studies in Innovation, Research and Education (NIFU). He has worked in the field of innovation and the economics of technological change since 1994, authoring more than 20 largely empirically oriented publications and leading or otherwise contributing to a wide range of policy-oriented projects for domestic as well as international agencies. He also holds administrative responsibility for innovation datasets at NIFU. His current work includes developing empirical approaches to study the socio-economic importance of emerging green technologies.

Antje Klitkou has a PhD from Humboldt University, Berlin (1993). She has worked at the Nordic Institute for Studies in Innovation, Research and Education (NIFU) (Oslo, Norway) since 2002, as Senior Researcher until 2014 and as Research Professor since 2014. She has research interests in research and innovation policy and the transition to a sustainable bioeconomy, energy and transport. She has been Project Coordinator for several international research projects, such as "SusValueWaste: Sustainable path creation for innovative value chains for organic waste products" (2015–19) on the transition to the bioeconomy, funded by the Research Council of Norway; "Technology opportunities in Nordic energy system transitions – TOP-NEST" (2011–15), funded by Nordic Energy Research; and "InnoDemo: Role of demonstration projects in innovation: Transition to sustainable energy and transport" (2013–15), funded by the Research Council of Norway. She has published several journal articles on bioeconomy topics.

Kari-Anne Lyng works as a Research Scientist at the Norwegian research institute Ostfold Research and is a specialist in life cycle assessment methodology. She holds an MSc in Mechanical Engineering and finalised her PhD at the Norwegian University of Life Sciences in 2018 with the title "Reduction of environmental impacts through optimisation of biogas value chains: Drivers, barriers and policy development".

Michael Spjelkavik Mark is Head of Research at the Nordic Institute for Studies in Innovation, Research and Education (NIFU). His main areas of work include impact assessment of R&D and innovation programmes, empirical analysis of R&D and innovation and university–industry collaboration. Previously he was Principal Economist at DAMVAD and Associated Partner as well as Principal Economist at Economics Norway.

Johanna Olofsson is a PhD student at the division of Environmental and Energy Systems Studies at the Faculty of Engineering, Lund University. Her PhD research explores the role of environmental assessments in identifying sustainable circular resource systems for organic residues.

Dorothy Sutherland Olsen received her PhD in 2013 from the faculty of Educational Science at the University of Oslo and, since 2010, has been a Senior Researcher at the Nordic Institute for Studies in Innovation, Research and Education (NIFU). Her current research is principally related to studies of adult learning and innovation. In the SusValueWaste project she has primarily worked on the interdisciplinary character of bioeconomy research and on the aquaculture case study. She is currently working on a project exploring the relationship between learning and innovation and leading a project on competence development among older workers in Norway. Dorothy has led many projects both in research and within business. In her research she primarily uses qualitative methods such as interviews, observations, document analysis, workshops and focus group interviews.

Kristoffer Rørstad is a Senior Adviser at the Nordic Institute for Studies in Innovation, Research and Education (NIFU), with a cand. scient. in chemistry from the University of Bergen (1998). Rørstad has 16 years of experience, including production of the national R&D statistics and mapping of R&D in thematic areas of high priority in Norway (climate and environmental research, agriculture and food and polar research). In addition, he has bibliometric competence and has conducted several bibliometric analyses related to different projects, resulting in a number of scientific articles.

Hilde Ness Sandvold is a PostDoc at Stavanger Business School, Centre for Innovation Research, University of Stavanger. She is an economist and holds a PhD in Industrial Economy. She has co-authored papers within productivity and efficiency analysis related to the seafood industry, and currently participates as a researcher in several ongoing externally funded projects (Norwegian Research Council and EU).

Lisa Scordato is a Senior Adviser at the Nordic Institute for Studies in Innovation, Research and Education (NIFU) and has worked there since 2007. She received her MSc in Political Science from the University of Linköping in Sweden (1999–2005). From 2004 to 2006 she worked as a policy analyst at the Swedish Institute for Growth Policy Studies, Brussels, Belgium. She has over 10 years of in-depth experience in the analysis of innovation and research policies and in studies of research and innovation policy in Norway, other Nordic countries and wider Europe. In the SusValueWaste project she has been the co-leader of the work package on bioeconomy policies and regulations. She has authored a number of scientific articles on the bioeconomy.

Eili Skrivervik is a PhD candidate at TIK (Centre for Technology, Innovation and Culture) at the University of Oslo. Prior to her PhD, she worked at the University of Oslo as a Research Assistant where she wrote a TIK working paper entitled "The bioeconomy and food waste: Insects' contribution" (2018) and a research dissemination article entitled "Innovation can contribute to less food waste" (Innovasjon kan føre til mindre matsvinn). She completed her Master's degree at TIK in the fall of 2017.

Louise Strange is a student assistant at the research group Climate Change and Sustainable Development, Department of Management Engineering, Technical University of Denmark. She has a Bachelor's degree in Geography and Geoinformatics from the University of Copenhagen and is pursuing an MSc degree in Agricultural Development at the Department of Food and Resource Economics, University of Copenhagen. The topic of her MSc thesis is the role of climate change perceptions and access to information and technology among smallholder coffee producers in the Central Highlands of Vietnam (forthcoming in 2019).

Nhat Strøm-Andersen has a Master's degree in Science in Innovation and Entrepreneurship from BI Norwegian Business School, 2013. Her academic interests have since further developed into the field of innovation studies with a focus on the transition towards a sustainable (bio)economy under her PhD study at the TIK (Centre for Technology, Innovation and Culture), University of Oslo. Her doctoral research aims to study actual and potential value chains and identify sustainable pathways of valorising organic by-product and side stream resources in the Norwegian food processing industry. The study examines why, how and under what condition(s) meat processing and dairy firms are likely to innovate with respect to the utilisation of animal and dairy by-products and side streams.

Julia Szulecka is a Postdoctoral Researcher at TIK (Centre for Technology, Innovation and Culture) at the University of Oslo. She holds a double PhD from TU Dresden and University of Padova, and an MSc from VU Amsterdam. She has also completed the International Course on Resource Efficiency, Cleaner Production and Waste Management at TU Dresden (CIPSEM). Her research interests include environmental politics, forest policy, sustainable use of waste resources and the governance of bioeconomy transitions. She has conducted fieldwork on different forest plantation management modes in Paraguay and Indonesia, and has been a visiting researcher at the Center for International Forestry Research (CIFOR) in Bogor, as well as the Universidad Nacional de Asuncion. Her research has appeared in *Forest Policy and Economics*, *Land Use Policy* and the *International Forestry Review*. Her current work focuses on food waste regulations in Norway and local innovation networks as drivers in the bioeconomy transition.

Anne Nygaard Tanner is an Assistant Professor and Group Leader of the Industrial Dynamics and Strategy Group at the Technical University of Denmark and holds a 20% Postdoctoral Researcher position at CIRCLE, Lund University. Her research is interdisciplinary and cuts across the fields of innovation studies, economic geography and institutional theory. She has been published in high-ranked ISI journals such as *Economic Geography*, *Journal of Economic Geography* and *Energy Policy*.

Acknowledgements

This book emerged out of the ambition to provide insights into how companies can valorise different residues and side streams, both inside and across a wide range of industry sectors, and how governments can foster the cascading and sustainable use of bio-resources. These insights have been developed in the context of a larger interdisciplinary research project under the title "Sustainable path creation for innovative value chains for organic waste products (SusValueWaste)" funded by the Research Council of Norway's Bionær programme (project number 244249). We want to thank the Research Council of Norway for funding the project and for making this book possible.

The project has been coordinated by the Nordic Institute for Studies in Innovation, Research and Education (NIFU), which co-funded this book. The other research partners were Østfold Research in Fredrikstad (Norway), TIK-Centre at the University of Oslo (Norway), the OREEC – Oslo Renewable Energy and Environment Cluster (Norway), the Norwegian Institute of Bioeconomy Research (NIBIO) in Ås (Norway), the Centre of Innovation Research at the University of Stavanger (Norway), the Technical University of Denmark (DTU) in Lyngby (Denmark), the Centre for Innovation, Research and Competence in the Learning Economy (CIRCLE) and the Faculty of Engineering, LTH at Lund University (Sweden). We want to thank all the researchers in our project for their discussion and development of the research ideas which contributed to this book.

Obviously, not all of the results from the project could be included in this book. Here we want to acknowledge the contributions from our former NIFU colleagues who helped make this project possible: Sverre Herstad (now at the University of Oslo), Svein Erik Moen (now at the Research Council of Norway) and Kyrre Lekve (now at the Simula Research Laboratory). We also want to acknowledge the contributions from our colleagues at the TIK-Centre who were unable to be part of this book project: our first PostDoc Valentina E. Tartiu (now at Unitelma-Sapienza, University of Rome in Italy), Håkon Endresen Normann, Olav Wicken and Fulvio Castellacci. We thank our colleagues at OREEC: Marianne Rist-Larsen Reime, Alexandra Maria Almasi and Mali Hole Skogen. Several of the book chapters have been made possible through discussion with colleagues who are not on the list of authors, such as Pål Børjesson from LTH and Ole Jørgen Hanssen from Østfold Research for Chapter 14 on life cycle assessment, Bjørn Langerud from NIBIO for Chapter 4 on new path development for forest-based value creation in Norway and Ragnar Tveterås from the University of Stavanger for Chapter 8 on new pathways for organic waste in land-based farming of salmon.

Especially we would like to thank the members of our reference group who accompanied our project with good advice and encouragement throughout its development. We would like to thank Marit Aursand from Sintef Ocean in Trondheim, Lars Coenen, Professor at the University of Melbourne, Australia, Gunnar Grini and Marit Holtermann Foss from the Federation of Norwegian Industries, Kari Kolstad, now Dean at the Norwegian University of Life Sciences, before working at the Norwegian Agriculture Authority, and Odd Jarle Skjelhaugen, Professor at the Norwegian University of Life Sciences.

Our project was based on collaboration with many companies and public actors engaged in different parts of the bioeconomy. Our thanks go to all of them, but especially we would like to thank the companies which were most central in our case studies. The order here is based on the order of the chapters: Treklyngen, Borregaard AS, Norske Skog Skogn, Biokraft AS, Oslo REN, Oslo Waste-to-Energy Agency, Lindum AS, Carlsberg Group, Danish Crown A/S, Norilia AS, Nortura AS, TINE AS and Arla Foods A/S. In addition, there should be mentioned many companies which have been interviewed, for example for the brewery chapter and the aquaculture chapter, and which are not named here specifically. We also would like to thank all the representatives from public agencies and public administration, such as Trond Einar Pedersen and Gudrun Langthaler (Research Council of Norway), Geir Oddson (Nordic Council of Ministers), Peter De Smedt (European Commission) and Thomas Malla (Ministry of Trade, Industry and Fisheries), and representatives of stakeholder organisations, such as Avfall Norge (Norwegian waste management and recycling association), the Norwegian Forestry Owner Association, Tekna – The Norwegian Society of Graduate Technical and Scientific Professionals, and the Federation of Norwegian Industries, who participated in our workshops and seminars and enriched our research with their comments and suggestions.

Our project interacted actively with other projects on the bioeconomy, such as the CYCLE project coordinated by SINTEF, the BioSmart project coordinated by Ruralis and the Bioeconomy-Region Interreg project coordinated by the Akerhus County Council, and the somehow broader Nordic project GONST – The Geography of Nordic Sustainability Transitions coordinated by Lund University. Our discussions with the participants in those projects were very inspiring. Special thanks go to Marit Aursand, Hilde Bjørkhaug and Monika Svanberg. We also want to thank our many international guests at our seminars, workshops and conferences, such as Rick Bosman from TU Delft in the Netherlands, Lea Fünfschilling from Lund

University in Sweden, Rob Weterings from TNO in the Netherlands, Charis Galanakis from ISEKI in Greece, Piergiuseppe Morone from Unitelma-Sapienza, University of Rome in Italy, Lene Lange from DTU, Nancy Bocken from TU Delft, Adrian Higson from Bio based Products, NNFCC in the United Kingdom, Kes McCormick from Lund University, Hilke Bos-Brouwers from Wageningen University in the Netherlands, Johanna Mossberg and Tomas Kåberger from Chalmers University of Technology in Gothenburg and Kean Birch from York University in Canada.

Last but not least, we would like to thank our anonymous reviewers for their encouraging reviews, and the academic publishing house Routledge for excellent guidance through the book writing process and enabling us to reach a wider audience with our research.

1 Introduction

Antje Klitkou, Arne Martin Fevolden and Marco Capasso

Organic waste and other residual materials from bio-based industries and households are of increasing value in today's economy. Substances that have long represented a cost to the economy are now becoming a valuable resource. Exploiting the full potential of these resources requires increased innovation and systemic change as well as better regulation and governance – or, in other words, a transition to a sustainable bioeconomy.

The transition to a sustainable bioeconomy has been on the agenda in policy, academia and business circles worldwide. Developing this sustainable bioeconomy is considered to be critical for several reasons: the need for the sustainable use of resources, the growing demand for food, materials and energy, and the need to decouple economic growth from environmental degradation. However, a sustainable bioeconomy will only emerge when certain challenges are addressed.

First, the entire economy must be involved in the transition process. The sustainable and circular bioeconomy will not only transform traditional bio-industries such as food production and forestry, it will also transform all sectors of the economy. Fossil resources must be replaced by renewable bio-resources in many industries and organic residuals and side-streams must be exploited for sustainable value creation. The transition requires a focus on the circularity of value creation: side-streams and former waste streams can become new input factors for new value circles. The valorisation of waste streams necessitates a higher degree of coordination along and across industries.

Second, a wide range of policy instruments must be employed. Traditional policy instruments, such as generic tax exemptions and R&D funding, are insufficient to foster such a comprehensive transition process and need to be complemented by other types of instruments, such as public procurement, new standards, regional specialisation strategies and entrepreneurship initiatives. In addition, policymakers must take into account the geographic embeddedness of the waste streams and the need for changes in the established rules of the game.

Third, researchers from diverse fields of study must be involved. The transition to a bioeconomy is a complex process and therefore interdisciplinary and transdisciplinary approaches need to be developed in order to facilitate the exchange of knowledge and experience across the established groups of actors and sectors. However, interdisciplinary and transdisciplinary collaborations are challenging, since partners work under different incentive structures and draw on different knowledge bases.

This book addresses these challenges through a holistic approach: (1) analyses of value chains crossing the established sector boundaries, (2) analyses of policy and governance perspective on the transition process and (3) interdisciplinary studies of the bioeconomy.

1.1 Framework

1.1.1 Background

Over the last decade, the notion of grand challenges has emerged as a central issue in policymaking and in academic discourse. The Lund Declaration (2009) stressed the urgency of pursuing solutions to the so-called grand societal challenges, such as climate change, food security, health, industrial restructuring and energy security. All these challenges are persistent problems which require long-term approaches and are highly complex, difficult to manage and characterised by uncertainties (Coenen, Hansen & Rekers, 2015; Schuitmaker, 2012). The concept of a bioeconomy has been introduced as an important pathway for addressing several of these challenges. Replacing fossil-based products with products based on organic waste resources is an important strategy not only for mitigating climate change, but also for fostering industrial restructuring, improving public health and ensuring food and energy security (Ollikainen, 2014; Pülzl, Kleinschmit & Arts, 2014; Richardson, 2012).

However, as Bugge et al. (2016) have pointed out, there seems to be little consensus about what a bioeconomy actually implies. Visions of the bioeconomy range from one that is very closely connected to the increasing use of biotechnology across sectors (e.g. Wield, 2013), to one where the focus is on the use of biological material (e.g. McCormick & Kautto, 2013). Others call for a shift towards locally embedded eco-economies, which use local good practice as the starting point (Marsden, 2012). Thus, describing the bioeconomy, it has been argued that "its meaning still seems in a flux" (Pfau, Hagens, Dankbaar & Smits, 2014; Pülzl et al., 2014, p. 386) and that the knowledge-based bioeconomy can be characterised as a "master narrative" (Levidow, Birch & Papaioannou, 2013, p. 95), which is open to very different interpretations (Bugge et al., 2016, p. 1f.). The different perspectives on the bioeconomy can roughly be aligned into three points of view: (1) the OECD's and the United States' focus on processes that convert raw material into value-added products using biotechnology and life sciences; (2) the European Union's emphasis on the use of biomass resources, such as biological resources and waste, as inputs for food, feed, energy and industrial products; and (3) environmental scientists' and NGOs' concentration on sustainability and planetary boundaries (Kleinschmit et al., 2014).

While the first European bioeconomy strategy had a focus on bioeconomy research and innovation to tackle the grand societal challenges (European Commission, 2012), the updated European bioeconomy strategy (European Commission, 2018) stresses the need for sustainability and circularity of the bioeconomy. A sustainable bioeconomy "can turn bio-waste, residues and discards into valuable resources and can create the innovations and incentives to help retailers and consumers cut food waste by 50% by 2030" (European Commission, 2018, p. 6).

The world's population is expected to increase from seven billion in 2012 to more than nine billion by 2050 (European Commission, 2012). This means that there will be an increased need for food, feed and many other bio-based materials. Reducing and preventing food waste is one important avenue to take. However, not all food waste can be avoided; therefore, we need to exploit this resource for other means of value creation.

Many authors emphasise the need to use new types of resource for producing food, feed and other bio-based materials. These resources require different technological pathways to the traditional bio-processing industry. Such pathways are provided, among others, by biological treatment (biogas production) and biorefining.

Biological treatment with anaerobic digestion is based on different types of feedstock, such as urban organic waste, food waste from the food processing industry and manure. One output is biogas, which can be used in transport as a replacement for fossil fuels. The other output is bio-digest, which can be used as a replacement for artificial fertiliser. This returns nutrients back into the soil. Lantz et al. have discussed the potential incentives and barriers for an expansion of biogas technology in the Swedish context, including the complete biogas chain from feedstock production to the final utilisation of biogas and the digested residues (Lantz, Svensson, Bjornsson & Borjesson, 2007). They distinguish between barriers to the production of biogas and barriers to the utilisation of biogas and digestate, and use a life cycle assessment (LCA) in order to estimate the potential for biogas production from waste resources found in different sectors and sources. Their scientific contribution resulted in a lively debate in Sweden about the agricultural use of sewage sludge from wastewater treatment plants; the debate in turn originated from frequent alarming reports of the possible presence of undesirable substances in the sludge. To ensure the quality of the digestate, a set of rules and voluntary agreements are used. Manure, being a by-product which does not require any additional handling by the farmer, is often considerably more easily available to the biogas producer, and its use is especially profitable if transportation costs are covered.

Another pathway is provided by biorefineries. Biorefineries can be classified in different ways (Parajuli et al., 2015) based on the types of raw material input used for the process, such as straw and stover from plant production, residues from food processing, sludge from wastewater treatment, residues from fish processing, aquaculture and residues from forestry and forest-based industries. A further classification is based on the applied technology: biochemical or thermochemical (based on gasification and/or pyrolysis). A third classification distinguishes between the main intermediate products produced in the biorefinery, such as syngas, sugar and lignin. Biorefineries must be optimised for the efficient use of bio-resources, energy use and recovery of valuable compounds, such as proteins and phosphorus. In the Scandinavian context, all these resource streams are valuable, but straw and stover might, due to the structure of the Danish agricultural sector, be more important for Denmark than for Norway and Sweden. Forestry residues have been exploited by biorefinery companies, such as Borregaard in Norway and Domsjö in Sweden. Biorefineries not only enable the replacement of fossil resources with renewable, organic resources in the production of materials and chemicals, but also allow for the production of new types of materials with different qualities to those of fossil-based materials.

When assessing the sustainability of biological treatment and biorefining *processes*, there are several elements to consider: (1) the mobilisation of waste and residue streams from the agricultural, forestry and food sectors; (2) technological options for converting biomass into biomaterials and bioenergy; and (3) the sustainability of bio-based products compared to traditional products (Kretschmer, Buckwell, Smith, Watkins & Allen, 2013). Food waste, crop and forest residues have significant potential as bio-resources since they offer a range of potential energy outputs from 1.55 to 5.56 EJ per year. The majority (over 90%) of this potential energy output is offered by crop and forest residues. The extent of biomass-based products on the market is influenced by three factors: feedstock availability and its price, market demand and investment decisions, which again are influenced by the maturity of the chosen technology and its economic viability.

When assessing the sustainability of bio-based *products*, Kretschmer et al. (2013) stress two issues for analysis: efficient use of biomass resources, incl. residues and waste, and greenhouse gas emission effects. LCAs can provide an evaluation of the sustainability of bio-based products. When analysing the effects of greenhouse gas emission, the consequences of diverting residues from previous uses (straw and forest residues) must be considered. While the replacement of fossil fuels by first-generation biofuels can be assessed as unsustainable, the massive deployment of advanced biofuels from forestry and agricultural residues can also have unintended environmental consequences and lead to new path dependencies. A resource-efficient use of biomass is not only related to replacing energy crops by using agricultural and forestry residues, but also to the *cascading use* of these bio-resources, which implies a shift from high volumes towards lower volumes, and from low added value to high added value.

1.1.2 Defining the concepts – waste valorisation, circularity, sustainable business models and the bioeconomy

This book explores how streams of organic waste and residues can be transformed into valuable products and thus foster a transition towards a sustainable and circular bioeconomy. When investigating this subject, some questions immediately arise: what is the bioeconomy, what is the circular bioeconomy and why is it sustainable? And, last but not least, what is organic waste and what are valorisation pathways?

A number of different definitions of "bioeconomy" exist (Bugge et al., 2016; Schmid, Padel & Levidov, 2012). In this book, we define bioeconomy as the set of economic activities related to the sustainable production and use of renewable biological feedstock and processes to generate economic outputs in the form of bio-based food, feed, energy, materials or chemicals. A bioeconomy is "sustainable" as far as it is maintaining our environment and protecting food quality and biodiversity. A "circular" bioeconomy means that the existing renewable bio-resources are used in an efficient way, which means that organic waste, co-products and by-products are treated as resources for the bioeconomy. Strategies to achieve this circularity include the following processes: prevention and reduction of organic waste streams, finding new highly valued bio-products based on the re-use of organic by-products, co-products, residues and waste streams, recycling of organic waste and residues, and recovering of the energy content of organic waste streams.

The cascading use of biomass and waste resources has become an important way in which to improve resource efficiency. This principle implies that burning such resources should be the last option, to be adopted only when no other use can be envisioned (de Besi & McCormick, 2015). Biomass resources have been extensively used for energy production, both for heat and power, as well as in relation to waste-to-energy processes. However, following the principles of cascading use of bio-resources, other renewable energy resources should instead be used for energy production (Knauf, 2015; Suominen, Kunttu, Jasinevicius, Tuomasjukka & Lindner, 2017).

These processes often cross existing sectoral borders in the bioeconomy: waste streams and by-products from agriculture or forestry might become a resource for aquaculture or biochemical industry, or vice versa. Urban organic waste can be transformed into biogas for transport and into fertiliser, replacing artificial fertiliser or peat-based compost. Different waste streams can be combined to produce new types of products, not just biogas and fertiliser, but also new feed sources. However, the expression "circular bioeconomy" can be seen as something of an idealised concept, since some materials will always be lost or degraded as they move along supply chains.

What is organic waste? There have been several attempts to define organic waste streams by providing important inclusion and exclusion criteria. The United Nations Statistics Division distinguishes *waste* from other residues in the following way: waste includes materials which are not prime products and the generator has no further use for these resources and discards them, or intends or is required to discard them. This means that the definition of waste is dynamic: (1) since the generator can change the production process and can introduce new processes which exploit the former waste streams, (2) the regulator might change the requirements for what should be discarded and (3) the generator might identify a demand for the resource from other firms and start trading the materials as a good. On the other hand, waste streams, by definition, exclude residuals which are directly recycled or reused at the place of generation, as well as waste materials which are directly discharged into ambient water or air. The latter means that resource streams which are discharged from fisheries and offshore aquaculture into the oceans are under-reported, which might contribute to the increased pollution of the oceans. In this book, we address the valorisation of both organic waste streams and side-streams. We distinguish between residues which have no economic value and side-streams which already have a value.

Waste valorisation means adding value to residues and side-streams through changes in markets and/or in the physical properties of these materials. Valorisation requires both technological and institutional innovation. When analysing valorisation pathways for organic waste and side-streams we can distinguish between different groups of technologies which are applied for organic waste valorisation: (1) more conventional technologies that have been used in the management and treatment of those streams of resources, such as animal feeding, composting, anaerobic digestion, incineration and landfill disposal, and (2) alternative, biorefinery technologies aiming at the extraction and recovery of high-value compounds and the production of chemicals, materials and fuels (Maina, Kachrimanidou & Koutinas, 2017). However, the choice of technology is not the only and most important dimension of valorisation. Valorisation pathways are the trajectories through which such values are created and distributed by and among actors from the private sector, policy, research, civil society and households. Valorisation pathways may even constitute so-called transitions pathways which involve changed technologies, institutions and regulations, infrastructures, production systems, business models and consumption patterns (Turnheim et al., 2015).

Sustainable business models address different ways in which firms can combine an improved customer value with societal, environmental and economic benefits (Boons, Montalvo, Quist & Wagner, 2013). They can target innovative value propositions, value creation and delivery, and mechanisms to capture value (Bocken, Short, Rana, & Evans, 2014, p. 43f.). As Lozano has pointed out, value includes flows of material resources and energy as inputs and products and services as outputs, but also economic value, human resources and, last but not least, environmental value (Lozano, 2018, p. 6). The concept of sustainable business models can be linked to the discourse about the sustainable transition of socio-technical systems (Boons et al., 2013; Geels, 2002; Schot & Steinmueller, 2018).

1.1.3 Methodological approaches

Different methodologies are applied throughout the book to analyse the emerging circular bioeconomy and the valorisation of waste streams. Given the breadth and diversity of the subject, the research goals have been pursued using a variety of research tools belonging to both the qualitative and quantitative spheres.

The studies of the sectors of origin of the residuals combine transition theory, innovation theory and global value chain analysis. Data sources are here principally represented by interviews, workshops, event history analyses and document analyses, all related to case studies of ongoing and emerging valorisation processes. Additional quantitative data, such as national and regional accounts of waste streams, complement the technical foundation of the study.

Cross-sectoral perspectives are centred on quantitative methods. An analysis of the curricula vitae of experts involved in Norwegian research and development projects, funded by the Research Council of Norway and explicitly addressing the bioeconomy, will shed light on the knowledge base of the Norwegian bioeconomy. In addition, an analysis of research and development (R&D) statistics, firm-level data and surveys will provide an insight into the actors and their roles in the bioeconomy.

The book's chapters, pertaining to scientific reviews and scoping reflections, are based on theoretical reasoning on the previous academic literature, supported by bibliometric inferences. When making policy suggestions, semi-structured interviews and expert workshops have been accompanied by document analyses of the regulatory framework for valorisation processes. LCA methods have also been explored, to allow for a more direct response to governance-related questions.

1.2 Important themes addressed in the book

In the following we explain some of the main thematic issues addressed in the book, such as circularity, regional embedding and geographical context of waste valorisation, resource ownership and inter-firm governance, and policy and regulations of waste valorisation.

1.2.1 Circularity across established sectors

When analysing processes in the bioeconomy, the literature has often applied a sectoral approach, i.e. analysing developments in forestry, agriculture, aquaculture, etc., and has largely focused on the primary production of organic products. Instead, when analysing the circular bioeconomy, we have decided to also consider value creation, which crosses the traditionally defined sectors. Cross-sectoral connections are difficult to establish, since the economic actors would need to understand the properties of waste streams in unfamiliar sectors, and to relate them to new processes or new products. Sometimes a bioeconomic improvement may even need the combination of several different waste streams.

Such cross-cutting relations across the sectors can emerge at several stages of the "value circle". We prefer to adopt the expression "value circle", rather than the more common "value chain", which is connected to a more traditional linear model of value creation: production of primary goods – processing – distribution – consumption – disposal. Instead, our focus here is on a more circular approach, where value creation can occur at any stage, and discarded products can still potentially be used for value creation.

Figure 1.1 attempts to visualise such an approach. This figure is generic and can be applied to many nature-based value creation processes.

A next step in the development is from individual value circles to a network of circles and sharing resources in cascades from high to lower resource qualities, culminating in industrial symbiosis. Industrial symbiosis requires coordination and co-location in a regional setting and aims at turning waste outputs from one production process into a feedstock for another production process. The circularity of the bioeconomy is strongly related to the *sustainable* valorisation of biological resources. However, the valorisation of residues is not automatically sustainable. Assessing the sustainability of such valorisation is one of the tasks of LCAs. Crossing sectoral boundaries for valorising biological resources can be challenging for performing LCAs, especially because the boundaries of the systems change if one compares a valorisation path inside one sector, such as wooden residues exploited for

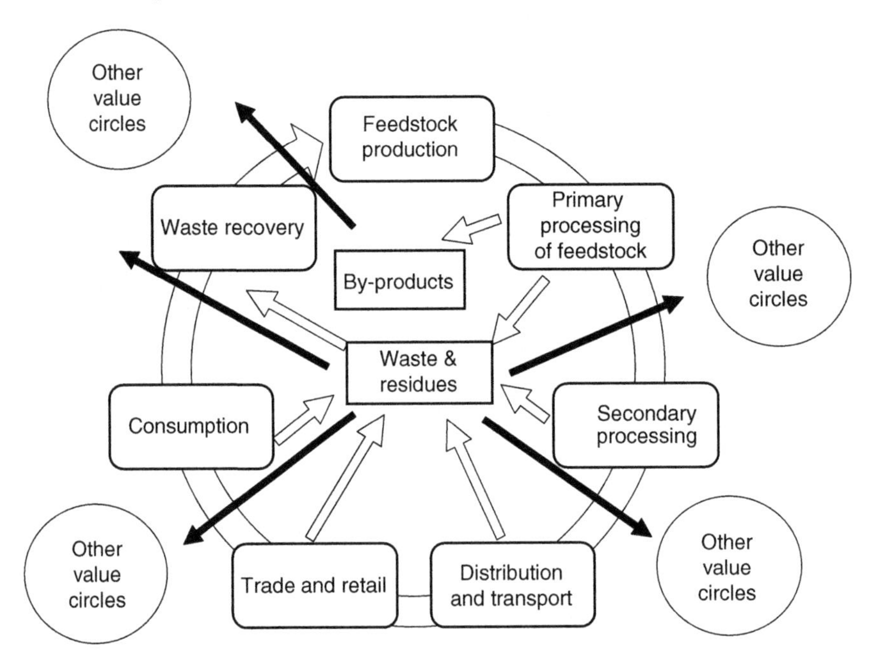

Figure 1.1 Value circles and valorisation of waste and residues.

wooden construction elements, with a path including different sectors, such as wooden residues used as feedstock for animals. However, LCAs focus on tracing the sustainability of new products across sectoral boundaries.

In this book we will study examples of value creation crossing sectoral boundaries. We will also discuss to what extent these are really more sustainable than established alternatives. Another important issue is how circularity across sector boundaries is dependent on cooperation with and knowledge of actors in other sectors. In particular, we target the need for addressing the specific geographic context to achieve such circularity across sector boundaries, as the next subsection, in particular, thematises.

1.2.2 Regional embedding and geographies of innovation

Making the economy "more circular", and adding value to resources that are currently unexploited, requires some form of innovation. The most direct and visible form of innovation would touch the existing value circles, and would prominently involve bridging across existing value circles: if some form of waste is generated at some level of one existing value circle, innovation may allow such waste to become an input within a different value circle.

In some cases, what is currently being wasted in a value circle could be technically ready to be an input in a different circle, but the bridge between the two circles cannot be easily crossed when the circles are not situated in the same geographic area. Innovation might then be required to allow easier transportation of waste: for instance, wood chips can be treated to not absorb humidity during transportation, and thus be introduced into geographically distant value circles.

In other cases, what is currently being wasted can become an input only after a chemical or mechanical transformation: for instance, food waste can be transformed into biofuel. If the type of waste cannot be transported easily or cheaply, the chemical/mechanical transformation (and the innovation connected to it) must occur in the same location where the waste is produced.

However, innovation may also occur on the consumption side. The creation of car models which employ biofuel is an example of the coordination of different value circles which intersect only at the consumption side. Innovations that are more consumption-based, and less technological, may involve the development of new business models. In these cases, the geographic dimension is also relevant. For instance, when local gardening is used to promote biofoods, trust may be required by the consumer to buy a product which looks or tastes differently than usual, and such trust needs to be created at a local level. Even more evident is the case of public procurement, where the consumption stimulated by the public sector, for instance in the case of a municipality acquiring buses fuelled by bio-methane, may have a politically clear local origin.

It should also be kept in mind that the innovation process itself can depend on locational attributes, and not only because of value circle considerations and institutional setting. Every location provides a particular knowledge set associated with the firms already present in the location. A progressive valorisation of waste is more likely when the current capabilities of a location allow it, and such capabilities, or "competency mix", are associated not only with the firm producing waste, but also with the other firms present in the same location. Knowledge from different economic sectors, all present in a location, could be recombined to generate the desired innovation. However, in order to assess whether a location can provide the right "embedding" for waste valorisation, it is important to understand at which geographical range the location should be considered. A new technology may relate to some competences built at country level, to some competences built at regional level, and to some competences built at firm level. For instance, exploiting knowledge from a technologically contiguous production process, currently present in the country but not in the region, may not be possible when the knowledge transmission requires a frequent interaction: in this case, a country-level analysis of the competence mix would fail in identifying the possibilities of a region. On the other hand, knowledge transmission requiring a less frequent interaction would be neglected by a region-level analysis of the competency mix which excludes knowledge spillovers from region to region. The competency mix which waste valorisation can draw upon can thus be assessed only by considering the recent economic history of the firm, of the region and of the country where waste valorisation is expected to occur.

To sum up, innovation can alter the ways in which value circles operate and intersect. The geographic scale on which the intersection takes place always plays an essential role in the innovation process. As a consequence, the geographic dimension constitutes an essential element to consider when planning new circular economies which valorise waste streams.

1.2.3 Resource ownership and interfirm governance structures

Innovative paths to the bioeconomy are dominated by uncertainty, stemming not only from the unpredictability of scientific and technological evolution, but also from the difficulty of value circles to maintain stability in the face of fluctuations in natural resource availability on the market. An innovative bioeconomic sector may struggle to grow and survive, especially when it needs a constant volume and quality of intermediate goods flowing from upstream producers, to justify a long-term investment. At the same time, ensuring that downstream buyers can guarantee a constant demand can be difficult when the reaction of final consumers to the new bioeconomic products is not predictable, or when the barrier to the entry of other firms into the innovative sector is low. A better coordination along the value circle can be achieved, for instance, by extending the firm boundaries through vertical integration, or by maintaining the existing firm boundaries while securing stability through long-term contracts. An alternative solution to value circle miscoordination may lie in sharing decisional power between firms that are directly connected within the value circle. Formally, an individual could have a seat on both the board of directors of the innovative bioeconomic firm and on the board of directors of an upstream or downstream firm. Such directorate interlocking could guarantee, also in the eyes of potential investors, that the flow of goods along the value circle is sufficiently stable.

Within the bioeconomy, the valorisation of waste streams necessitates an even higher degree of coordination along and across value circles. This is due to the intrinsic properties of waste: both its amount and its production timing are not functional to its future use, but instead depend on the needs of the value circle from which it originates. As a consequence, its availability does not obey the market forces which usually modulate, at least partially, the supply of products in relation to their demand. There are three possible solutions to this "waste puzzle". The first one is technical: altering waste in order not only to make it useful for an economic purpose, but also to improve its properties for storage and transportation so that production amount and timing become less important. The second solution is still technical but goes one step further: processing waste in order to increase its value, up to a point where the demand for it has a similar importance for the firm as the demand for the main product associated with it; at this point, the firm will have an incentive to coordinate the production of the main product and of the "former" waste. The third solution is organisational: if the firm from which the waste originates has a connection, in terms of ownership or of management, to a potential downstream sector for the waste, then the actors in the potential downstream sector can be reassured that the flow of waste will be sufficiently constant, and therefore the incentives for long-term investments will be sufficient to allow waste valorisation in the downstream sector.

1.2.4 Policy and regulation of waste valorisation

Policymakers want to influence the bioeconomy for a wide range of reasons. Among others, they want to ensure that the food we eat is safe and that the methods used to produce it do not harm the environment; they want to make sure that biological resources are used in a way that generates new jobs and economic growth, especially in economically disadvantaged areas; and they want to make sure that biological residual resources (such as organic household waste, wooden residues, brewers' spent grain, animal by-products) are used in a sustainable manner to produce new products and new sources of energy. To accomplish these ambitions, policymakers have a large arsenal of different mechanisms at their disposal, ranging from bans and prohibitions to subsidies, tariffs and quotas and from innovative public procurement to government strategies. The combination of multiple goals and multiple mechanisms makes regulation of the bioeconomy challenging. However, there are also other factors that complicate policymaking.

Policymaking and the regulation of waste streams take place at different levels. Some policies and regulations are developed and implemented at the local level by counties and municipalities. At the local level, policies and regulations are often strongly connected to services that the counties and municipalities are responsible for providing the citizens and might involve such things as encouraging the citizens to sort their waste to facilitate biological treatment and demanding that bus companies which provide public transportation services use biofuels. At a national level, the policies are more broad-based and can affect entire sectors of the bioeconomy. Ministries are sometimes dedicated to the regulation of specific sectors, such as the Ministry of Trade, Industry and Fisheries and the Ministry of Agriculture and Food in Norway. At the supranational level, there are policies that apply for multiple countries and that each country has agreed to adhere to by signing a specific agreement (e.g. the Paris Agreement) or joining a supranational organisation (e.g. the EU or WTO). There might be different levels of compliance with supranational regulations, since these regulations do not necessarily enjoy the same level of popular support as national policies and the supranational organisation might be unable to effectively enforce the regulations. In practice, many sectors of the bioeconomy are at the same time subject to regulation at several levels.

The bioeconomy consists of actors with varying and often conflicting agendas. Some actors own bio-resources and want to exploit these resources commercially. Other actors possess strong technological capabilities and want to introduce new bio-based products or processes. In addition, yet other actors want to protect the environment and mitigate climate change. All these actors want to influence policymaking to further their own agenda. Sometimes, it is possible for policymakers to find solutions that accommodate the wishes of all these actors. Other times, it is impossible for policymakers to find solutions that fit everyone, and they must choose which agenda they want to support. In these cases, they often have to balance between sustainability and economic growth as central principles of the bioeconomy.

In this book, we will try to answer several important questions related to policy and regulation. We want to look at what are the most important policies for improving the sustainability and economic viability of the bioeconomy, what are the main pitfalls that policymakers should be aware of when they regulate the bioeconomy and what kinds of conflicting agendas policymakers face when they want to introduce new regulation.

1.3 An overview of the book

This book is organised into four parts, each exploring different aspects related to the valorisation of organic waste, circularity and the bioeconomy.

Part I discusses the main concepts and approaches applied in the book. Chapter 2 improves our understanding of the bioeconomy by exploring the origins, uptake and use of the term "bioeconomy" in academic literature. It reviews the literature and identifies three kinds of vision for the bioeconomy: the bio-technology vision, the bio-resource vision and the bio-ecology vision. Each of these visions has different origins, represents different sectors and emphasises different underlying values, directions and drivers in the bioeconomy. Together they illustrate the multifaceted nature of this sector of the economy. Chapter 3 develops a theoretical framework for understanding the patterns, drivers and contexts of organic waste valorisation within and across bio-based sectors. The chapter discusses and connects four central concepts into a coherent framework – governance, value creation and capture, technological change, and spatial relatedness and dynamics. These concepts and the related theory serve as the basis for the analysis carried out in the next part.

Part II presents empirical case studies of organic waste valorisation in six bio-based sectors – forestry (Chapter 4), urban waste management (Chapter 5), brewing (Chapter 6), meat processing (Chapter 7), aquaculture (Chapter 8) and dairy (Chapter 9). These sectors are not only important from an environmental and climate perspective, but also in terms of food security, rural employment and economic growth. Although Part II presents six sector studies, all the case studies are cross-sectoral. The sector under consideration is only the starting point of a longer valorisation pathway that spans several sectors. For instance, the case study on urban waste management begins with an analysis of organic household waste, but then focuses on how the waste is valorised into biomethane fuel for buses and fertilisers for farmers. All the case studies draw on the theoretical framework developed in Chapter 3, but emphasise different aspects of this framework.

Part III investigates aspects of the bioeconomy that span across several sectors such as knowledge flow and innovation activity. Chapter 10 investigates what kind of knowledge is required in the bioeconomy. The chapter analyses the CVs of scientific personnel who have received public funding through bioeconomy-focused research programmes in Norway in order to explore which disciplines and institutions are most relevant to the bioeconomy. Chapter 11 investigates activity in the Norwegian bioeconomy by combining and analysing a range of data sources, including patent data and project data, as well as the results from two recent surveys. The chapter aims to provide an overview of the population and actions of actors in the Norwegian bioeconomy.

Part IV investigates valorisation pathways and the bioeconomy from a policy perspective. Chapter 12 explores whether there are conflicting interests and policy rationales shaping the bioeconomy. The chapter analyses submissions to a public hearing on the development of a bioeconomy strategy in Norway to map the actors involved in shaping the new bioeconomy and to analyse their positions within this emerging field. Chapter 13 explores how policies on food waste are developed and implemented. It investigates this topic by analysing the introduction of voluntary targets for food waste reduction in Norway. Chapter 14 presents the LCA method and its use as a policy tool in relation to waste management. The chapter discusses how policymaking can be enriched by LCA, but also raises the potential pitfalls of blindly relying on LCA results. Finally, Chapter 15 concludes the part and the book by articulating suggestions and recommendations for policymakers and companies concerned about the valorisation of various waste streams.

References

Bocken, N. M. P., Short, S. W., Rana, P. & Evans, S. (2014). A literature and practice review to develop sustainable business model archetypes. *Journal of Cleaner Production, 65*, 42–56. doi:10.1016/j.jclepro.2013.11.039.

Boons, F., Montalvo, C., Quist, J. & Wagner, M. (2013). Sustainable innovation, business models and economic performance: An overview. *Journal of Cleaner Production, 45*, 1–8. doi:10.1016/j.jclepro.2012.08.013.

Bugge, M. M., Hansen, T. & Klitkou, A. (2016). What is the bioeconomy? A review of the literature. *Sustainability, 8*(7), 1–22. doi:10.3390/su8070691.

Coenen, L., Hansen, T. & Rekers, J. V. (2015). Innovation policy for grand challenges: An economic geography perspective. *Geography Compass, 9*(9), 483–496. doi:10.1111/gec3.12231.

de Besi, M., & McCormick, K. (2015). Towards a bioeconomy in Europe: National, regional and industrial strategies. *Sustainability*, 7(8), 10461–10478. doi:10.3390/su70810461.

European Commission. (2012). *Innovating for sustainable growth: A bioeconomy for Europe*. Brussels: European Commission, Retrieved from http://ec.europa.eu/research/bioeconomy/pdf/official-strategy_en.pdf.

European Commission. (2018). *A sustainable bioeconomy for Europe: Strengthening the connection between economy, society and the environment. Updated Bioeconomy Strategy*. Brussels: European Commission, Retrieved from https://ec.europa.eu/research/bioeconomy/pdf/ec_bioeconomy_strategy_2018.pdf.

Geels, F. W. (2002). Technological transitions as evolutionary reconfiguration processes: A multi-level perspective and a case-study. *Research Policy, 31*(8–9), 1257–1274.

Kleinschmit, D., Lindstad, B. H., Thorsen, B. J., Toppinen, A., Roos, A. & Baardsen, S. (2014). Shades of green: A social scientific view on bioeconomy in the forest sector. *Scandinavian Journal of Forest Research, 29*(4), 402–410. doi:10.1080/02827581.2014.921722.

Knauf, M. (2015). Waste hierarchy revisited: An evaluation of waste wood recycling in the context of EU energy policy and the European market. *Forest Policy and Economics, 54*, 58–60. doi:10.1016/j.forpol.2014.12.003.

Kretschmer, B., Buckwell, A., Smith, C., Watkins, E. & Allen, B. (2013). *Recycling agricultural, forestry and food wastes and residues for sustainable bioenergy and biomaterials*. Retrieved from Brussels: 978-92-823-4738-6.

Lantz, M., Svensson, M., Bjornsson, L. & Borjesson, P. (2007). The prospects for an expansion of biogas systems in Sweden: Incentives, barriers and potentials. *Energy Policy, 35*(3), 1830–1843. doi:10.1016/j.enpol.2006.05.017.

Levidow, L., Birch, K. & Papaioannou, T. (2013). Divergent paradigms of European agro-food innovation: The knowledge-based bio-economy (KBBE) as an R&D agenda. *Science, Technology & Human Values, 38*(1), 94–125. doi:10.1177/0162243912438143.

Lozano, R. (2018). Sustainable business models: Providing a more holistic perspective. *Business Strategy and the Environment*, 8. doi:10.1002/bse.2059.

Lund Declaration. (2009). *Europe must focus on the grand challenges of our time*. Retrieved from https://era.gv.at/object/document/130.

Maina, S., Kachrimanidou, V. & Koutinas, A. (2017). A roadmap towards a circular and sustainable bioeconomy through waste valorization. *Current Opinion in Green and Sustainable Chemistry, 8*, 18–23. doi:10.1016/j.cogsc.2017.07.007.

Marsden, T. (2012). Towards a real sustainable agri-food security and food policy: Beyond the ecological fallacies? *Political Quarterly*, *83*(1), 139–145. doi:10.1111/j.1467-923X.2012.02242.x.

McCormick, K., & Kautto, N. (2013). The bioeconomy in Europe: An overview. *Sustainability*, *5*(6), 2589–2608.

Ollikainen, M. (2014). Forestry in bioeconomy: Smart green growth for the humankind. *Scandinavian Journal of Forest Research*, *29*(4), 360–366. doi:10.1080/02827581.2014.926392.

Parajuli, R., Dalgaard, T., Jorgensen, U., Adamsen, A. P. S., Knudsen, M. T., Birkved, M., ... Schjorring, J. K. (2015). Biorefining in the prevailing energy and materials crisis: A review of sustainable pathways for biorefinery value chains and sustainability assessment methodologies. *Renewable & Sustainable Energy Reviews*, *43*, 244–263. doi:10.1016/j.rser.2014.11.041.

Pfau, S. F., Hagens, J. E., Dankbaar, B. & Smits, A. J. M. (2014). Visions of sustainability in bioeconomy research. *Sustainability*, *6*(3), 1222–1249. doi:10.3390/su6031222.

Pülzl, H., Kleinschmit, D. & Arts, B. (2014). Bioeconomy: An emerging meta-discourse affecting forest discourses? *Scandinavian Journal of Forest Research*, *29*(4), 386–393. doi:10.1080/02827581.2014.920044.

Richardson, B. (2012). From a fossil-fuel to a biobased economy: The politics of industrial biotechnology. *Environment and Planning C: Government and Policy*, *30*(2), 282–296.

Schmid, O., Padel, S. & Levidov, L. (2012). The bio-economy concept and knowledge base in a public goods and farmer perspective. *Bio-based and Applied Economics*, *1*(1), 47–63.

Schot, J., & Steinmueller, W. E. (2018). Three frames for innovation policy: R&D, systems of innovation and transformative change. *Research Policy*, *47*(9), 1554–1567.

Schuitmaker, T. J. (2012). Identifying and unravelling persistent problems. *Technological Forecasting and Social Change*, *79*(6), 1021–1031. doi:10.1016/j.techfore.2011.11.008.

Suominen, T., Kunttu, J., Jasinevicius, G., Tuomasjukka, D. & Lindner, M. (2017). Trade-offs in sustainability impacts of introducing cascade use of wood. *Scandinavian Journal of Forest Research*, *32*(7), 588–597. doi:10.1080/02827581.2017.1342859.

Turnheim, B., Berkhout, F., Geels, F. W., Hof, A., McMeekin, A., Nykvist, B. & van Vuuren, D. (2015). Evaluating sustainability transitions pathways: Bridging analytical approaches to address governance challenges. *Global Environmental Change*, *35*(November), 239–253. doi:10.1016/j.gloenvcha.2015.08.010.

Wield, D. (2013). Bioeconomy and the global economy: Industrial policies and bio-innovation. *Technology Analysis & Strategic Management*, *25*(10), 1209–1221. doi:10.1080/09537325.2013.843664.

Part I

Perspectives on the bioeconomy

2 What is the bioeconomy?

Markus M. Bugge, Teis Hansen and Antje Klitkou

2.1 Introduction[1]

The notion of grand challenges has over the last decade emerged as a central issue in policymaking and – increasingly – academia. In a European context, the Lund Declaration [1] stressed the urgency of pursuing solutions to problems in diverse fields such as climate change, food security, health, industrial restructuring and energy security. A key common denominator for these grand challenges is that they can be characterised as persistent problems, which are highly complex, open-ended and characterised by uncertainty in terms of how they can be addressed and solved – a partial solution may result in further problems at a later point in time due to feedback effects (Coenen, Hansen & Rekers, 2015; Schuitmaker, 2012; Upham, Klitkou & Olsen, 2016).

Still, despite these uncertainties, the concept of a bioeconomy has been introduced as an important part of the solution to several of these challenges. Moving from fossil-based to bio-based products and energy is important from a climate change perspective, but it is also suggested that a transition to a bioeconomy will address issues related to food security, health, industrial restructuring and energy security (Ollikainen, 2014; Pülzl, Kleinschmit & Arts, 2014; Richardson, 2012).

However, despite the key role attributed to the bioeconomy in addressing these grand challenges, there seems to be little consensus concerning what a bioeconomy actually implies. For instance, the conceptualisations of the bioeconomy range from one that is closely connected to the increasing use of bio-technology across sectors, e.g. Wield (2013), to one where the focus is on the use of biological material, e.g. McCormick and Kautto (2013). Thus, describing the bioeconomy, it has been argued that "its meaning still seems in a flux" (Pülzl et al., 2014, p. 386) and that the bioeconomy can be characterised as a "master narrative" (Levidow, Birch & Papaioannou, 2013, p. 95), which is open to very different interpretations.

With this in mind, the aim of this chapter is to provide an enhanced understanding of the notion of the bioeconomy. Arguably, this is important if the transition to the bioeconomy is indeed a key element in targeting a number of central grand challenges. Specifically, the chapter seeks to explore the origins, uptake and contents of the term "bioeconomy" in the academic literature. First, this includes a bibliometric analysis of peer-reviewed articles on the topic (Section 2.3), which identifies central organisations, countries and scientific fields. A main result is that the bioeconomy concept has been taken up in multiple scientific fields. Consequently, in Section 2.4 we review literature on the bioeconomy in order to examine the differences in the understanding of the bioeconomy concept that are put forward in the academic literature. Specifically, we focus on the implications regarding overall aims and objectives, value creation, drivers and mediators of innovation, and spatial focus. Before proceeding to the analysis, the following section presents the methodology.

2.2 Methodology

2.2.1 Bibliometric analysis

The bibliometric analysis is based on a literature retrieval of relevant scientific articles indexed in a recognised scientific article database, the Core Collection of Web of Science. The delimitation of a sample can be defined by the chosen publishing period, the geographical location of the authors, the selection of research areas, the selection of a journal sample or the selection of keywords. For the purpose of this study, we analysed the literature indexed during the last decade, from 2005 to 2014. We did not include 2015 to allow the papers published in the last year to gather citations in 2015. Since we decided to analyse the existing scientific literature about the bioeconomy, we chose to take a global approach and to include all research domains. (Furthermore, there is significant overlap in the research carried out on the bioeconomy between the human, social, natural and technical research domains. For example, ethical aspects of the development of the bioeconomy are often covered by journals categorised as humanities, so this research domain is included as well.)

The following keywords and their variants were selected: bioeconomy, bio-based economy, bio-based industry, circular economy and bio*, bio-based society, bio-based products, and bio-based knowledge economy (variations are created by hyphens and truncation). A list of calculated indicators is provided in Appendix A. In the analysis of most active organisations and their collaboration in terms of co-publishing we used fraction counts and not absolute counts to achieve a more accurate picture of the position of the different organisations.

Social network analysis (SNA) techniques were applied to measure different types of centrality in the networks, such as degree centrality and betweenness centrality. While degree centrality is defined as the number of links that a node has (Borgatti, 2005), betweenness centrality is defined as the number of times a node acts as a bridge along the shortest path between two other nodes (Freeman, 1977). Both indicators are calculated with the help of UCINET 6 developed by Borgatti, Everett and Freeman (2002) and network graphs were created with NetDraw developed by Borgatti (2002). The network graphs were based on degree centrality measures. The structure of the identified network was analysed by identifying cliques. A clique is a subset of the network in which the nodes are more closely and intensely tied to each other than they are tied to other members of the network.

2.2.2 Literature review

The literature review aims to examine differences in the understanding of the bioeconomy concept. It is based on a subset of the papers included in the bibliometric analysis. The main inclusion criterion was that papers had to include a discussion of the bioeconomy. Importantly, the resulting bioeconomy visions described in Section 2.4 should not be understood as visions promoted by the academic writers, but as bioeconomy visions that result from academic analysis of the actions of policymakers, industry actors, etc.

In order to improve our understanding of the underpinnings and conditions for the emergence of the bioeconomy, we included papers that concentrated on conceptual aspects such as innovation and value creation, driving forces, governance and the spatial focus of the bioeconomy. We thus excluded papers that primarily discussed technical issues. The review consisted of a screening of the abstracts of 110 papers. From these we made a discretionary selection of 65 papers that were considered relevant to the analysis.

These papers were then read by between two and four persons in order to enhance reliability. The content of the papers was summarised in a database, considering aspects such as research objectives, methods, scope regarding geography and industry sector, and main conclusions. Differing opinions concerning individual articles were resolved in discussions. The database provided the point of departure for identifying papers containing relevant content on bioeconomy aims and objectives, value creation processes, drivers and mediators of innovation, or spatial focus. These papers were then re-read and synthesised into the analysis presented in Section 2.4.

2.3 Bibliometric analysis of scientific literature on the bioeconomy

We identified 453 papers for the period 2005 to 2014. Figure 2.1 shows that the topic has gained increasing attention in the scientific discourse.

The total number of citations achieved by the whole sample was 9207, but the distribution of citations is skewed (see Table 2.1). The three most-cited papers received 18% of all citations. The 15 most-cited papers received 41% of the citations. Forty-one papers received one citation, and 55 papers received no citations.

It is more interesting to look at the average number of citations per year than the total number of citations because older papers will by default tend to achieve more citations than the most recent papers. Still, the results do not

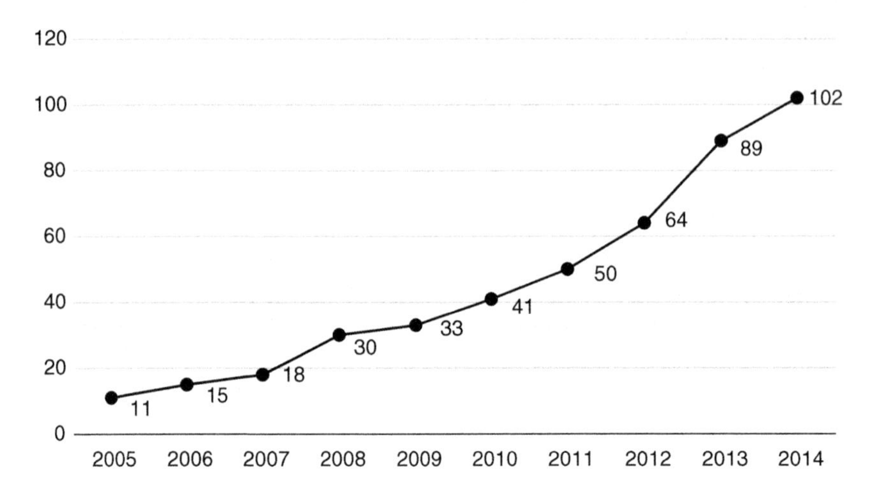

Figure 2.1 Number of papers per year (n = 453 papers).

Table 2.1 The 10 most-cited papers (491 citations) and the 10 papers with the most citations per year

Most-cited papers		*Papers with most citations per year*	
Reference	*Number of citations*	*Reference*	*Average number of citations per year*
(Bozell & Petersen, 2010)	760	(Bozell & Petersen, 2010)	127
(Zhang, Himmel & Mielenz, 2006)	509	(Zhang et al., 2006)	51
(Lee, Doherty, Linhardt & Dordick, 2009)	351	(Lee et al., 2009)	50
(Bordes, Pollet & Averous, 2009)	344	(Bordes et al., 2009)	49
(Graham, Nelson, Sheehan, Perlack & Wright, 2007)	234	(Dusselier, Van Wouwe, Dewaele, Makshina & Sels, 2013)	37
(Li, Wang & Zhao, 2008)	230	(Horn, Vaaje-Kolstad, Westereng & Eijsink, 2012)	36
(FitzPatrick, Champagne, Cunningham & Whitney, 2010)	211	(FitzPatrick et al., 2010)	35
(Carvalheiro, Duarte & Girio, 2008)	209	(Burrell, 1991)	35

Note
Citation data retrieved 23 February 2016. There can be some delay in the indexing process. Therefore, the number of citations for papers published towards the end of 2014 may be underestimated.

differ much across the two different ways of calculating citations. The data fit with Bradford's law of scattering, which means that the most significant articles in a given field of investigation are found within a relatively small core cluster of journal publications and a large group of articles does not get any citations (Vaaje-Kolstad et al., 2010).

The analysis of the journals revealed that this topic has been pursued in a large number of journals: the 453 papers were published in 222 journals; 149 of the journals had just one paper on this topic. Table 2.2 shows the journals with more than seven articles, the number of achieved citations and their share of citations of the total number of citations. It seems that no journal has positioned itself as the central journal for academic debate on the bioeconomy.

The 453 articles were authored by 1487 researchers. Most of the researchers (89% or 1324) had only one paper in the sample. Five researchers had more than four papers in the sample (Table 2.3).

Where do these researchers come from? An analysis of the 992 addresses listed in the database provided two types of information: the origin of country and the organisation. Two hundred and seven articles listed only one address

Table 2.2 Journals with more than seven articles ($n = 117$) – number of articles, sum and share of citations per journal (total $n = 9{,}207$ citations)

Journal	*Number of papers*	*Share of papers (%)*	*Number of citations*	*Share of all citations (%)*
Biofuels Bioproducts & Biorefining – Biofpr	27	6.0	244	2.7
Biomass & Bioenergy	18	4.0	251	2.7
Journal of the American Oil Chemists Society	15	3.3	202	2.2
Journal of Cleaner Production	12	2.6	204	2.2
International Journal of Life Cycle Assessment	10	2.2	164	1.8
International Sugar Journal	10	2.2	30	0.3
Bioresource Technology	9	2.0	361	3.9
Applied Microbiology and Bio-Technology	8	1.8	249	2.7
Scandinavian Journal of Forest Research	8	1.8	14	0.2
Sum	117	25.8	1,719	18.7

Note
See Appendix B for more details.

Table 2.3 The five most prominent authors, with more than four papers

Author	*Number of articles*
Sanders, J. P. M.	8
Zhang, Y. H. P.	6
Birch, K.	5
Montoneri, E.	5
Patel, M. K.	5

and four articles did not list any address. Therefore, we have a sample of 449 papers for the analysis of organisational affiliation. For all articles, the shares of the addresses have been calculated to get fractional counts (Table 2.4). The most important countries in the total sample are the United States, the Netherlands and the United Kingdom.

The authors listed organisational affiliations to 459 organisations in the 449 papers. We calculated fractions of addresses and standardised the types of organisations (Table 2.5). Most of the papers (73%) have listed a university address, 13% listed a research institute address, 6% a company, 1% an international organisation and 6% a public agency.

The most prominent organisations measured in numbers of papers and in degree centrality in the co-authorship network (see Table 2.6) are mainly universities. However, the U.S. Department of Agriculture has the central position in the network when measuring betweenness centrality. That means that the ministry is important for bridging distant networks of expertise. Higher values of degree centrality in Table 2.6 indicate the centrality of the respective organisation in the network, while higher values for betweenness centrality show the bridging function of the respective organisation. Some of the most important universities in the United States (Michigan State University and the University of Florida) achieve high values for degree

Table 2.4 The 10 countries with the most articles, based on address fraction counts

Country	*Number of papers*
United States	116
Netherlands	45
United Kingdom	43
Germany	27
Canada	22
Belgium	21
Italy	20
People's Republic of China	19
Australia	18
Sweden	14

Table 2.5 Types of organisation by number of papers, and their share of the total number of papers ($n = 449$ papers)

Type of organisation	*Number of papers*	*Share (%)*
Higher education institution	327.3	72.9
Research institute	57.6	12.8
Company	26.6	5.9
Public agency	25.0	5.6
International organisation	6.3	1.4
Science agency	4.0	0.9
Cluster organisation	2.3	0.5

Table 2.6 The 10 most prominent organisations in terms of number of papers ($n = 99$, fraction counts) and Freeman's degree centrality in co-authorship networks; values for Freeman's betweenness centrality are added

Organisation	*Number of papers*	*Degree centrality*	*Betweenness centrality*
Wageningen University & Research Centre	19.2	8.200	9,471.480
Iowa State University	17.6	1.861	1,529.762
U.S. Department of Agriculture	15.4	3.242	11,896.121
University of Ghent	12.0	3.003	9,493.600
University of Utrecht	7.2	2.000	1,145.533
University of York	5.8	1.833	933.000
Lund University	5.8	0.833	235.000
Michigan State University	5.5	0.867	0.000
University of Florida	5.3	0.333	0.000
Cardiff University	4.8	0.833	1,782.586

Note
Degree centrality is defined as the number of links that a node has (Borgatti, 2005), while betweenness centrality is defined as the number of times a node acts as a bridge along the shortest path between two other nodes (Freeman, 1977). Centrality measures for degree centrality and betweenness centrality have been calculated with UCINET 6.

centrality, but low values for betweenness centrality because they do not function as connectors between important subnetworks. The measurements of degree centrality in the co-authorship network show that the research field consists of a core of networked organisations and a surrounding plethora of many smaller sub-networks of organisations to which the researchers are affiliated. We identified 179 cliques with at least two nodes and 79 cliques with at least three nodes. The biggest sub-network consists of 237 nodes.

The surrounding plethora of small-sized sub-networks is dominated by higher education organisations. The main sub-network shows not only universities but also companies and other types of actors placed centrally in the network and a geographical clustering of collaboration. Notably, a number of geographical clusters can be identified: (a) an U.S. cluster with a central position around the U.S. Department of Agriculture and other U.S. actors, whether universities, public agencies or companies; (b) a western and central European cluster with the central position of University Wageningen in the Netherlands, ETH in Switzerland and the University of Ghent in Belgium; (c) a small Canadian-French cluster around the University of Toronto; (d) a small Scandinavian cluster; and (e) a small South American cluster with Universidad Estadual Campinas in Brazil. Other regions are less centrally positioned in the network and are more linked to the outer borders of some of these clusters, such as East Asian actors to the U.S. cluster.

In order to get an idea of where the bioeconomy is discussed, we identified the main scientific fields in the sample. Papers are mostly listed under several categories. Therefore, weighted counts have been applied. The sample included 99 Web of Science categories, which represents a very dispersed distribution. There are 249 categories applied in the database, but for many categories this is just a very minor topic so far. Most important are three categories belonging to the natural sciences and technological sciences: biotechnology & applied microbiology, energy & fuels, and environmental sciences. Social science studies are less visible in the sample. The 15 most prominent categories are summarised in Figure 2.2 and the complete overview is listed in a table in Appendix C.

In summary, the bibliometric analysis highlights that bioeconomy research has become more visible over the last few years. Almost three-fourths of the papers are co-authored by researchers affiliated to a higher education institution, while researchers from private firms are much less visible. The research community is still rather fragmented, with a core of European and American regional clusters most active and networked in the field. Conversely, organisations from other parts of the world are much less connected to the network of bioeconomy research. Topic-wise, the research field appears fragmented, dispersed over many fields of science. It is, however, dominated by natural and engineering sciences, while the social sciences are less visible.

2.4 Bioeconomy visions

Considering the many origins and the wide diffusion of the bioeconomy concept across multiple scientific fields, the aim of this section is to examine

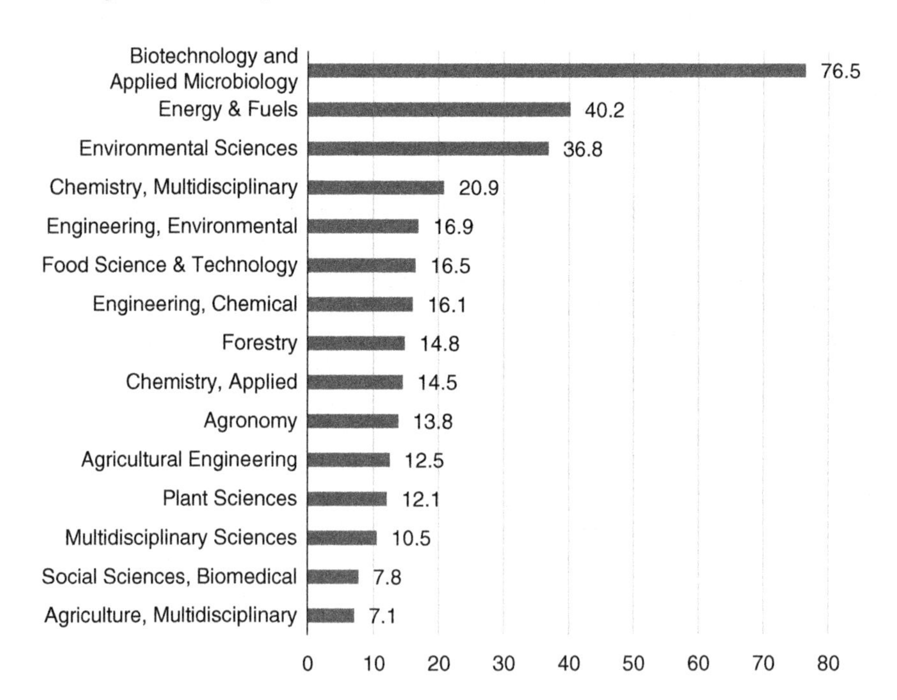

Figure 2.2 Share of Web of Science categories, based on weighted counts ($n = 453$).

differences in the understanding of this concept, which are put forward in the academic literature. Broadly speaking, we find that it is possible to distinguish between three ideal type visions of what a bioeconomy constitutes (see also Levidow et al., 2013; Staffas, Gustavsson & McCormick, 2013). Reflecting on the importance of bioeconomy research in the fields of natural and engineering science, it is perhaps not surprising that at least the first two visions appear to be significantly influenced by a technical perspective:

1 A *bio-technology vision* that emphasises the importance of bio-technology research and application and commercialisation of bio-technology in different sectors.
2 A *bio-resource vision* that focuses on the role of research, development and demonstration (RD & D) related to biological raw materials in sectors such as agriculture, marine, forestry and bioenergy, as well as on the establishment of new value chains. Whereas the bio-technology vision takes a point of departure in the potential applicability of science, the bio-resource vision emphasises the potentials in upgrading and conversion of the biological raw materials.
3 A *bio-ecology vision* that highlights the importance of ecological processes that optimise the use of energy and nutrients, promote biodiversity and avoid monocultures and soil degradation. While the previous two visions are technology-focused and give a central role to RD & D in globalised systems, this vision emphasises the potential for regionally concentrated circular and integrated processes and systems.

Importantly, these visions should not be considered completely distinct from each other, but rather as ideal type visions of the bioeconomy. Thus, while certain actors are predominantly associated with the different visions such as the OECD (the bio-technology vision), the European Commission (the bio-resource vision) and the European Technology Platform TP Organics (the bio-ecology vision) (Levidow et al., 2013; Staffas et al., 2013), it is also highlighted that the visions interrelate. For example, initial policy work in the European Commission was significantly influenced by existing work on the bio-technology vision (Richardson, 2012). (Similarly, individual papers included in the bibliometric analysis (Section 2.3) may often not subscribe to a single understanding of the bioeconomy concept; however, the aim in this part of the analysis is not to classify all bioeconomy papers according to the different visions, but rather to identify the key interpretations of the bioeconomy concept, which are put forward in the academic literature.)

In the following, we identify key features of the three bioeconomy visions, focusing specifically on implications in terms of overall aims and objectives, value creation, drivers and mediators of innovation, and spatial focus. This is summarised in Table 2.7.

Table 2.7 Key characteristics of the bioeconomy visions

	The bio-technology vision	*The bio-resource vision*	*The bio-ecology vision*
Aims and objectives	Economic growth and job creation	Economic growth and sustainability	Sustainability, biodiversity, conservation of ecosystems, avoiding soil degradation
Value creation	Application of biotechnology, commercialisation of research and technology	Conversion and upgrading of bio-resources (process oriented)	Development of integrated production systems and high-quality products with territorial identity
Drivers and mediators of innovation	R & D, patents, TTOs, research councils and funders (science push, linear model)	Interdisciplinary, optimisation of land use, including degraded land in the production of biofuels, use and availability of bio-resources, waste management, engineering, science and market (interactive and networked production mode)	Identification of favourable organic agro-ecological practices, ethics, risk, transdisciplinary sustainability, ecological interactions, re-use and recycling of waste, land use (circular and self-sustained production mode)
Spatial focus	Global clusters/ central regions	Rural/peripheral regions	Rural/peripheral regions

2.4.1 The bio-technology vision

The primary *aims and objectives* in the bio-technology vision relate to economic growth and job creation (Pollack, 2012; Staffas et al., 2013). Thus, while positive effects on climate change and environmental aspects are assumed, economic growth is clearly prioritised above sustainability. Therefore, feedback effects following from the use of bio-technology are most often ignored (Richardson, 2012). Similarly, risks and ethical concerns are subordinate priorities to economic growth (Hilgartner, 2007).

Value creation is linked to the application of biotechnologies in various sectors, as well as to the commercialisation of research and technology. It is expected that economic growth will follow from capitalising on biotechnologies, and intermediaries (such as bio-technology news providers) between bio-technology research firms and investors play an important role in stimulating economic growth around the bioeconomy (Morrison & Cornips, 2012). Consequently, investments in research and innovation, which will result in the production of scientific knowledge, are an absolutely central aspect in this version of the bioeconomy. Research starts from processes operating at the molecular level and products and production processes are subsequently constructed. In principle, this allows the transformation of biomass into a very wide spectrum of marketable products (Hansen, 2014).

Related to *drivers and mediators of innovation*, the implicit understanding of innovation processes in the bio-technology vision is in many ways similar to the so-called linear model of innovation, where innovation processes are assumed to start with scientific research, which is then followed by product development, production and marketing (Bush (1945); see Hansen and Winther (2011) for a summary of critiques of this model). Thus, close interaction between universities and industry is needed in the process in order to ensure that relevant research is indeed commercialised (Zilberman, Kim, Kirschner, Kaplan & Reeves, 2013). In this bioeconomy vision technological progress will solve resource shortages, and resource scarcity is therefore not a central parameter to analyse (McCormick & Kautto, 2013; Staffas et al., 2013). Similarly, it seems to be more or less implicitly assumed that waste will not be a key issue since bio-technology production processes will result in little or no waste. Since the starting point is at the molecular level, processes can in principle be designed to result in very little waste. Biotechnologies may also help transform organic waste into new end-products (Richardson, 2012). It is also suggested that the wide possibilities for application of biotechnology lead to a blurring of boundaries between traditional industries once the technologies approach the stage of commercialisation (Boehlje & Bröring, 2011; Wield, 2013). Since research is a central component in this vision, research councils and other research funding bodies become central actors in translating the visions of the bioeconomy into the actual development of the field itself (Kearnes, 2013). Related to the prominent role ascribed to research, some contributions in the literature focus upon issues of governance of research, such as the history of research policies for the bioeconomy (Aguilar, Magnien & Thomas, 2013).

In terms of *spatial focus*, the bio-technology vision of the bioeconomy is expected to lead to a concentration of growth in a limited number of regions globally that host a combination of large pharmaceutical firms, small biotech firms and venture capital (Cooke, 2007a, 2009). Also regions specialised in high-quality public research related to bio-technology may benefit in developmental terms (Birch, 2009). It is furthermore suggested that connections between these global bio-technology centres are very important for innovation in the bioeconomy and that certain regions in emerging and developing economies may also take advantage of the bioeconomy (Cooke, 2006; Wield, 2013). As a consequence of the focus on global competition in the bioeconomy, the notion of governance of innovation also constitutes a central feature in some of the research underpinning such a vision (Hogarth & Salter, 2010; Rosemann, 2014). Associated with the geographies of the bioeconomy, it is also pointed out how value creation in the bioeconomy comprises both a material component associated with bio-resources, but nonetheless also an immaterial component in terms of knowledge and an ability to develop new knowledge (Birch, 2012). Other parts of this literature revolve around issues such as the conditions for and strategies applied in building a bio-economy in various emerging economies (Chen & Gottweis, 2013; Hsieh & Lofgren, 2009; Salter, 2009; Salter, Cooper, Dickins & Cardo, 2007; Salter, Cooper & Dickins, 2006; Waldby, 2009).

2.4.2 The bio-resource vision

In the bio-resource vision the overall *aims and objectives* relate to both economic growth and sustainability. There is an expectation that bio-innovations will provide both economic growth and environmental sustainability (Levidow et al., 2013). Whereas economic growth in the bio-technology vision would follow from capitalising on biotechnologies, capitalising on bio-resources is expected to drive economic growth in the bio-resource vision. While it is often assumed that effects in terms of environmental sustainability will also be positive, the main focus is on technological development of new bio-based products, and much less on environmental protection (Duchesne & Wetzel, 2003). Thus, quite paradoxically, the climate change effects of the transition to a bioeconomy are rarely assessed, and the sustainability aspect receives relatively limited attention from policymakers (Ollikainen, 2014; Staffas et al., 2013). Notably, this weak integration of sustainability aspects in bioeconomy policies is despite the fact that academics frequently question the positive sustainability effects of the bioeconomy (Pfau, Hagens, Dankbaar & Smits, 2014). Ponte (2009) argues that processes and procedures associated with standard setting in the bioeconomy become more important than outcomes in terms of sustainable development. The bioeconomy discourse may in fact lead to a decreasing emphasis on issues such as deforestation and loss of biological diversity (Pülzl et al., 2014).

In terms of *value creation*, the bio-resource vision highlights the processing and conversion of bio-resources into new products. Related to the use and availability of bio-resources, waste management also takes up a more prominent position in the bio-resource vision. Minimising organic waste production along the value chain is a central concern, and waste production, which cannot be avoided, is an important input to renewable energy production (European Commission, 2012). The concept of cascading use of biomass is central in this regard since it highlights the efforts to maximise the efficiency of biomass use (Keegan, Kretschmer, Elbersen & Panoutsou, 2013). Finally, it is also argued that processing of waste that allows recycling by converting it to fertilisers is central to allow large-scale biofuel production (Mathews, 2009).

In relation to *drivers and mediators of innovation*, and as a natural consequence of the prime focus on bio-resources, the issue of land use constitutes a more explicit element than in the bio-technology vision. An important driver in the bio-resource vision is thus to improve land productivity (Levidow et al., 2013; Mathews, 2009) and to include degraded land in the production of biofuels (Mathews, 2009). However, there is often little discussion of the implications for changes between different types of land use such as forestry and agriculture on other aspects such as climate change (Ollikainen, 2014). Additionally, while considerations concerning the use and availability of bio-resources are prominent, the relation between the use of bio-resources and the use of other resources and products (such as water, fertilisers and pesticides) is rarely considered (Staffas et al., 2013).

Indeed, similar to the bio-technology vision, the bio-resource vision also highlights the role of research and innovation activities as an important driver for value creation. However, while the former takes a narrower point of departure in bio-technology research, the latter emphasises the importance of research in multiple fields, which are in different ways related to biological materials. Consequently, research and innovation efforts often involve collaboration between actors with dissimilar competences, and the importance of research on issues such as consumer preferences is also stressed (Levidow et al., 2013). Innovation is also understood to require collaboration across sectors, e.g. that firms from the forestry industry engage closely with downstream actors (Kleinschmit et al., 2014). According to McCormick and Kautto (2013), the importance of cross-sectoral collaborations for bioeconomy innovation is also frequently underlined in bioeconomy policies. Thus, in summary, the drivers of innovation underlying value creation in the bio-resource vision are less linear than in the bio-technology vision, as cross-sectoral collaborations and interaction with customers are emphasised.

In terms of *spatial focus*, the bio-resource vision emphasises the significant potential for stimulating development in rural settings. It is argued that plants producing new bio-products will positively influence employment in rural locations and will most likely be less footloose than other forms of economic activities due to the importance of natural resources as key location factors (Low & Isserman, 2009). Thus, the bio-resource bioeconomy opens up for a revived rural development driven by diversification into higher value-added products (Horlings & Marsden, 2014). Still, while localised competencies related to cultivating and processing of the biological material are central to this development, this will in most cases need to be complemented with externally located knowledge (Albert, 2007).

2.4.3 The bio-ecology vision

The *aims and objectives* of the bio-ecology vision are primarily concerned with sustainability. While economic growth and employment creation is a main concern in the bio-technology and bio-resource visions, these aspects are clearly secondary to sustainability concerns in the bio-ecology vision (Levidow et al., 2013). Reflecting the focus on and concern for sustainability, the literature on the bioeconomy also contains tensions and critical voices to the focus on economic growth and commercialisation in the bio-technology and in the bio-resource visions. In the literature on health there are several contributions that criticise the commercialisation of bio-resources in areas such as trade in various forms of human tissues (examples of such criticism include questioning trade in cord-blood (Brown, 2013; Brown, Machin & McLeod, 2011; Martin, Brown & Turner, 2008; Waldby & Cooper, 2010), oocytes (Gupta, 2012; Haimes, 2013; Waldby, 2008), foetal tissue (Kent, 2008), stem cells (Fannin, 2013), femoral head (Hoeyer, 2009) and blood (Mumtaz, Bowen & Mumtaz, 2012; Schwarz, 2009)). Examples of topics that are discussed are the ethics of commercialisation of bioresources (Bahadur & Morrison, 2010), safety in blood supply (Mumtaz et al., 2012), inequalities in access to bio-resources (Davies, 2006) and the moral dilemmas of surrogacy (Gupta, 2012).

Regarding *value creation*, the bio-ecology vision emphasises the promotion of biodiversity, conservation of ecosystems, the ability to provide ecosystem services and prevention of soil degradation (Levidow et al., 2013; McCormick & Kautto, 2013). Moreover, it is emphasised that energy production from bio-waste only takes place at the very end of the chain, after reuse and recycling. Also, the use of own waste as well as waste from urban areas is important to reduce or even eliminate the need for external inputs to bioproduct production facilities (Levidow et al., 2013; McCormick & Kautto, 2013). In this sense this vision emphasises a circular and self-sustained production mode.

With reference to the underlying *drivers and mediators of innovation*, the bio-ecological vision of the bioeconomy highlights the identification of favourable organic bio-ecological practices (Marsden, 2012; Siegmeier & Möller, 2013) and ecological interactions related to the re-use and recycling of waste and efficiency in land use. A related key topic is bio-ecological engineering techniques that aim to "design agricultural systems that require as few agrochemicals and energy inputs as possible, instead relying on ecological interactions between biological components to enable agricultural systems to boost their own soil fertility, productivity and crop protection" (Levidow et al., 2013, pp. 98–99).

Whereas the two other bioeconomy visions place emphasis on the role of technically focused research and innovation activities, this is not the case in the bio-ecology vision. In fact, certain technologies such as genetically modified crops are ruled out in the bio-ecology vision. This does not imply that research and innovation activities are deemed unimportant, but rather that they have different foci. For instance, Albrecht, Gottschick, Schorling and Stint (2012) call for greater emphasis in research on transdisciplinary sustainability topics related to e.g. cultivation potentials of sustainable biomass, global fair trade and wider participation in discussions and decisions on transition processes. Finally, calls are made for research that takes the global scale as the point of departure and accounts for the negative consequences of the competing bioeconomy visions (Hansen, 2014).

In terms of *spatial focus*, the bio-ecology vision emphasises the opportunities for rural and peripheral regions in a similar way to the bio-resource vision. It is suggested that rural growth opportunities may result from a focus on high-quality products with territorial identity (Levidow et al., 2013). However, while the importance of external linkages is stressed in the bio-resource vision, the bio-ecology vision calls for development of locally embedded economies, i.e. "place-based agri-ecological systems" (Marsden, 2012, p. 140), as a central part of the efforts to ensure a sustainable bioeconomy.

2.5 Findings and concluding remarks

Based on a review of the research literature, this chapter has documented the scope, origins and reach of the notion of the bioeconomy. Moreover, the chapter has sought to deepen our understanding of the notion of the bioeconomy through the identification of three different visions of the bioeconomy. In sum, the chapter has sought to map the diverse grounds and perspectives in this field.

While the transition to the bioeconomy is often argued to play a key role in targeting grand challenges such as climate change, food security, health, industrial restructuring and energy security, the chapter has shown that the bioeconomy constitutes a young research field, although it is likely that the research covered in this analysis probably has been involved in related domains before, or in similar research under different headings, such as biotechnology. As opposed to former research on biotechnology as such, the more recent research on the bioeconomy seems to refer to a broader concept that encompasses several sectors spanning from health and the chemical industry, to agriculture, forestry and bioenergy. The chapter has shown how a range of different disciplines are involved in the knowledge production underpinning the emergence of the bioeconomy. This breadth reflects the generic characteristic and nature of the notion of the bioeconomy. However, among the variety of disciplines researching the bioeconomy, natural and engineering sciences take up the most central role.

With this in mind, it is perhaps not surprising that the literature review identified three visions of the bioeconomy, of which at least the first two appear to be significantly influenced by an engineering and natural sciences perspective. The bio-technology vision emphasises the importance of biotechnology research and the application and commercialisation of biotechnology in different sectors of the economy. The bio-resource vision focuses on processing and upgrading of biological raw materials, as well as on the establishment of new value chains. Finally, the bio-ecology vision highlights sustainability and ecological processes that optimise the use of energy and nutrients, promote biodiversity and avoid monocultures and soil degradation.

The perception of a bioeconomy also contains different objectives in terms of a focus on reducing waste-streams of bio-resources on the one hand, and developing new products and economic value chains based on existing waste-streams from bio-resources on the other. To the degree that there emerge new economic value chains surrounding biowaste, this may constitute a disincentive to reduce the amount of biowaste in the first place. These two objectives may thus constitute contrasting rationalities. Such opposing rationales reflect the diversity among policy areas involved and highlight the difficulty of speaking of horizontal policies across sectors or domains. However, at the same time, given the emphasis on engineering and the natural sciences, the bio-technology vision and the bio-resources vision overlap to some extent and may represent complementary strategies in terms of the possibility of applying biotechnology to bio-resources. In this sense it may be a viable strategy for countries and regions to possess both localised bio-resources and the technology to refine and upgrade these. Instead of exporting bio-resources for upgrading elsewhere, domestic upgrading would ensure a higher value creation locally, in addition to expected synergies in terms of research and innovation.

Given the main emphasis on natural and engineering sciences in much bioeconomy research, an important topic for future studies is the connection between the bioeconomy and its wider societal and economic implications. The notion of the bioeconomy is often seen to cover a wide range of industries that are very different in terms of technological advancement and value chains. Moreover, the emergence of a bioeconomy is expected to imply the implementation and application of generic biotechnologies into several other sectors and domains. Such application of biotechnology in different existing industry sectors may serve to redefine how these sectors operate and what they produce. Thus, further research into the position of the bioeconomy in societal and economic development strategies following the principles of regional and context-sensitive smart specialisation (Morgan, 2015) or constructed regional advantage (Cooke, 2007b) is welcome.

Thus, whether and how the transition to a bioeconomy will indeed contribute to addressing key grand challenges remains to be seen. Quite paradoxically, while the master narrative surrounding the bioeconomy stresses these particular aspects, consequences in terms of e.g. environmental protection and climate change effects are rarely assessed (Duchesne & Wetzel, 2003; Ollikainen, 2014). This may be attributed to the dominance of natural and engineering science research, which often focuses on narrow aspects of the bioeconomy rather than the wider, systemic consequences. Thus, additional bioeconomy research in non-technical fields is arguably important in order to provide a more profound understanding of the socioeconomic aspects of the bioeconomy and thereby its potential for addressing the grand challenges of our time.

As an attempt to answer the question "What is the bioeconomy?" posed in the title, this chapter has shown that the notion of the bioeconomy is multifaceted: in breadth, e.g. in terms of origins and sectors represented; and in depth, i.e. in terms of rationales or visions of the underlying values, direction and drivers of the bioeconomy. The chapter has shown how these different visions seem to co-exist in the research literature, and how they bear implications for objectives, value creation, drivers of innovation and spatial focus. Still, although we must remember that bioeconomy is a broad (and deep) term covering many sectors and meanings, it seems possible to distil a joint interest in "an exploration and exploitation of bio-resources". Such an interest may imply different ways of applying biotechnology to bio-resources and various forms of harvesting new bioproducts. Nonetheless, it may also foster an improved understanding of the ecosystems in which we live and possibilities in terms of new and sustainable solutions and the knowledge and technologies underpinning these.

Appendix A

The following indicators have been calculated in the bibliometric analysis:

- Number of papers per year
- Total number of citations: we obtained the citation data in February 2016
- Citations per paper
- Average number of citations of each paper per year since publishing
- Number of papers per journal
- Citations of papers per journal
- Number of papers per author
- Affiliation of authors based on fraction counts: papers per country and per organisation
- Organisational affiliation of the authors distinguishing between types of organisations: higher education institutions, research institutes, companies, public agencies, international organisations, science agencies and cluster organisations
- Centrality of organisations measured in number of papers based on fraction counts, and SNA centrality measures, such as degree centrality and betweenness centrality
- Distribution of scientific field based on the categories of the database Web of Science as an indicator for the scientific field and based on fraction counts.

Appendix B

Table 2.A1 Alphabetic list of journals with number of papers, share of papers, number of citations and share of citations

Journal	*Number of papers*	*Share of papers (%)*	*Number of citations*	*Share of citations (%)*
Acs Sustainable Chemistry & Engineering	2	0.4	9	0.10
Advances in Agronomy	1	0.2	0	0.00
Advances in Applied Microbiology	1	0.2	19	0.21
African Journal of Biotechnology	1	0.2	45	0.49
Agricultural Economics	7	1.5	16	0.17
Agriculture Ecosystems & Environment	1	0.2	26	0.28
Agrociencia	1	0.2	0	0.00
Agronomy Journal	1	0.2	234	2.54
Aiche Journal	1	0.2	28	0.30
Analytica Chimica Acta	1	0.2	54	0.59
Annual Review of Chemical and Biomolecular Engineering	1	0.2	4	0.04
Antioxidants & Redox Signaling	1	0.2	6	0.07
Applied and Environmental Microbiology	2	0.4	44	0.48
Applied Biochemistry and Biotechnology	1	0.2	11	0.12
Applied Microbiology and Biotechnology	8	1.8	249	2.70
Applied Radiation and Isotopes	1	0.2	0	0.00
Applied Soft Computing	1	0.2	6	0.07
Area	1	0.2	11	0.12
Asian Journal of Chemistry	2	0.4	1	0.01
Australian Forestry	1	0.2	1	0.01
Australian Health Review	1	0.2	2	0.02
Biocatalysis and Biotransformation	1	0.2	2	0.02
Biochimie	1	0.2	67	0.73
Biocontrol	1	0.2	1	0.01
Bioenergy Research	2	0.4	19	0.21
Bioethics	1	0.2	7	0.08

Biofuels bioproducts & Biorefining – biofpr	27	6.0	244	2.65
Biomacromolecules	2	0.4	34	0.37
Biomass & Bioenergy	18	4.0	251	2.73
Bioresource Technology	9	2.0	361	3.92
Biosocieties	3	0.7	17	0.18
Biotechnology & Biotechnological Equipment	1	0.2	0	0.00
Biotechnology Advances	5	1.1	571	6.20
Biotechnology and Bioengineering	1	0.2	351	3.81
Biotechnology and Genetic Engineering Reviews, vol 26	1	0.2	0	0.00
Biotechnology for Biofuels	3	0.7	151	1.64
Biotechnology in China II: Chemicals, Energy and Environment	2	0.4	12	0.13
Biotechnology Journal	2	0.4	16	0.17
Body & Society	2	0.4	7	0.08
Botanical Journal of the Linnean Society	1	0.2	16	0.17
Brazilian Journal of Chemical Engineering	1	0.2	5	0.05
Carbohydrate Polymers	2	0.4	8	0.09
Carbohydrate Research	1	0.2	35	0.38
Catalysis Letters	1	0.2	18	0.20
Catalysis Today	1	0.2	52	0.56
Cellulose	2	0.4	5	0.05
Chemical and Biochemical Engineering Quarterly	1	0.2	5	0.05
Chemical Communications	1	0.2	9	0.10
Chemical Engineering & Technology	1	0.2	7	0.08
Chemical Engineering Journal	1	0.2	7	0.08
Chemical Engineering Progress	2	0.4	7	0.08
Chemical Society Reviews	1	0.2	36	0.39
Chemistry – A European Journal	1	0.2	71	0.77
Chemsuschem	2	0.4	73	0.79
Chimica Oggi – Chemistry Today	3	0.7	0	0.00
Critical Reviews in Biotechnology	1	0.2	0	0.00
Critical Reviews in Environmental Science and Technology	1	0.2	4	0.04

continued

Table 2.A1 Continued

Journal	*Number of papers*	*Share of papers (%)*	*Number of citations*	*Share of citations (%)*
Croatian Medical Journal	1	0.2	2	0.02
Crop Science	3	0.7	54	0.59
Current Microbiology	1	0.2	28	0.30
Current Opinion in Chemical Engineering	2	0.4	7	0.08
Current Opinion in Environmental Sustainability	6	1.3	54	0.59
Current Opinion in Solid State & Materials Science	1	0.2	2	0.02
Current Organic Chemistry	1	0.2	2	0.02
Defence Science Journal	1	0.2	2	0.02
Drewno	2	0.4	0	0.00
Ecology and Evolution	1	0.2	0	0.00
Ecology and Society	1	0.2	8	0.09
Economic Development Quarterly	1	0.2	25	0.27
Educational Philosophy and Theory	1	0.2	4	0.04
Energies	2	0.4	10	0.11
Energy	1	0.2	4	0.04
Energy & Environmental Science	2	0.4	167	1.81
Energy & Fuels	1	0.2	59	0.64
Energy Conversion and Management	2	0.4	21	0.23
Energy Policy	5	1.1	50	0.54
Energy Sources Part A – Recovery Utilization and Environmental Effects	1	0.2	0	0.00
Engineering in Life Sciences	3	0.7	9	0.10
Environment and Planning A	1	0.2	4	0.04
Environment and Planning C – Government and Policy	1	0.2	9	0.10
Environmental Engineering and Management Journal	1	0.2	2	0.02
Environmental Progress & Sustainable Energy	1	0.2	13	0.14
Environmental Science & Technology	3	0.7	49	0.53
Enzyme and Microbial Technology	1	0.2	6	0.07
European Journal of Agronomy	1	0.2	16	0.17

European Journal of Lipid Science and Technology	2	0.4	4	0.04
European Planning Studies	1	0.2	14	0.15
European Urban and Regional Studies	1	0.2	7	0.08
Feedstocks for the Future: Renewables for the Production of Chemicals and Materials	1	0.2	4	0.04
Feminist Theory	1	0.2	29	0.31
Fems Yeast Research	1	0.2	71	0.77
Food Security	1	0.2	12	0.13
Forest Products Journal	1	0.2	9	0.10
Forestry Chronicle	3	0.7	20	0.22
Fresenius Environmental Bulletin	1	0.2	0	0.00
Frontiers in Plant Science	1	0.2	11	0.12
Fuel	1	0.2	2	0.02
Fuel Processing Technology	1	0.2	16	0.17
Functional Plant Biology	1	0.2	53	0.58
Future Trends in Biotechnology	1	0.2	13	0.14
Futures	1	0.2	9	0.10
Genome Medicine	1	0.2	13	0.14
Geoforum	1	0.2	7	0.08
Geopolitics	1	0.2	4	0.04
Global Change Biology Bioenergy	1	0.2	0	0.00
Green Chemistry	3	0.7	1,056	11.47
Health Policy and Planning	1	0.2	2	0.02
Hortscience	1	0.2	0	0.00
Human Reproduction	1	0.2	9	0.10
Ices Journal of Marine Science	1	0.2	4	0.04
Industrial & Engineering Chemistry Research	3	0.7	36	0.39
Industrial Crops and Products	5	1.1	149	1.62
Interface Focus	1	0.2	12	0.13
International Affairs	1	0.2	7	0.08
International Food and Agribusiness Management Review	1	0.2	5	0.05
International Journal of Feminist Approaches to Bioethics	1	0.2	10	0.11

continued

Table 2.A1 Continued

Journal	*Number of papers*	*Share of papers (%)*	*Number of citations*	*Share of citations (%)*
International Journal of Life Cycle Assessment	10	2.2	164	1.78
International Sugar Journal	10	2.2	30	0.33
Invasive Plant Science and Management	1	0.2	5	0.05
J-for-Journal of Science & Technology for Forest Products and Processes	1	0.2	0	0.00
Journal of Agricultural & Environmental Ethics	1	0.2	10	0.11
Journal of Agricultural and Food Chemistry	5	1.1	117	1.27
Journal of Analytical and Applied Pyrolysis	2	0.4	9	0.10
Journal of Applied Polymer Science	1	0.2	1	0.01
Journal of Bacteriology	2	0.4	32	0.35
Journal of Biobased Materials and Bioenergy	3	0.7	51	0.55
Journal of Biotechnology	7	1.5	146	1.59
Journal of Chemical Technology and Biotechnology	1	0.2	3	0.03
Journal of Cleaner Production	12	2.6	204	2.22
Journal of Environmental Health	1	0.2	3	0.03
Journal of Environmental Management	1	0.2	13	0.14
Journal of Ethnopharmacology	1	0.2	23	0.25
Journal of Green Building	1	0.2	3	0.03
Journal of Industrial Ecology	3	0.7	60	0.65
Journal of Industrial Microbiology & Biotechnology	1	0.2	56	0.61
Journal of Integrative Environmental Sciences	2	0.4	2	0.02
Journal of Magnetic Resonance	1	0.2	12	0.13
Journal of Maps	1	0.2	6	0.07
Journal of Medicinal Plants Research	1	0.2	3	0.03
Journal of Nanoscience and Nanotechnology	1	0.2	133	1.44
Journal of Peasant Studies	1	0.2	70	0.76
Journal of Photochemistry and Photobiology A – Chemistry	1	0.2	16	0.17
Journal of Proteomics	1	0.2	6	0.07
Journal of Scientific & Industrial Research	1	0.2	209	2.27

Journal of Surfactants and Detergents	1	0.2	14	0.15
Journal of the American Leather Chemists Association	3	0.7	10	0.11
Journal of the American Oil Chemists Society	15	3.3	202	2.19
Journal of the Chemical Society of Pakistan	1	0.2	16	0.17
Journal of the Chilean Chemical Society	1	0.2	15	0.16
Journal of the Science of Food and Agriculture	1	0.2	10	0.11
Jove-Journal of Visualized Experiments	1	0.2	0	0.00
Landbauforschung	1	0.2	2	0.02
Life Science Journal-Acta Zhengzhou University Overseas Edition	1	0.2	0	0.00
Macromolecular Bioscience	1	0.2	109	1.18
Medical Journal of Australia	1	0.2	2	0.02
Metabolic Engineering	2	0.4	21	0.23
Microbial Biotechnology	1	0.2	18	0.20
Microbial Cell Factories	1	0.2	0	0.00
Microbiology – SGM	1	0.2	22	0.24
Molecular Crystals and Liquid Crystals	1	0.2	3	0.03
Mrs Bulletin	1	0.2	5	0.05
New Biotechnology	2	0.4	20	0.22
New Genetics and Society	5	1.1	79	0.86
New Medit	1	0.2	0	0.00
New Phytologist	3	0.7	74	0.80
OMICS – A Journal of Integrative Biology	1	0.2	6	0.07
Organic & Biomolecular Chemistry	1	0.2	35	0.38
Philosophical Transactions of the Royal Society B – Biological Sciences	1	0.2	1	0.01
Phytochemistry	1	0.2	87	0.94
Plant Biotechnology Journal	2	0.4	22	0.24
Plant Cell	1	0.2	62	0.67
Plant Cell Tissue and Organ Culture	1	0.2	0	0.00
Plant Journal	1	0.2	143	1.55
Plant Science	1	0.2	19	0.21
Plos One	4	0.9	28	0.30

continued

Table 2.A1 Continued

Journal	*Number of papers*	*Share of papers (%)*	*Number of citations*	*Share of citations (%)*
Political Quarterly	1	0.2	6	0.07
Polymers	1	0.2	0	0.00
Precision Agriculture	1	0.2	19	0.21
Proceedings of the National Academy of Sciences of the United States of America	2	0.4	80	0.87
Process Biochemistry	1	0.2	67	0.73
Process safety and Environmental Protection	1	0.2	16	0.17
Progress in Polymer Science	1	0.2	344	3.74
Pulp & Paper – Canada	1	0.2	0	0.00
Pure and Applied Chemistry	3	0.7	3	0.03
Radiocarbon	2	0.4	2	0.02
Regenerative Medicine	3	0.7	48	0.52
Renewable & Sustainable Energy Reviews	7	1.5	76	0.83
Resources Conservation and Recycling	2	0.4	37	0.40
Risk Analysis	1	0.2	8	0.09
Romanian Biotechnological Letters	2	0.4	0	0.00
RSC Advances	2	0.4	44	0.48
Scandinavian Journal of Forest Research	8	1.8	14	0.15
Science	1	0.2	207	2.25
Science and Public Policy	2	0.4	6	0.07
Science as Culture	2	0.4	3	0.03
Science Technology & Human Values	4	0.9	39	0.42
Small-Scale Forestry	1	0.2	7	0.08
Social Science & Medicine	5	1.1	50	0.54
Sociology of Health & Illness	2	0.4	9	0.10
Southern Journal of Applied Forestry	1	0.2	1	0.01
Spanish Journal of Agricultural Research	1	0.2	0	0.00
Springerplus	1	0.2	6	0.07
Studies in Informatics and Control	1	0.2	0	0.00

Sustainability	4	0.9	43	0.47
Sustainability Science	1	0.2	8	0.09
Technological and Economic Development of Economy	1	0.2	0	0.00
Technology Analysis & Strategic Management	1	0.2	1	0.01
Tijdschrift Voor Economische en Sociale Geografie	1	0.2	4	0.04
Topia-Canadian Journal of Cultural Studies	1	0.2	0	0.00
Transactions of the Asabe	1	0.2	0	0.00
Transactions of the Institute of British Geographers	1	0.2	18	0.20
Transgenic Research	2	0.4	37	0.40
Transnational Environmental Law	1	0.2	1	0.01
Transportation Research Record	3	0.7	4	0.04
Tree Genetics & Genomes	1	0.2	3	0.03
Trends in Biotechnology	2	0.4	104	1.13
Trends in Microbiology	1	0.2	12	0.13
Trends in Plant Science	1	0.2	7	0.08
Tribology & Lubrication Technology	1	0.2	0	0.00
Waste and Biomass Valorization	1	0.2	1	0.01
Water Air and Soil Pollution	1	0.2	9	0.10
Water Research	2	0.4	21	0.23

Appendix C

Table 2.A2 List of Web of Science categories, sorted by number of papers and share of papers

Web of Science category	*Number of papers*	*Share (%)*
Biotechnology & Applied Microbiology	76.5	16.9
Energy & Fuels	40.2	8.9
Environmental Sciences	36.8	8.1
Chemistry, Multidisciplinary	20.9	4.6
Engineering, Environmental	16.9	3.7
Food Science & Technology	16.5	3.6
Engineering, Chemical	16.1	3.6
Forestry	14.8	3.3
Chemistry, Applied	14.5	3.2
Agronomy	13.8	3.1
Agricultural Engineering	12.5	2.8
Plant Sciences	12.1	2.7
Multidisciplinary Sciences	10.5	2.3
Social Sciences, Biomedical	7.8	1.7
Agriculture, Multidisciplinary	7.1	1.6
Microbiology	6.8	1.5
Biochemistry & Molecular Biology	6.7	1.5
Environmental Studies	6.4	1.4
Economics	5.8	1.3
Polymer Science	5.3	1.2
Social Issues	5.3	1.2
Geography	5.3	1.2
Materials Science, Paper, & Wood	5.2	1.1
Agricultural Economics & Policy	4.5	1.0
Chemistry, Organic	4.0	0.9
Biochemical Research Methods	4.0	0.9
Public, Environmental, & Occupational Health	4.0	0.9
Genetics & Heredity	3.3	0.7
Biology	3.0	0.7
Chemistry, Physical	2.7	0.6
Sociology	2.7	0.6
History & Philosophy of Science	2.3	0.5
Planning & Development	2.3	0.5
Materials Science, Textiles	2.2	0.5
Chemistry, Analytical	2.0	0.4
Cultural Studies	2.0	0.4
Geochemistry & Geophysics	2.0	0.4
Medicine, General & Internal	2.0	0.4
Ecology	1.8	0.4
Cell & Tissue Engineering	1.5	0.3
Engineering, Biomedical	1.5	0.3
Political Science	1.5	0.3
Horticulture	1.3	0.3
Materials Science, Biomaterials	1.3	0.3
Spectroscopy	1.3	0.3
Women's Studies	1.3	0.3

Table 2.A2 Continued

Web of Science category	*Number of papers*	*Share (%)*
Chemistry, Medicinal	1.3	0.3
Management	1.2	0.3
Public Administration	1.2	0.3
Urban Studies	1.1	0.2
Materials Science, Multidisciplinary	1.0	0.2
Physics, Applied	1.0	0.2
Architecture	1.0	0.2
Crystallography	1.0	0.2
Education & Educational Research	1.0	0.2
Engineering, Civil	1.0	0.2
Engineering, Mechanical	1.0	0.2
Engineering, Multidisciplinary	1.0	0.2
Entomology	1.0	0.2
Health Care Sciences & Services	1.0	0.2
Health Policy & Services	1.0	0.2
International Relations	1.0	0.2
Nutrition & Dietetics	1.0	0.2
Thermodynamics	1.0	0.2
Transportation	1.0	0.2
Transportation Science & Technology	1.0	0.2
Water Resources	1.0	0.2
Ethics	0.8	0.2
Physics, Condensed Matter	0.5	0.1
Anthropology	0.5	0.1
Automation & Control Systems	0.5	0.1
Computer Science, Artificial Intelligence	0.5	0.1
Computer Science, Interdisciplinary Applications	0.5	0.1
Endocrinology & Metabolism	0.5	0.1
Geography, Physical	0.5	0.1
Law	0.5	0.1
Mechanics	0.5	0.1
Obstetrics & Gynaecology	0.5	0.1
Operations Research & Management Science	0.5	0.1
Physics, Nuclear	0.5	0.1
Reproductive Biology	0.5	0.1
Soil Science	0.5	0.1
Cell Biology	0.3	0.1
Chemistry, Inorganic & Nuclear	0.3	0.1
Fisheries	0.3	0.1
Marine & Freshwater Biology	0.3	0.1
Mathematics, Interdisciplinary Applications	0.3	0.1
Meteorology & Atmospheric Sciences	0.3	0.1
Mycology	0.3	0.1
Nuclear Science & Technology	0.3	0.1
Oceanography	0.3	0.1
Physics, Atomic, Molecular, & Chemical	0.3	0.1
Radiology, Nuclear Medicine, & Medical Imaging	0.3	0.1
Social Sciences, Mathematical Methods	0.3	0.1

continued

Table 2.A2 Continued

Web of Science category	*Number of papers*	*Share (%)*
Engineering, Industrial	0.3	0.1
Integrative & Complementary Medicine	0.3	0.1
Medical Ethics	0.3	0.1
Pharmacology & Pharmacy	0.3	0.1
Nanoscience & Nanotechnology	0.2	0.04

Note

1 This chapter is an adaption of Bugge, Hansen and Klitkou, "What is the bioeconomy? A review of the literature", published in *Sustainability* in 2016.

References

Aguilar, A., Magnien, E. & Thomas, D. (2013). Thirty years of European biotechnology programmes: From biomolecular engineering to the bioeconomy. *New Biotechnology*, *30*(5), 410–425. doi:10.1016/j.nbt.2012.11.014.

Albert, S. (2007). Transition to a bio-economy: A community development strategy discussion. *Journal of Rural and Community Development*, *2*(2), 64–83.

Albrecht, S., Gottschick, M., Schorling, M. & Stint, S. (2012). Bio-economy at a crossroads: Way forward to sustainable production and consumption or industrialization of biomass? *GAIA – Ecological Perspectives for Science and Society*, *21*(1), 33–37.

Bahadur, G., & Morrison, M. (2010). Patenting human pluripotent cells: Balancing commercial, academic and ethical interests. *Human Reproduction*, *25*(1), 14–21.

Birch, K. (2009). The knowledge–space dynamic in the UK bioeconomy. *Area*, *41*(3), 273–284. doi:10.1111/j.1475-4762.2008.00864.x.

Birch, K. (2012). Knowledge, place, and power: Geographies of value in the bioeconomy. *New Genetics and Society*, *31*(2), 183–201. doi:10.1080/14636778.2012.662051.

Boehlje, M., & Bröring, S. (2011). The increasing multifunctionality of agricultural raw materials: Three dilemmas for innovation and adoption. *International Food and Agribusiness Management Review*, *14*(2), 1–16.

Bordes, P., Pollet, E. & Averous, L. (2009). Nano-biocomposites: Biodegradable polyester/nanoclay systems. *Progress in Polymer Science*, *34*(2), 125–155. doi:10.1016/j.progpolymsci.2008.10.002.

Borgatti, S. P. (2002). *NetDraw: Graph Visualization Software*. Cambridge, MA: Analytic Technologies.

Borgatti, S. P. (2005). Centrality and network flow. *Social Networks*, *27*, 55–71.

Borgatti, S. P., Everett, M. G. & Freeman, L. C. (2002). *Ucinet for Windows: Software for Social Network Analysis*. Cambridge, MA: Analytic Technologies.

Bozell, J. J., & Petersen, G. R. (2010). Technology development for the production of biobased products from biorefinery carbohydrates: The US Department of Energy's "Top 10" revisited. *Green Chemistry*, *12*(4), 539–554. doi:10.1039/b922014c.

Brown, N. (2013). Contradictions of value: Between use and exchange in cord blood bioeconomy. *Sociology of Health & Illness*, *35*(1), 97–112.

Brown, N., Machin, L. & McLeod, D. (2011). Immunitary bioeconomy: The economisation of life in the international cord blood market. *Social Science & Medicine*, *72*(7), 1115–1122.

Bugge, M. M., Hansen, T. & Klitkou, A. (2016). What is the bioeconomy? A review of the literature. *Sustainability*, *8*(7), 1–22. doi:10.3390/su8070691.

Burrell, Q. L. (1991). The Bradford distribution and the Gini Index. *Scientometrics*, *21*(2), 181–194. doi:10.1007/bf02017568.

Bush, V. (1945). *Science: The Endless Frontier*. Washington, DC: United States Government Printing Office.

Carvalheiro, F., Duarte, L. C. & Girio, F. M. (2008). Hemicellulose biorefineries: A review on biomass pretreatments. *Journal of Scientific & Industrial Research*, *67*(11), 849–864.

Chen, H. D., & Gottweis, H. (2013). Stem cell treatments in China: Rethinking the patient role in the global bio-economy. *Bioethics*, *27*(4), 194–207.

Coenen, L., Hansen, T. & Rekers, J. V. (2015). Innovation policy for grand challenges: An economic geography perspective. *Geography Compass*, *9*(9), 483–496. doi:10.1111/gec3.12231.

Cooke, P. (2006). Global bioregional networks: A new economic geography of bioscientific knowledge. *European Planning Studies*, *14*(9), 1265–1285. doi:10.1080/09654310600933348.

Cooke, P. (2007a). *Growth Cultures: The Global Bioeconomy and Its Bioregions*. Abingdon: Routledge.

Cooke, P. (2007b). To construct regional advantage from innovation systems first build policy platforms. *European Planning Studies*, *15*(2), 179–194.

Cooke, P. (2009). The economic geography of knowledge flow hierarchies among internationally networked medical bioclusters: A scientometric analysis. *Tijdschrift Voor Economische en Sociale Geografie*, *100*(3), 332–347.

Davies, G. (2006). Patterning the geographies of organ transplantation: Corporeality, generosity and justice. *Transactions of the Institute of British Geographers*, *31*(3), 257–271.

Duchesne, L. C., & Wetzel, S. (2003). The bioeconomy and the forestry sector: Changing markets and new opportunities. *The Forestry Chronicle*, *79*(5), 860–864. doi:10.5558/tfc79860-5.

Dusselier, M., Van Wouwe, P., Dewaele, A., Makshina, E. & Sels, B. F. (2013). Lactic acid as a platform chemical in the biobased economy: The role of chemocatalysis. *Energy & Environmental Science*, *6*(5), 1415–1442. doi:10.1039/c3ee00069a.

European Commission. (2012). *Innovating for Sustainable Growth: A Bioeconomy for Europe*. Brussels: European Commission Retrieved from http://ec.europa.eu/research/bioeconomy/pdf/official-strategy_en.pdf.

Fannin, M. (2013). The hoarding economy of endometrial stem cell storage. *Body & Society*, *19*(4), 32–60.

FitzPatrick, M., Champagne, P., Cunningham, M. F. & Whitney, R. A. (2010). A biorefinery processing perspective: Treatment of lignocellulosic materials for the production of value-added products. *Bioresource Technology*, *101*(23), 8915–8922. doi:10.1016/j.biortech.2010.06.125.

Freeman, L. C. (1977). A set of measures of centrality based on betweenness. *Sociometry*, *40*(1), 35–41.

Graham, R. L., Nelson, R., Sheehan, J., Perlack, R. D. & Wright, L. L. (2007). Current and potential US corn stover supplies. *Agronomy Journal*, *99*(1), 1–11. doi:10.2134/argonj2005.0222.

Gupta, J. A. (2012). Reproductive biocrossings: Indian egg donors and surrogates in the globalized fertility market. *International Journal of Feminist Approaches to Bioethics, 5*(1), 25–51.

Haimes, E. (2013). Juggling on a rollercoaster? Gains, loss and uncertainties in IVF patients' accounts of volunteering for a UK egg sharing for research scheme. *Social Science & Medicine, 86*, 45–51.

Hansen, J. (2014). The Danish biofuel debate: Coupling scientific and politico-economic claims. *Science as Culture, 23*(1), 73–97. doi:10.1080/09505431.2013.808619.

Hansen, T., & Winther, L. (2011). Innovation, regional development and relations between high- and low-tech industries. *European Urban and Regional Studies, 18*(3), 321–339.

Hilgartner, S. (2007). Making the bioeconomy measurable: Politics of an emerging anticipatory machinery. *BioSocieties, 2*(3), 382–386. doi:10.1017/S1745855207005819.

Hoeyer, K. (2009). Tradable body parts? How bone and recycled prosthetic devices acquire a price without forming a "market". *BioSocieties, 4*(2–3), 239–256.

Hogarth, S., & Salter, B. (2010). Regenerative medicine in Europe: Global competition and innovation governance. *Regenerative Medicine, 5*(6), 971–985.

Horlings, L. G., & Marsden, T. K. (2014). Exploring the "New Rural Paradigm" in Europe: Eco-economic strategies as a counterforce to the global competitiveness agenda. *European Urban and Regional Studies, 21*(1), 4–20. doi:10.1177/0969776412441934.

Horn, S. J., Vaaje-Kolstad, G., Westereng, B. & Eijsink, V. G. H. (2012). Novel enzymes for the degradation of cellulose. *Biotechnology for Biofuels, 5*. doi:10.1186/1754-6834-5-45.

Hsieh, C. R., & Lofgren, H. (2009). Biopharmaceutical innovation and industrial developments in South Korea, Singapore and Taiwan. *Australian Health Review, 33*(2), 245–257.

Kearnes, M. (2013). Performing synthetic worlds: Situating the bioeconomy. *Science and Public Policy, 40*(4), 453–465. doi:10.1093/scipol/sct052.

Keegan, D., Kretschmer, B., Elbersen, B. & Panoutsou, C. (2013). Cascading use: A systematic approach to biomass beyond the energy sector. *Biofuels, Bioproducts and Biorefining*, 7(2), 193–206.

Kent, J. (2008). The fetal tissue economy: From the abortion clinic to the stem cell laboratory. *Social Science & Medicine, 67*(11), 1747–1756.

Kleinschmit, D., Lindstad, B. H., Thorsen, B. J., Toppinen, A., Roos, A. & Baardsen, S. (2014). Shades of green: A social scientific view on bioeconomy in the forest sector. *Scandinavian Journal of Forest Research, 29*(4), 402–410. doi:10.1080/02827581.2014.921722.

Lee, S. H., Doherty, T. V., Linhardt, R. J. & Dordick, J. S. (2009). Ionic liquid-mediated selective extraction of lignin from wood leading to enhanced enzymatic cellulose hydrolysis. *Biotechnology and Bioengineering, 102*(5), 1368–1376. doi:10.1002/bit.22179.

Levidow, L., Birch, K. & Papaioannou, T. (2013). Divergent paradigms of European agro-food innovation: The knowledge-based bio-economy (KBBE) as an R&D agenda. *Science, Technology & Human Values, 38*(1), 94–125. doi:10.1177/0162243912438143.

Li, C. Z., Wang, Q. & Zhao, Z. K. (2008). Acid in ionic liquid: An efficient system for hydrolysis of lignocellulose. *Green Chemistry, 10*(2), 177–182. doi:10.1039/b711512a.

Low, S. A., & Isserman, A. M. (2009). Ethanol and the local economy: Industry trends, location factors, economic impacts, and risks. *Economic Development Quarterly, 23*(1), 71–88. doi:10.1177/0891242408329485.

Marsden, T. (2012). Towards a real sustainable agri-food security and food policy: Beyond the ecological fallacies? *The Political Quarterly, 83*(1), 139–145. doi:10.1111/j.1467-923X.2012.02242.x.

Martin, P., Brown, N. & Turner, A. (2008). Capitalizing hope: The commercial development of umbilical cord blood stem cell banking. *New Genetics and Society, 27*(2), 127–143.

Mathews, J. A. (2009). From the petroeconomy to the bioeconomy: Integrating bioenergy production with agricultural demands. *Biofuels, Bioproducts and Biorefining, 3*(6), 613–632. doi:10.1002/bbb.181.

McCormick, K., & Kautto, N. (2013). The bioeconomy in Europe: An overview. *Sustainability, 5*(6), 2589. doi:10.3390/su5062589.

Morgan, K. (2015). Smart specialisation: Opportunities and challenges for regional innovation policy. *Regional Studies, 49*(3), 480–482.

Morrison, M., & Cornips, L. (2012). Exploring the role of dedicated online biotechnology news providers in the innovation economy. *Science, Technology & Human Values, 37*(3), 262–285. doi:10.1177/0162243911420581.

Mumtaz, Z., Bowen, S. & Mumtaz, R. (2012). Meanings of blood, bleeding and blood donations in Pakistan: Implications for national vs global safe blood supply policies. *Health Policy and Planning, 27*(2), 147–155.

Ollikainen, M. (2014). Forestry in bioeconomy: Smart green growth for the humankind. *Scandinavian Journal of Forest Research, 29*(4), 360–366. doi:10.1080/02827581.2014.926392.

Pfau, S. F., Hagens, J. E., Dankbaar, B. & Smits, A. J. M. (2014). Visions of sustainability in bioeconomy research. *Sustainability, 6*(3), 1222–1249.

Pollack, A. (2012, 26 April 2012). White House promotes a bioeconomy. *New York Times*. Retrieved from www.nytimes.com/2012/04/26/business/energy-environment/white-house-promotes-a-bioeconomy.html?_r=0 (accessed on 9 March 2016).

Ponte, S. (2009). From fishery to fork: Food safety and sustainability in the "virtual" knowledge-based bio-economy (KBBE). *Science as Culture, 18*(4), 483–495. doi:10.1080/09505430902873983.

Pülzl, H., Kleinschmit, D. & Arts, B. (2014). Bioeconomy: An emerging meta-discourse affecting forest discourses? *Scandinavian Journal of Forest Research, 29*(4), 386–393. doi:10.1080/02827581.2014.920044.

Richardson, B. (2012). From a fossil-fuel to a biobased economy: The politics of industrial biotechnology. *Environment and Planning C: Government and Policy, 30*(2), 282–296.

Rosemann, A. (2014). Standardization as situation-specific achievement: Regulatory diversity and the production of value in intercontinental collaborations in stem cell medicine. *Social Science & Medicine, 122*, 72–80. doi:10.1016/j.socscimed.2014.10.018.

Salter, B. (2009). State strategies and the geopolitics of the global knowledge economy: China, India and the case of regenerative medicine. *Geopolitics, 14*(1), 47–78.

Salter, B., Cooper, M. & Dickins, A. (2006). China and the global stem cell bioeconomy: An emerging political strategy? *Regenerative Medicine*, *1*(5), 671–683.

Salter, B., Cooper, M., Dickins, A. & Cardo, V. (2007). Stem cell science in India: Emerging economies and the politics of globalization. *Regenerative Medicine*, *2*(1), 75–89.

Schuitmaker, T. J. (2012). Identifying and unravelling persistent problems. *Technological Forecasting and Social Change*, *79*(6), 1021–1031. doi:10.1016/j.techfore.2011.11.008.

Schwarz, M. T. (2009). Emplacement and contamination: Mediation of Navajo identity through excorporated blood. *Body & Society*, *15*(2), 145–168.

Siegmeier, T., & Möller, D. (2013). Mapping research at the intersection of organic farming and bioenergy: A scientometric review. *Renewable and Sustainable Energy Reviews*, *25*, 197–204. doi:10.1016/j.rser.2013.04.025.

Staffas, L., Gustavsson, M. & McCormick, K. (2013). Strategies and policies for the bioeconomy and bio-based economy: An analysis of official national approaches. *Sustainability*, *5*(6), 2751.

Upham, P., Klitkou, A. & Olsen, D. S. (2016). Using transition management concepts for the evaluation of intersecting policy domains ("grand challenges"): The case of Swedish, Norwegian and UK biofuel policy. *International Journal of Foresight and Innovation Policy*, *11*(1–3), 73–95. doi:10.1504/IJFIP.2016.078326.

Vaaje-Kolstad, G., Westereng, B., Horn, S. J., Liu, Z. L., Zhai, H., Sorlie, M. & Eijsink, V. G. H. (2010). An oxidative enzyme boosting the enzymatic conversion of recalcitrant polysaccharides. *Science*, *330*(6001), 219–222. doi:10.1126/science.1192231.

Waldby, C. (2008). Oocyte markets: Women's reproductive work in embryonic stem cell research. *New Genetics and Society*, *27*(1), 19–31.

Waldby, C. (2009). Biobanking in Singapore: Post-developmental state, experimental population. *New Genetics and Society*, *28*(3), 253–265.

Waldby, C., & Cooper, M. (2010). From reproductive work to regenerative labour: The female body and stem cell industries. *Feminist Theory*, *11*(1), 3–22.

Wield, D. (2013). Bioeconomy and the global economy: Industrial policies and bioinnovation. *Technology Analysis & Strategic Management*, *25*(10), 1209–1221. doi:10.1080/09537325.2013.843664.

Zhang, Y. H. P., Himmel, M. E. & Mielenz, J. R. (2006). Outlook for cellulase improvement: Screening and selection strategies. *Biotechnology Advances*, *24*(5), 452–481. doi:10.1016/j.biotechadv.2006.03.003.

Zilberman, D., Kim, E., Kirschner, S., Kaplan, S. & Reeves, J. (2013). Technology and the future bioeconomy. *Agricultural Economics*, *44*(s1), 95–102. doi:10.1111/agec.12054.

3 Theoretical perspectives on innovation for waste valorisation in the bioeconomy

Markus M. Bugge, Simon Bolwig, Teis Hansen and Anne Nygaard Tanner

3.1 Introduction

This book is anchored in a systemic and evolutionary understanding of how society evolves through technological development and innovation that are socially embedded and conditioned by actors, networks and institutions. This chapter outlines the conceptual framework for the empirical case studies in the book, which will present the innovative dynamics of turning waste into value in breweries, forestry-based industry, meat production, dairy production and urban waste management. The chapter starts by introducing *the notion of the circular bioeconomy* and interpreting it as an ongoing and broader transition in society. Here we account for the generic and pervasive nature of the bioeconomy as well as the benefits and overall objectives associated with the current transition towards a more circular economy. Following this introduction, *waste is presented as a potentially valuable resource within the bioeconomy*.

We define the term *waste* as "unwanted or unusable material, substances, or by-products" that are "eliminated or discarded as no longer useful or required after the completion of a process" (*Oxford Dictionaries*, 2018). In economic terms, waste is "unwanted material left over from a production process, or output which has no marketable value" (*Business Dictionary*, 2018), implying that the nature of the market (including firms, value chains, infrastructures, consumers, etc.) and not just a material's physical properties determine whether a material is considered waste. Therefore, the elimination of waste in the ideal circular bioeconomy involves changes in both the properties of materials and markets. A distinction can also be made between residues (with no use or market value) and side-streams or by-products (with a value). It follows that *waste valorisation* means adding value to residues, side-streams and by-products through changes in markets and/or in the physical properties of these substances, involving both technological and institutional innovation. Finally, *valorisation pathways* are the trajectories through which such values are created and distributed by and among actors from the private sector, policy, research, civil society and households. At a large spatial and temporal scale, such valorisation pathways may constitute so-called *transition pathways*, defined as:

> patterns of changes in socio-technical systems unfolding over time that lead to new ways of achieving specific societal functions. Transitions pathways involve varying degrees of reconfiguration across technologies, supporting infrastructures, business models and production systems, as well as the preferences and behaviour of consumers.
>
> (Turnheim et al., 2015)

In this chapter, we initially describe the concept of a circular bioeconomy (section 3.2), before focusing on the role of waste in the bioeconomy (section 3.3). Here, we introduce key concepts such as the waste pyramid and the cascading use principle. In section 3.4 we focus on barriers to waste valorisation and specify lock-in mechanisms that may hinder the transition towards a circular bioeconomy. Conversely, section 3.5 describes drivers for innovation in waste valorisation. Here, we introduce three generations of innovation policies, which reflect different perspectives on the nature and dynamics of innovation: science-driven innovation, systems of innovation and socio-technical transitions. These perspectives on innovation are then applied to the waste pyramid in order to distinguish between improving an existing system and replacing it with another system higher in the pyramid. In section 3.6 we discuss the roles of policy and governance in order to understand how to avoid or overcome the barriers and challenges associated with the shift towards the bioeconomy. Finally, section 3.7 summarises the chapter.

3.2 The circular bioeconomy

Parallel to the emergence of information technologies in the 1970s and 1980s and their subsequent application into the information society, the development of biotechnologies over the last few decades has been an important driver in the transformation of the economy and society towards the bioeconomy. This trend is influenced by, and indeed an important part of, the growing societal emphasis on sustainable development. The emergence and development of the bioeconomy create a potential for a return to (more) circular modes of production and consumption. This implies taking an intersectoral perspective on different industrial activities, where the rest materials from one industry process are utilised as an input in another. The notion of industrial symbiosis encompasses such cross-industry integration, and is based on the co-location and coordination of different industrial activities, which facilitates the exploitation of side-streams and residues.

The bioeconomy is a very broad concept, which encompasses multiple actors and resources and spans several sectors from health and the chemical industry to agriculture, fishery and aquaculture, dairy, slaughterhouses, breweries, forestry and energy. This breadth reflects the generic nature of the notion of the bioeconomy, and some of its potential transformational power. The transition to the bioeconomy is often argued to play a key role in targeting grand challenges such as climate change, food security and renewable energies. The bioeconomy might signal a shift (or return) to a circular economy and a society that replaces fossil fuels with renewable energy sources. Yet the broad coverage of the term also means that there are diverging perspectives on the bioeconomy. While some argue that we need to use more pesticides and precision fertilisation in agriculture, others prescribe so-called "no till" and biodiversity to avoid diseases and soil degradation. Rather than a lack of knowledge of how to run the bioeconomy, there is a large variety of contrasting recipes for how to arrive at more circular and sustainable modes of production and consumption. As a consequence, one challenge is to also make sense of these different views and perspectives. In Chapter 2, we outlined three visions of the bioeconomy, which represent one way to handle this breadth in perspectives (Bugge, Hansen & Klitkou, 2016).

3.3 The roles of waste in the bioeconomy

A major strategy in the transition towards the bioeconomy is an improved exploitation of organic residues – previously referred to as waste – and side-streams from industrial production and household consumption. This implies creating a circular economy in which the outputs from one value chain are used as inputs in another. Hence, what has been formerly regarded as waste in one sector is now turned into a resource for another sector, representing a smarter and more sustainable way of organising and exploiting limited energy and resources.

Figure 3.1 below presents the waste pyramid, which hierarchically ranks different waste treatment options according to their level of sustainability; waste disposal and energy recovery are the least favoured options, while recycling, reuse and prevention are the more favoured and sustainable options. The latter preferred types are usually more resource- and energy-efficient, although there can be trade-offs between resource and energy savings, and they often, though not always, involve lower greenhouse gas emissions. It is, however, important to carefully assess the multiple life cycle impacts for specific options rather than assuming higher or lower general sustainability based on the pyramid's categories (see Chapter 14). Moreover, a specific option may encompass several categories, for example the treatment of waste in a biogas plant involves both recovery (of energy) and recycling (use of the digestate as fertiliser).

The waste pyramid illustrates how side-streams and residues may be processed and utilised in different ways (European Commission, 2008). Historically, waste disposal in landfills has gradually been replaced by innovative and potentially more sustainable forms of waste management, focusing first on energy recovery, and then on recycling, reuse, minimisation and, ultimately, waste prevention. In this book, we conceptualise each of these forms of management as *integrated socio-technical systems of production and consumption* consisting of key elements, i.e. actors, capabilities, networks, institutions and infrastructures. The composition and characteristics of these elements condition the system's innovative abilities.

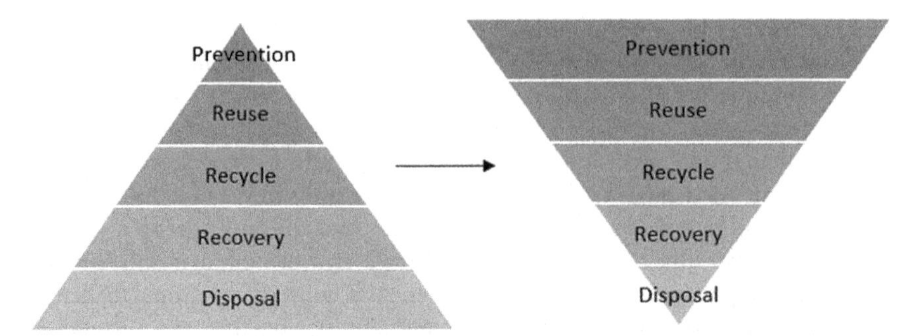

Figure 3.1 The waste pyramid. Innovation in waste systems implies turning the pyramid on its head so that less or no waste is disposed of (hence it is no longer waste), illustrated by a smaller area in the right-hand pyramid, and more waste (or resources) is prevented or reused, illustrated by a larger area in the right-hand pyramid.

In regard to the point made earlier that the bioeconomy consists of diverging perspectives, it is here interesting to note that waste plays various roles and is assigned different values at each level of the waste pyramid depending on the industry and country in focus.

The different socio-technical waste systems represented in the pyramid can co-exist in a given country or region; below we outline each system in turn. In a landfill system, waste has no value but is rather a cost in terms of transport and storage. In a recovery system, waste is an energy resource that can be exploited through incineration, e.g. in district heating or combined heat and power plants. In a recycling system, waste is a potential input to the production of various products such as biofertiliser, animal fodder, nutrition products or recycled materials such as paper, plastics, glass, metals and textiles. In systems of reuse and prevention, waste is avoided altogether, and the food or other biomass retains much of its initial value. One example is a restaurant owned by the student association at the campus of the University of Oslo serving cheap gourmet food which is close to its expiration date (see Chapter 13).

In a circular bioeconomy, the waste pyramid is further substantiated by a *cascading use principle*. Cascading use has been defined as "the efficient utilisation of resources by using residues and recycled materials for material use to extend total biomass availability within a given system" (European Commission, 2016). In general, cascading utilisation refers to a principle of multiple uses of biomass resources by using residues, recycling resources or recovering resources after consumption. Focus can either be on extending the timespan during which resources stay in the system (cascading-in-time) or on maximising the added value of resources (cascading-in-value) (Olsson et al., 2016). Cascading-in-time (Figure 3.2a) builds on the idea "that resources should be re-used sequentially in the order of the specific resource quality of each stage" (ibid., p. 7). Wood is often used as an example to illustrate the cascading-in-time principle (European Commission, 2016; Vis, Reumerman & Gärtner,

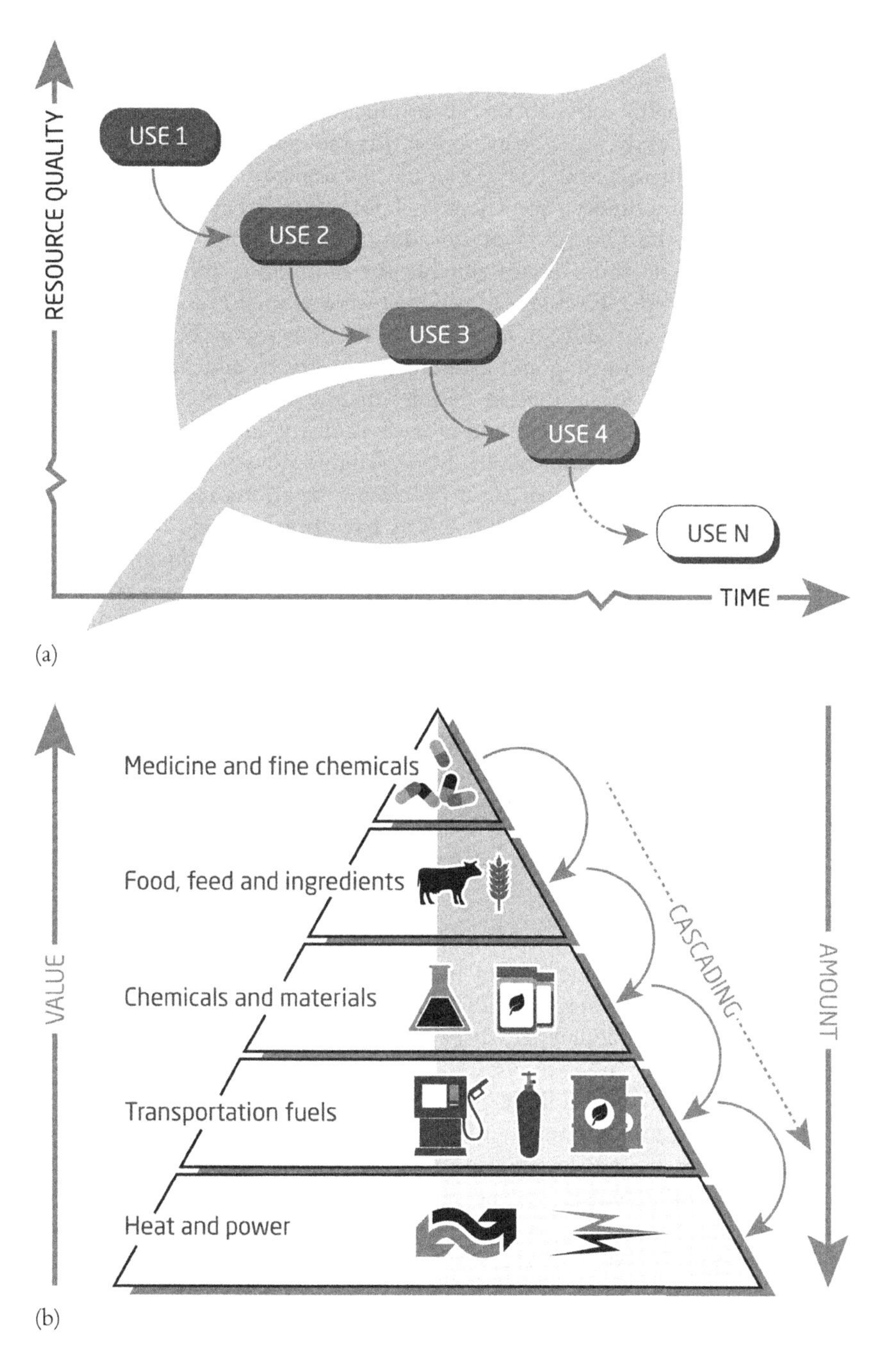

Figure 3.2 (a) (top): The cascade-in-time chain (adapted from Sirkin & Houten, 1994); (b) (bottom): The cascade-in-value concept (adapted from Olsson et al., 2016).

2014), where freshly harvested wood should first be used to veneer wood products, then particle-based products, then fibre-based products, then, finally, biofuels or incineration for heat or power. The cascading-in-value principle (Figure 3.2b) refers to the co-production of several different bio-products simultaneously, as is characteristic for bio-refineries. Hence, wood can also be used in this principle as cascade fractionation of valuable wood composites in a biorefinery. See Chapter 4 for more details. There can be conflicts between the two forms of cascading. For example, wood that has been reused multiple times is not suitable for biorefining processes, which normally rely on whole, fresh logs to produce various chemicals.

An example of a cascading-in-value use of biomass resources is the efforts made by the meat processing and rendering industry to add more value to animal by-products. During the last decade, incumbent meat processors and rendering companies have established new divisions or subsidiaries dedicated to preventing or utilising side-streams from slaughterhouses, and in so doing have increased the value of their meat processing by selling new products to new markets. At the slaughterhouses, efforts have been focused on reducing the volume of side-streams by utilising a higher proportion of the animal, for example by the export of pig ears, snouts, hooves or gallstones to Asian markets. Other companies have focused on developing new products, such as functional ingredients for the food industry, pet food, animal feed for mink production or, finally, biodiesel. All these initiatives aim to increase the overall value of meat processing and reduce the amount of animal by-products to be used in the lower part of the waste pyramid. See Chapter 7 for more details on this case.

3.4 Path dependence and barriers to waste valorisation

3.4.1 Path dependence

Path dependence is the tendency of institutions or technologies to become committed to develop in certain ways because of their structural properties or their beliefs and values (Greener, 2017). Path dependence is important for understanding changes in waste systems as it suggests the existence of mechanisms that (under certain conditions) can cause some technologies, behaviours or policies to persist or dominate even if "superior" alternatives exist. While such lock-in mechanisms cause considerable inertia in waste systems, a change in path-dependent systems is still possible through feedback mechanisms, as highlighted in, for example, studies of strategic niche management (Kemp, Schot & Hoogma, 1998).

Moving from established value chains in specific industries and sectors to new ones that are cross-sectoral and interwoven through the waste streams involved may require fundamental changes. These changes may take the form of new technologies, altered production modes, infrastructures, logistics and new consumer practices and habits. It can be challenging to change a system that has existed for a long period of time, as routinised practices tend to become institutionalised both socially and materially over time. Indeed, for innovation collaborations, there is little empirical evidence that the bioeconomy entails intensified cross-industry collaboration (Bauer, Hansen & Hellsmark, 2018). This form of path dependence and lock-in (see below) within given socio-technical systems may well prevent or slow down innovation and change (David, 1985; Martin & Sunley, 2006). The policies and institutions that have emerged to serve and support practices of a given production regime also hold implications for innovation, as existing institutions often reflect the interests and perspectives of the actors that make up these systems. Consequently, changing established practices and waste systems may be challenging due to the existing incentive structures and institutional set-up.

Contributions within evolutionary economic geography have supplemented such a systemic view on path dependence with a historical and evolutionary approach to studying patterns of regional path dependence. The argument here is that the portfolios and competencies of existing industrial actors will often condition the future paths and scenarios for a given sector or region (Boschma, 2015; Boschma & Frenken, 2006, 2011a, 2011b). To summarise, path dependence may be caused by several factors, both tangible and intangible. In this regard, investments in heavy physical infrastructures and material equipment may be a barrier to change. Repeated social practice and habits may likewise cause segmented cultures and values, which may also serve to prevent or slow down change and innovation. In practice, different combinations of tangible and intangible factors are likely to restrict or limit change and innovation.

In addition to possessing different roles and being subject to valorisation relative to various waste systems, the practices within a given system might constitute barriers towards moving upwards in the waste pyramid. For example, improving and optimising a recycling system might become dependent on generating waste as an input to this system, which would therefore create no incentives for aiming at a waste prevention system. In this sense the established actors, practices, institutions and infrastructures of existing waste valorisation systems might be barriers to a more sustainable system change.

3.4.2 Lock-in mechanisms

Klitkou, Bolwig, Hansen and Wessberg (2015) have developed an analytical framework for systematically studying the role of lock-in mechanisms in transition processes. They understand lock-in mechanisms as "mechanisms, which reinforce a certain pathway of economic, technological, industrial and institutional development and can lead to path dependence" (ibid., p. 23). Klitkou et al. (2015) observed that there could be interactions between lock-in mechanisms, such as between learning effects, network externalities and technological interrelatedness, which are reinforcing each other, while other interactions could have weakening effects. These mechanisms also have different functions in the different stages of path development. Only in the last stages will the process get locked in and become path-dependent (Sydow, Schreyögg & Koch, 2009). Moreover, non-predictability and coexistence of several outcomes normally characterise the start of the process, while inflexibility and inefficiency are typical for the later stages. Table 3.1 summarises the mechanisms discussed by Klitkou et al. (2015), including key literary sources.

The influences of learning effects, and economies of scale and scope, are evident in the forestry industry. Pulp and paper firms continue to prioritise incremental improvements relating to existing processes and products, which

Table 3.1 Lock-in mechanisms potentially affecting waste valorisation

Lock-in mechanism	*Description*	*Key sources*
Learning effects	Specialisation leads to increasing returns from learning in relation to existing products and production processes.	Arthur (1990); Cimoli (1994)
Economies of scale	Earlier investment in production equipment leads to increasing returns from additional built-up and further investments in this production system.	Hughes (1983, 1987)
Economies of scope	Existing product specialisations may guide diversification into new product groups due to potential cost efficiencies.	Panzar and Willig (1981)
Technological interrelatedness	Existing technologies lead to favourable conditions for development of technologies with complementarities.	Van den Bergh and Oosterhuis (2008); Boschma and Frenken (2011b)
Network externalities	Uptake of existing technologies leads to de facto standard setting due to institutionalised use patterns.	Katz and Shapiro (1986); David (1985)
Informational increasing returns	Uptake of existing technologies leads to increasing knowledge and attention about them, further stimulating their diffusion.	Van den Bergh and Oosterhuis (2008)
Collective action	Uptake of existing technologies leads to development of societal norms, customs and formal regulations, which further stimulates their diffusion.	Foxon (2002)
Institutional learning effects	Existing formal institutions limit the possibilities for establishing new policies, which are not aligned with them.	Foxon (2002)
Differentiation of power and institutions	Incumbent organisations may exercise power to prevent institutional change to their disadvantage; change of institutions is hampered by institutional complementarities.	Foxon (2002); Ostrom, Schroeder and Wynne (1993)

have long been their central profit-generating activities. The firms' core competencies closely relate to these activities, making it hard to change technologies (Laestadius, 2000). Moreover, the capital intensity of the industry implies that firms have made large investments in existing equipment. Consequently, few commercial-scale investments target the conversion of side-streams into new high-value products, which require obtaining knowledge about new markets and techniques. Furthermore, adding new technologies to an existing production system in a mill is highly complicated due to the economic importance of avoiding pauses in the production process (Bauer, Coenen, Hansen, McCormick & Palgan, 2017; Hansen & Coenen, 2017). However, in some cases bottlenecks in the production process may be overcome by extracting components (e.g. lignin). Subsequently, these substances may form the basis of new product lines. This underlines the importance of considering economies of scope in moving up the waste pyramid (Gregg et al., 2017). For lignin, this allows moving into a variety of new products from binders to fuels and speciality chemicals, rather than simply recovering the energy for use in the production process. See Chapter 4 for more details on this case.

In urban waste systems, path dependence is created by large investments in technological and physical infrastructure. In the municipality of Oslo, investments in an optical sorting plant and a biogas plant constitute an advanced system for managing organic household waste. Organic waste is sorted by the households in plastic bags with different colours, collected at the kerbside and sorted optically at the sorting plant. The waste is then treated mechanically and chemically and used for producing biofertiliser and biogas. The biofertiliser is sold to regional farms and the biogas is used for public bus transport. It exemplifies a circular system for waste recycling. Yet it is also a system that depends on constant flows of organic waste (Uyarra & Gee, 2013), and which may create disincentives for reducing or preventing waste generation in the first place (Bulkeley & Gregson, 2009; Mourad, 2016). Therefore, investments in one system of waste treatment create path dependence where economies of scale (e.g. investments in infrastructure) and scope (e.g. optical sorting of multiple waste fractions) are mechanisms that prevent leaps up the waste pyramid. See Chapter 5 for more details on this case.

The dairy sector provides a third example of how lock-in mechanisms can influence and reinforce innovation and value chain development in the bioeconomy. The Danish dairy cooperative Arla Foods is one of the largest dairy companies in the world. Because of a series of mergers and acquisitions, as well as specialisation in whey, over the past few decades, Arla Foods benefits from economies of scale, economies of scope and learning effects. The subsidiary Arla Foods Ingredients was created to find solutions to whey processing and utilisation at a time when new regulations restricted the disposal of whey as waste. This move not only created a long-term learning effect through a niche specialisation in whey handling and processing, but also expanded the product range of the company. Today, Arla supplies protein-based food ingredients within six product categories: paediatric nutrition, sports nutrition, medical nutrition, health foods, bakery and dairy. Through investments in state-of-the-art production equipment as well as in research and development in the context of globalisation, Arla has achieved significant market power, further increasing its ability to exploit the advantages of economies of scope and scale. See Chapter 9 for more details on this case.

3.5 Drivers of innovation in waste valorisation

To better understand the drivers and challenges associated with enabling a shift towards a circular bioeconomy, we here discuss the literature on innovation and innovation policies. Schot and Steinmueller (2018) distinguish between three generations of innovation policies in terms of different perspectives on what constitutes the main drivers of innovation and with respective implications for innovation policies. In the 1960s, innovation was primarily seen to emerge from research and scientific discovery. This view of innovation was very much oriented around technological development and scientific discovery and a belief in the commercialisation of new technologies and scientific breakthroughs. However, from the 1990s this view was supplemented by a more pronounced systemic understanding of how innovation occurs through impulses from user needs in the market and through the interplay and collaboration between various types of actors (Edquist, 1997; Lundvall, 1992).

Such a systemic understanding of innovation also implies supplementing the supply-oriented focus on the role of science with demand as also determining and conditioning innovation. The scholarly tradition on systems of innovation has shown how innovation should not be understood as isolated phenomena, but rather as being the output from collaboration and interactive learning across diverse types of actors that possess various and complementary capabilities. Moreover, it has illustrated how the spatial embeddedness and context for the industry actors, such as networks, institutions, infrastructures and policy frameworks, may also strongly affect innovation performance in firms, sectors and regions. This has been elaborated in the economic geography literature on path development, which specifies how regional characteristics condition future development opportunities, outlines the various stages in path development processes and considers the role of agency (Martin, 2010; Simmie, 2012). Each sector within the bioeconomy has traditionally consisted of established value chains and industrial processes where different inputs and resources – labour, investments, biological resources, technologies, infrastructures, policies and management – determine how value is created in the respective sectors. These value chains can thus be viewed as conditioned by their surrounding systems of innovation.

From the 2000s, the systemic understanding of innovation was complemented by research on socio-technical transitions, which has served to reconsider and broaden conventional innovation theories and policies by focusing on how entire systems may need to change more fundamentally. Particularly important for this focus on transformative change has been the multi-level perspective, which sees systemic transitions as co-evolutionary processes that unfold through an interplay between three interrelated analytical levels; regimes, niches and landscapes (Geels, 2002, 2004, 2005; Geels & Schot, 2007; Schot & Geels, 2008). A *regime* refers to an existing dominant system of production and consumption, *niches* are the locus for disruptive innovation and *landscapes* are understood as contextual factors conditioning regimes and niches. This tradition represents an important discontinuity in the main object of study from "innovations" to "transitions in socio-technical systems". Whereas the systems of innovation tradition were primarily driven by a technological and economic logic, the turn to socio-technical transitions has introduced a stronger sense of society beyond the economy and technological development.

Our discussion of these three perspectives reveals that opinions differ regarding what actors are important for driving innovation forward (Schot & Steinmueller, 2018). In a *traditional understanding of innovation processes*, the focus is on universities and research institutes as well as private firms, which are seen as central to making scientific discoveries and their commercialisation through the introduction of technical innovations in the market.

In the context of the waste pyramid, this implies a focus on improving technologies in the lower part of the pyramid, i.e. developing new and improved recovery and recycling processes. Examples are technologies that improve the efficiency in the recovery process, leading to a higher production of electricity and heat, or developments in recycling technologies that improve the quality of the sorting or reduce the need for other inputs such as electricity and labour.

The *innovation system perspective* broadens out the types of actors seen as important in innovation processes. Firms, universities and other knowledge institutions are still considered to play a key role, but inputs from users and customers are also seen to provide important inputs. Furthermore, public sector actors are attributed a central position, not least as the systemic emphasis underlines the importance of intermediaries (Kivimaa, Boon, Hyysalo & Klerk, 2018). The latter are organisations with a focus on connecting and brokering between other actors in the system, e.g. technology transfer offices and cluster and network organisations. Many intermediaries are public or quasi-public bodies, but are increasingly also established by private interest organisations or as independent private enterprises.

The systems perspective on innovation in waste prevention and handling also implies a focus on technological innovations in the bottom part of the pyramid. Yet, unlike the science-driven model of innovation, it will assign a stronger prominence to the interactions between various actors in different parts of the value chain, e.g. between actors in recycling and energy recovery, or between goods producers and recycling firms. Therefore, innovation is not perceived as the result of activities taking place within specific firms and organisations but rather as caused by their collaboration and interactions.

The *transformative change perspective* further broadens the actors considered important to innovation – and transition – processes. This includes groups traditionally considered outsiders to innovation processes such as civil society groups and interest organisations (Coenen, Hansen & Rekers, 2015; Geels & Raven, 2006). The emphasis on entire socio-technical systems implies that the development and production of technical artefacts are seen as closely connected to their use and associated social practices. Thus, users and consumers are not only regarded as input providers to innovation processes, but as important agents that may preserve or challenge regimes.

In a transformative change perspective on waste, innovation efforts are beneficial when they contribute to transitioning the waste system towards an increasing emphasis on the upper parts of the waste pyramid, i.e. recycling and especially preparing for reuse and prevention. Hence, this perspective gives less attention to incremental improvements of process technologies in the lower parts of the pyramid. Indeed, improvements in energy recovery technologies may be viewed as counterproductive to transformative change since they disincentivise efforts and investments in developing the higher parts of the waste pyramid.

3.6 Governance for waste valorisation

Reflecting the three generations of innovation perspectives introduced in the previous section, the role and scope of policy and governance of innovation have steadily developed and expanded over the last 50 years (Schot & Steinmueller, 2018). This trend represents a move from an initial emphasis on the role of new technologies themselves to the role of a range of other social, geographical, institutional and organisational factors affecting innovation.

An important distinction between the three generations of innovation policy discussed above is that the first two have a generic focus on innovation and growth, whereas the overall objectives and primary goals of the third generation are solutions to specific societal challenges (Schot & Steinmueller, 2018). In a linear model of innovation (first-generation innovation policy), one would invest in R&D to develop technologies that could help exploit waste in new ways. An innovation systems perspective (second-generation innovation policy) would develop innovative and cost-effective systems of waste collection and treatment across public, private and civic sectors, enabling a cost-effective exploitation of all possible forms of rest-products from consumption. A transitions perspective (third-generation innovation policy) would, however, put the social values of sustainability upfront and let these guide the search for more sustainable consumption in the first place (e.g. eco-designs such as reducing portions or organic packaging). Instead of aiming for optimising and greening existing value chains, a transitions mode would typically question the existence of the value chain altogether. In this sense, a transitions mode of innovation takes a broader perspective on the entire value chain.

Exemplifying this trend, and showing how diverse actor groups are crucial to innovation, Fagerberg (2017) has examined the main drivers behind Danish wind power, the German *Energiewende* and Norwegian electromobility. It is concluded that the social drivers have been more important than the technologies themselves, which had often been around for decades. Instead, the forces that seemed the most powerful in determining the pace and scope of these socio-technical transitions were those associated with the practices and interests of (local) user groups.

Thus, although there has been a continuous expansion in terms of the roles of policy and high-level governance in arranging for systemic innovation and system change, this does not eliminate the need for governance at the micro-level. Here, governance can be in the form of developing and renewing the competencies, routines, value chains and business models of individual organisations and within the boundaries of specific sectors. This illustrates how governance of waste may be diverse and manifold, depending on the case and context. In the subsequent chapters, we present case studies on innovation in waste valorisation in various industry sectors such as forestry, aquaculture, breweries, dairies and slaughterhouses. We also present a case study on urban waste systems, which supplements the production focus in the industry cases with a focus on the public sector and the consumer side of waste. So, depending on the case and context in question, various forms and levels of governance for waste valorisation are actualised.

Below we discuss how directionality towards specific societal goals and missions is often a result of multiple initiatives and practices that co-evolve at different levels and across various types of societal sectors and actors.

3.6.1 Directionality through international regulations

Amidst the widespread agreement on the need to include diverse types of actors in the governance of innovation, the literature on socio-technical transitions further argues that a strong element of priorities and directionality is required to accomplish certain missions or to arrive at more profound system changes – so-called socio-technical transitions (Mazzucato, 2017; Schot & Steinmueller, 2018; Shove & Walker, 2007; Smith & Raven, 2012; Smith, Stirling & Berkhout, 2005; Weber & Rohracher, 2012). Regarding organic waste, the UN sustainable development goals (United Nations, 2015), the Paris Agreement and the EU landfill ban in 2009 are central landscape elements that frame and guide international development in this area (see Chapter 5 on urban waste management and Chapter 13 on multi-level governance of food waste).

3.6.2 Directionality through national regulations

Directionality may also be set at national levels. In Norway, the recent *Industry agreement on the reduction of food waste* (Regjeringen, 2017) represents a similar sense of directionality, which is likely to guide innovations and behaviours in the years to come. The agreement aims to reduce food waste by 50% by 2030, and illustrates the importance of not underestimating the role of the private and civic sectors in the innovative dynamics towards more sustainable waste systems.

Indicators and performance measurement systems also often operate at national levels. In order to facilitate transition, Huguenin and Jeannerat (2017) suggest replacing "innovation" with "valuation" to ensure that the solutions address the most pressing societal questions. They propose to focus on the purpose behind developments in the economy and society, e.g. reducing greenhouse gas emissions, rather than on the factors contributing to these developments, e.g. strengthening R&D budgets or university-industry collaboration. Such an approach, they argue, would give innovative work a clear societal direction and "mission" and thus better facilitate addressing important issues in the first place. Similarly, Papargyropoulou et al. (2014) have applied the waste pyramid to food waste and call for a holistic approach to food waste that takes into account all production and consumption activities in global food value chains, i.e. agriculture, food processing and manufacturing, retail and consumption. Such a broad lens would also favour adding social and cultural aspects to the usual suspects of technology and economy. Preventing food waste could involve changes in technology and infrastructure in harvesting, storage, transport and distribution, and also in consumer-related issues such as the promotion of eco-designs and eco-labels, re-sizing of products and portions, and taxation of non-sustainable packaging (European Commission, 2008). Having food waste prevention as the overall concern represents an important objective and capability. Supporting this broader perspective on organic waste valorisation might be a way to avoid or overcome the risk of causing lock-ins or path dependencies when introducing solutions, practices or systems of production and consumption in the lower parts of the waste pyramid. See Chapter 13 on multi-level governance and food waste.

3.6.3 Directionality through industrial practices

Nonetheless, most learning, innovation and development work is anchored and embedded in existing and localised organisations, incentive structures and value chains. When individual organisations apply the transition agenda in their daily operations there may well be unresolved issues in terms of how to interpret a given challenge or task as falling under either "business as usual" or as "here there is reason to rethink the way we do things". We know that the two options imply and involve fundamentally different actors, approaches and resources. Therefore, this points back to the importance of innovation policies and governance in terms of giving direction, articulating demands, mobilising relevant stakeholders and arranging for joint reflexivity and learning (Weber & Rohracher, 2012).

The literature on private forms of governance within firms and along value chains offers insights into the dynamics of firms and industries that can augment second- and third-generation perspectives on innovation.

Along value chains, global value chain (GVC) scholars highlight how firms acquire capabilities and access new market segments ("upgrade") through participation in specific value chains, where learning from downstream firms is seen as a central upgrading mechanism (Bolwig, Ponte, du Toit, Riisgaard & Halberg, 2010; Gereffi & Lee, 2016). Value chain governance is the process by which so-called "lead firms" organise activities with the purpose of achieving a certain functional division of labour within a chain. It involves setting the terms of chain membership, such as prices or the compliance with technical, environmental and legal standards. It also includes the way in which such market requirements are implemented along the chain, and how they affect chain participation for firms, the re-allocation of value-adding activities and the distribution of costs and benefits (Gibbon, Bair & Ponte, 2008). In the context of waste valorisation, the GVC perspective and the governance mechanisms just mentioned suggest that the capabilities and incentives of innovation are strongly influenced by the nature of the inter-firm linkages and power relationships in specific markets. Yet, similarly to the trend within innovation studies, recent GVC literature highlights that a broader range of actors, such as governments, standard-setters and NGOs, can yield significant influence on value chain governance, especially in emerging industries such as renewable energy (Nygaard & Bolwig, 2018; Ponte & Sturgeon, 2013).

At the firm level, scholars have long studied the links between private sustainability measures or corporate social responsibility (CSR) on the one hand and the competitive advantage to companies on the other. See Chapter 6 on brewing. In the brewing industry, CSR efforts include the sustainable use of organic residues, reduced water consumption, waste water management, more efficient energy use and diminished CO_2 emissions, sustainable packaging and responsible drinking. To pursue a competitive advantage, companies must choose between product differentiation and low costs in terms of cost leadership (Porter, 1985). CSR serves as a means of product differentiation by functioning as a co-specialised asset that makes other assets more valuable (McWilliams & Siegel, 2011). Most evident here is the effect of CSR on reputation or branding (Roberts & Dowling, 2002). Branding and reputation are hard-to-get resources that cannot be imitated and thus serve as entry barriers to competitors (Reinhardt, 1998). Hence CSR can serve as a means for obtaining a sustainable competitive advantage (McWilliams & Siegel, 2011). In this context, a review of 200 studies by Clark et al. (2015) found a positive association between companies' sustainability measures and their economic performance in terms of the cost of capital, operational performance and stock price, although the direction of causality is ambiguous. Despite such benefits, Whelan and Fink (2016) observe that sustainability and broader CSR measures are only rarely placed at the core of a business's strategies.

3.7 Summary

In this chapter we have discussed the notion of the circular bioeconomy, the drivers and barriers for adding value to waste and thereby creating a more sustainable bioeconomy, and the special role of governance including innovation policy in developing the bioeconomy. There are many views on the bioeconomy, which is an emerging area for research, policy and economic activity. Our focus has been on the role of innovation in waste valorisation, not only technological but also social and institutional innovation.

We have discussed the *waste pyramid* that illustrates the hierarchy of alternative forms of waste management in terms of resource efficiency and sustainability, and the associated notion of cascading use of biological resources. We conceptualised the alternative forms of waste management as *integrated socio-technical systems of production and consumption.* This concept provided a gateway into the studies on socio-technical transitions, innovation and governance that we claim are central for analysing the patterns and dynamics of waste valorisation.

The *drivers of innovation in waste valorisation* were approached through a discussion of three generational perspectives on innovation and innovation policy. Today, many scholars have come to understand innovation in the bioeconomy as *transformative change*. This perspective focuses on the upper parts of the waste pyramid (recycle, reuse and prevention), emphasises entire socio-technical systems and not only considers companies, researchers and policy makers as agents, but also intermediary organisations, users and consumers. Hence, analysis of the dynamics of bio-economic value chains should consider not only the firm actors handling the products and technologies, but also the broader institutional, economic and social context of production and trade.

The barriers to innovation in the bioeconomy were discussed through the concepts of *path dependence and lock-in mechanisms*. Economic, institutional and social mechanisms may cause inertia in waste systems and constrain an upward movement in the waste pyramid, and there are important regional and sectoral dimensions of path dependence arising from the characteristics of specific industries. The Swedish pulp and paper industry is an example thereof. However, under the right conditions, the same mechanisms, e.g. learning effects and economies of scale, may bring waste systems onto a more sustainable path, as illustrated by the case of whey valorisation in Arla Foods.

Governance is essential to understanding and enhancing waste valorisation. The concept not only comprises public policy, but also private governance by firms within value chains, as well as the activities of civil society organisations, industry organisations and consumer groups. Another key insight is that the governance of production, consumption and innovation is not an abstract process but is instead deeply rooted in existing value chains, organisations and localities. Firm efforts to gain sustainable competitive advantage through product differentiation (including CSR) or cost reductions can spur innovations in waste valorisation, while the incentives and capabilities to undertake innovation not only originate within the firm but also at the level of the value chain or "value network" comprising various non-commercial actors.

Finally, recent contributions to the transitions literature have emphasised the need for a much stronger "directionality" in the governance of sustainable innovation as well as a focus on social values as the key driver of sustainable development. For example, a holistic approach to food waste should first consider social values related to food production and consumption, including ecology, health and food waste prevention, as well as processes and impacts along entire global value chains. Such a broader perspective on waste systems, we argue, would help overcome transition issues related to path dependence and lock-in, thereby paving the way for the circular bioeconomy.

Acknowledgements

This work was supported by the Research Council of Norway (grant number 244249). Tobias Pape Thomsen created the artwork.

References

Bauer, F., Hansen, T. & Hellsmark, H. (2018). Innovation in the bioeconomy: Dynamics of biorefinery innovation networks. *Technology Analysis & Strategic Management, 30*, 935–947. doi:10.1080/09537325.2018.1425386.

Bauer, F., Coenen, L., Hansen, T., McCormick, K. & Palgan, Y. V. (2017). Technological innovation systems for biorefineries: A review of the literature. *Biofuels, Bioproducts and Biorefining, 11*(3), 534–548. doi:10.1002/bbb.1767.

Bolwig, S., Ponte, S., du Toit, A., Riisgaard, L. & Halberg, N. (2010). Integrating poverty and environmental concerns into value chain analysis: A conceptual framework. *Development Policy Review, 28*(28), 173–194.

Boschma, R. A. (2015). Towards an evolutionary perspective on regional resilience. *Regional Studies, 49*(5), 733–751. doi:10.1080/00343404.2014.959481.

Boschma, R. A., & Frenken, K. (2006). Why is economic geography not an evolutionary science? Towards an evolutionary economic geography. *Journal of Economic Geography, 6*(3), 273.

Boschma, R. A., & Frenken, K. (2011a). The emerging empirics of evolutionary economic geography. *Journal of Economic Geography, 11*(2), 295–307.

Boschma, R. A., & Frenken, K. (2011b). Technological relatedness and regional branching. In H. Bathelt, M. Feldman & D. F. Kogler (Eds.), *Beyond Territory: Dynamic Geographies of Knowledge Creation, Diffusion, and Innovation* (pp. 64–81). London and New York: Routledge.

Bugge, M., Hansen, T. & Klitkou, A. (2016). What is the bioeconomy? A review of the literature. *Sustainability, 8*(691), 1–22.

Bulkeley, H., & Gregson, N. (2009). Crossing the threshold: Municipal waste policy and household waste generation. *Environment and Planning A, 41*, 929–945.

Business Dictionary. (2018). Waste. www.businessdictionary.com.

Clark, G. L., Feiner, A. & Viehs, M. (2015). *From the Stockholder to the Stakeholder: How Sustainability Can Drive Financial Outperformance*. Retrieved from https://arabesque.com/research/From_the_stockholder_to_the_stakeholder_web.pdf.

Coenen, L., Hansen, T. & Rekers, J. V. (2015). Innovation policy for grand challenges: An economic geography perspective. *Geography Compass*, *9*(9), 483–496. doi:10.1111/gec3.12231.

David, P. A. (1985). Clio and the economics of QWERTY. *American Economic Review*, *75*(2), 332–337.

Edquist, C. (1997). *Systems of Innovation: Technologies, Institutions and Organizations*. London: Pinter.

European Commission. (2008). *Directive 2008/98/EC of the European Parliament and of the Council of 19 November 2008 on Waste and Repealing Certain Directives*. Official Journal of the European Union.

European Commission. (2016). *Study on the Optimised Cascading Use of Wood*. Brussels: European Commission.

Fagerberg, J. (2017). *Mission (Im)possible? The Role of Innovation (and Innovation Policy) in Supporting Structural Change and Sustainability Transitions*. TIK Working Papers on Innovation Studies. TIK Centre, University of Oslo. Retrieved from http://ideas.repec.org/s/tik/inowpp.html.

Geels, F. W. (2002). Technological transitions as evolutionary reconfiguration processes: A multi-level perspective and a case-study. *Research Policy*, *31*(8–9), 1257–1274. doi:10.1016/S0048-7333(02)00062-8.

Geels, F. W. (2004). From sectoral systems of innovation to socio-technical systems: Insights about dynamics and change from sociology and institutional theory. *Research Policy*, *33*(6–7), 897–920.

Geels, F. W. (2005). Processes and patterns in transitions and system innovations: Refining the co-evolutionary multi-level perspective. *Technological Forecasting & Social Change*, *72*(6), 681–696.

Geels, F. W., & Raven, R. (2006). Non-linearity and expectations in niche-development trajectories: Ups and downs in Dutch biogas development (1973–2003). *Technology Analysis & Strategic Management*, *18*(3–4), 375–392. doi:10.1080/09537320600777143.

Geels, F. W., & Schot, J. (2007). Typology of sociotechnical transition pathways. *Research Policy*, *36*(3), 399–417.

Gereffi, G., & Lee, J. (2016). Economic and social upgrading in global value chains and industrial clusters: Why governance matters. *Journal of Business Ethics*, *133*, 25–38.

Gibbon, P., Bair, J. & Ponte, S. (2008). Governing global value chains: An introduction. *Economy and Society*, *37*(3), 315–338.

Greener, I. (2017). Path dependence. In *Encyclopædia Britannica*. Retrieved from: www.britannica.com/topic/path-dependence.

Gregg, J., Bolwig, S., Hansen, T., Solér, O., Ben Amer-Allam, S., Pladevall Viladecans, J., … Fevolden, A. (2017). Value chain structures that define European cellulosic ethanol production. *Sustainability*, *9*(1), 118.

Hansen, T., & Coenen, L. (2017). Unpacking resource mobilisation by incumbents for biorefineries: The role of micro-level factors for technological innovation system weaknesses. *Technology Analysis & Strategic Management*, *29*(5), 500–513. doi:10.1080/09537325.2016.1249838.

Huguenin, A., & Jeannerat, H. (2017). Creating change through pilot and demonstration projects: Towards a valuation policy approach. *Research Policy*, *46*(3), 624–635.

Kivimaa, P., Boon, W., Hyysalo, S. & Klerk, L. (2018). Towards a typology of intermediaries in sustainability transitions: A systematic review and a research agenda. *Research Policy*, In Press. doi:doi.org/10.1016/j.respol.2018.10.006.

Klitkou, A., Bolwig, S., Hansen, T. & Wessberg, N. (2015). The role of lock-in mechanisms in transition processes: The case of energy for road transport. *Environmental Innovation and Societal Transitions, 16*(September), 22–37.

Laestadius, S. (2000). Biotechnology and the potential for a radical shift of technology in forest industry. *Technology Analysis & Strategic Management, 12*(2), 193–212.

Lundvall, B. A. (1992). *National Systems of Innovation: Towards a Theory of Innovation and Interactive Learning* (B.-Å. Lundvall Ed.). London: Pinter.

Martin, R. (2010). Roepke Lecture in economic geography: Rethinking regional path dependence: Beyond lock-in to evolution. *Economic Geography, 86*(1), 1–27. doi:10.1111/j.1944-8287.2009.01056.x.

Martin, R., & Sunley, P. (2006). Path dependence and regional economic evolution. *Journal of Economic Geography, 6*(4), 395–437.

Mazzucato, M. (2017). *Mission-Oriented Innovation Policy: Challenges and Opportunities* (RSA Action and Research Centre Ed.). London, UK: UCL Institute for Innovation and Public Purpose.

McWilliams, A., & Siegel, D. (2011). Creating and capturing value: Strategic corporate social responsibility, resource-based theory, and sustainable competitive advantage. *Journal of Management, 37*(5), 1480–1495.

Mourad, M. (2016). Recycling, recovering and preventing "food waste": Competing solutions for food systems sustainability in the United States and France. *Journal of Cleaner Production, 126*(10 July), 461–477.

Nygaard, I., & Bolwig, S. (2018). The rise and fall of foreign private investment in the jatropha biofuel value chain in Ghana. *Environmental Science & Policy, 84*(June), 224–234. doi:10.1016/j.envsci.2017.08.007.

Olsson, O., Bruce, L., Hektor, B., Roos, A., Guisson, R., Lamers, P. & Thrän, D. (2016). *Cascading of Woody Biomass: Definitions, Policies and Effects on International Trade.* Retrieved from www.ieabioenergy.com/publications/cascading-of-woody-biomass-definitions-policies-and-effects-on-international-trade.

Oxford Dictionaries. (2018). Waste. www.oxforddictionaries.com.

Papargyropoulou, E., Lozano, R., Steinberger, J. K., Wright, N. & Ujang, Z. b. (2014). The food waste hierarchy as a framework for the management of food surplus and food waste. *Journal of Cleaner Production, 76*(August), 106–115.

Ponte, S., & Sturgeon, T. (2013). Explaining governance in global value chains: A modular theory-building effort. *Review of International Political Economy, 21*(1), 1–29. doi:10.1080/09692290.2013.809596.

Porter, M. E. (1985). *The Competitive Advantage: Creating and Sustaining Superior Performance.* New York: Free Press.

Regjeringen. (2017). Industry agreement on reduction of food waste. Retrieved from www.regjeringen.no/contentassets/1c911e254aa0470692bc311789a8f1cd/industry-agreement-on-reduction-of-food-waste_norway.pdf.

Reinhardt, F. L. (1998). Environmental product differentiation: Implication for corporate strategy. *California Management Review, 40*(4), 43–73.

Roberts, P. W., & Dowling, G. R. (2002). Corporate reputation and sustained superior financial performance. *Strategic Management Journal, 23*(12), 1077–1093.

Schot, J., & Geels, F. W. (2008). Strategic niche management and sustainable innovation journeys: Theory, findings, research agenda, and policy. *Technology Analysis & Strategic Management, 20*(5), 537–554.

Schot, J., & Steinmueller, W. E. (2018). Three frames for innovation policy: R&D, systems of innovation and transformative change. *Research Policy, 47*(9), 1554–1567.

Shove, E., & Walker, G. (2007). CAUTION! Transitions ahead: Politics, practice, and sustainable transition management. *Environment and Planning A*, *39*(4), 763–770.

Simmie, J. (2012). Path dependence and new technological path creation in the Danish wind power industry. *European Planning Studies*, *20*(5), 753–772. doi:10.1080/09654313.2012.667924.

Smith, A., & Raven, R. (2012). What is protective space? Reconsidering niches in transitions to sustainability. *Research Policy*, *41*(6), 1025–1036.

Smith, A., Stirling, A. & Berkhout, F. (2005). The governance of sustainable socio-technical transitions. *Research Policy*, *34*(10), 1491–1510.

Sydow, J., Schreyögg, G. & Koch, J. (2009). Organizational path dependence: Opening the black box. *Academy of Management Review*, *34*(4), 689–709.

Turnheim, B., Berkhout, F., Geels, F. W., Hof, A., McMeekin, A., Nykvist, B. & van Vuuren, D. (2015). Evaluating sustainability transitions pathways: Bridging analytical approaches to address governance challenges. *Global Environmental Change*, *35*(November), 239–253. doi:10.1016/j.gloenvcha.2015.08.010.

United Nations. (2015). *Resolution adopted by the General Assembly on 25 September 2015: Transforming our world: The 2030 Agenda for Sustainable Development* Retrieved from: www.un.org/ga/search/view_doc.asp?symbol=A/RES/70/1&Lang=E (accessed 23.01.2018).

Uyarra, E., & Gee, S. (2013). Transforming urban waste into sustainable material and energy usage: The case of Greater Manchester (UK). *Journal of Cleaner Production*, *50*(1), 101–110. doi:10.1016/j.jclepro.2012.11.046.

Vis, M. V., Reumerman, P. & Gärtner, S. (2014). *Cascading in the Wood Sector: Final Report*. Retrieved from Enchede, The Netherlands.

Weber, K. M., & Rohracher, H. (2012). Legitimizing research, technology and innovation policies for transformative change: Combining insights from innovation systems and multi-level perspective in a comprehensive "failures" framework. *Research Policy*, *41*(6), 1037–1047.

Whelan, T., & Fink, C. (2016). Sustainability: The comprehensive business case for sustainability. *Harvard Business Review*, *21*(October).

Part II

Sector studies

4 New path development for forest-based value creation in Norway

Antje Klitkou, Marco Capasso, Teis Hansen and Julia Szulecka

4.1 Introduction

This chapter focuses on path development in the forest-based industries of Norway, based on the valorisation of side-streams and residues.

Forest-based value creation is one of the main avenues for the emerging bioeconomy in Norway and the neighbouring Nordic countries Sweden and Finland. Historically, the forestry sector has been an important part of the Norwegian economy, both in terms of GDP and employment. The sector contributed 10.4% of Norwegian GDP in 1845 (Grytten, 2004, p. 254). After the discovery of oil and natural gas on the Norwegian shelf, the importance of the forestry sector has diminished. And over the last decade this negative trend has multiplied.

For decades, the forestry-based industry in Norway has specialised in pulp and paper production, especially newsprint paper. Therefore, huge volumes of forest resources, including residues and side-streams, have been poured into this industry. However, due to changing global market conditions – the massive deployment of the Internet reducing demand for newspapers – and the rise of competitors in other parts of the world, European pulp and paper production has declined significantly in the last decade (Karltorp & Sandén, 2012). This development has had a tremendous impact on the market possibilities of forestry residues, since they can now be processed for pulp and paper to a much lesser degree. Nevertheless, other valorisation pathways exist besides pulp and paper production, such as the wooden construction industry, wooden furniture manufacturing, bioenergy – including solid bioenergy and liquid biofuels – and the production of lignocellulosic chemicals and materials. We want to explore how important these pathways could be in valorising forestry residues and side-streams.

This chapter analyses and compares new path development processes in three Norwegian regions specialising in forest-based value creation. These developments take place in different regional contexts and take different directions regarding choice of technology and the success of these developments. We can distinguish between three pathways and compare three empirical cases: (1) replacing pulp and paper production with an integrated biorefinery which produces chemicals and materials; (2) integrating pulp and paper production with liquefied biogas production; and (3) developing an industrial cooperation of different firms in order to replace pulp and paper production with new forest-based products from logs and residuals, such as bioethanol, biochar, wooden construction materials, etc. In this chapter we will answer the following research questions:

1 What are the main new pathways for forest-based value creation in Norway?
2 How did these new pathways emerge and how is valorisation of side-streams and residues accomplished?

The chapter is structured as follows: after the introduction comes a short section on the valorisation of side-stream and residues in forest-based industries. The third section discusses the conceptual framework for this chapter. The fourth section explains the methodology and data sources applied and gives an account of the three empirical cases as well as applying the conceptual framework; and the final section discusses the results in light of the research questions.

4.2 Forest-based value creation with a focus on the valorisation of side-streams and residues

When exploring new possibilities of value creation in forest-based industries it is necessary to understand the different types of residues and side-streams and how they can be valorised. We have to distinguish between three main groups of residues in forestry-based value chains:

1 *Primary residues:* leftovers from cultivation, harvesting or logging activities from trees within and outside forests;
2 *Secondary residues:* wood processing residues and side-streams, such as sawdust, bark, black liquor;
3 *Tertiary residues:* used wood (in household-waste, end-of-life wood from industrial and trade use, discarded furniture, demolition wood, etc.) considered to be organic waste.

In this chapter we will mainly focus on secondary residues and side-streams, as these are especially important from a regional development perspective.

The Food and Agriculture Organization (FAO) has assessed the shares of residue in the valorisation process of a tree as follows (FAO, 1990):

The harvesting and logging of a tree lead to the following residues being left in the forest:

- Tops, branches and foliage: 23%
- Stumps (excluding roots): 10%
- Sawdust: 5%

Operations at saw mills result in the following products and residues:

- Slabs, edgings and off-cut: 17%
- Sawdust and fines: 7.5%
- Various losses: 4%
- Bark: 5.5%
- Sawn timber: 28%

This gives a high potential for value creation based on these residues. Forestry residues are expensive to collect and to transport, particularly in high-cost Norwegian society, where the forestry sector is struggling to stay competitive in the global market (Talbot & Astrup, 2014). If the bark, leaves and thinnings are left behind in the forest, the nutrients in the soil will not be depleted (FAO, 1990). However, often the bark will first be removed at the plant, following which it will become a residue and will be used as a fuel for other operations.

We want to highlight opportunities for the valorisation of side-streams and residues resulting from the manufacturing of wooden construction materials and furniture, bioenergy production (solid and liquid), manufacturing of pulp and paper and manufacturing of lignocellulosic chemicals, lignin-based products, fibres and other material.

Manufacturing of wooden construction materials and furniture

The value chain starts with the processing of the logs harvested in the forest to produce sawn wood. Only about one-third of a tree becomes sawn wood, leaving a vast amount of residue (Parikka, 2004), e.g. sawdust and fines. The efficiency of particular sawmills depends on various factors (e.g. wood properties, types of operations, machinery) and can be measured using special formulas, e.g. lumber recovery factor (LRF) (Keegan, Morgan, Blatner & Daniels, 2010). Sawmill residues can be used individually or combined as mulch, firewood, hog fuel, animal bedding, for use in particle or strand boards and for pulp recovery (Krigstin, Hayashi, Tchórzewski & Wetzel, 2012). They are particularly attractive to panel and pulp manufacturers, and can be upgraded into various wood-based materials. Markets and the valorisation of the products depend on the sawmill location and the local forest industries.

Bioenergy production

We can distinguish between solid bioenergy, liquid and gaseous bioenergy. As examples of solid bioenergy there is a possible valorisation pathway for sawdust, involving the production of wooden briquettes from sawdust. Edgings and slabs from sawmills can be used for fire wood. More technologically advanced is the pathway leading to the production of advanced pellets from sawdust, etc. using patented steam explosion technology.

There exist different approaches to producing advanced biofuels from wooden material, including residues, but these approaches are mostly integrated into the production of other products in biorefineries (Gregg et al., 2017). Mainly we distinguish between an *anaerobic digestion pathway* to produce biogas, a thermo-chemical pathway for biodiesel or bio-oil and a *biochemical pathway* to produce bioethanol (Fevolden & Klitkou, 2016). The *thermo-chemical approach* involves heating the biomass either with oxygen (gasification for producing syngas and later through a Fischer-Tropsch process-created biodiesel) or without oxygen (pyrolysis for producing bio-oils).

Manufacturing of pulp and paper

The amount of waste in the pulp and paper industry is substantial: around 40–50 kg of dry sludge is generated during the production of one tonne of paper while as much as 300 kg results from one tonne of recycled paper (Najpai, 2015). The composition of waste from pulp and paper mills depends on the final products, production methods and equipment. Waste from mechanical pulping includes rejects (bark and wood residues, sand), ash from energy production, green liquor sludge, dregs and lime mud, primary and biological sludge and chemical flocculation sludge. Papermaking using virgin fibres results in waste in the form of rejects from stock preparation and sludge from water treatments (sludge from chemical pre-treatment, from clarification, biological treatment and chemical flocculation). Papermaking from recovered paper requires many cleaning processes, resulting in more waste, especially deinking sludge composed of cellulose fibres, printing inks and mineral components (Monte, Fuente, Blanco & Negro, 2009). Most of these wastes can be used and valorised, largely eliminating the use of landfills.

One of the most common waste treatment methods in the European pulp and paper industry is the incineration of residues (rejects and sludge) by power and steam generation (Monte et al., 2009; Oral et al., 2005). Other thermal processes, such as pyrolysis, steam reforming, wet oxidation or gasification, are also possible but the technologies for sludge application are still being improved. In the cement industry, both material and energy residues from pulp and paper production can be used to improve products and production processes. Wastes and sludge can be used as soil improvers, through anaerobic digestion converted to biogas and humus (Monte et al., 2009). Other interesting valorisation pathways are cat litter and other absorbents from dried sludge, pesticide/fertiliser carriers or conversion to fuel components (Najpai, 2015). Research on waste from the pulp and paper industry confirms useful elements for both value-added products and industry. While some producers 'capitalise on these opportunities', current best practices are still 'far from gaining the maximum value from paper resources' (CEPI, 2013).

Manufacturing of lignocellulosic products in integrated biorefineries

In integrated biorefineries the whole tree is processed and no off-cuts, sawdust, etc. are lost. The bark is used for heating purposes. Energy produced in one operation is reused in other operations, which means that an integrated biorefinery should co-locate a number of plants to enable the symbiotic exploitation of side-streams and residues in the most cost-effective way. Integrated biorefineries produce a wide spectrum of products such as fuels, platform chemicals and materials of various types including plastics and textiles (Bauer, Coenen, Hansen, McCormick & Palgan, 2017). An economic risk analysis of different biorefinery concepts is in favour of upgrading bioethanol to higher value-added chemicals (Cheali, Posada, Gernaey & Sin, 2016).

One of the main issues when processing lignocellulosic materials in biorefineries is how to handle lignin in the production process. Lignin originally appeared as the main residue of paper production and represents *c.*30% of dry mass of wood. Lignin needs to be removed from the pulp to get a better quality of paper. Traditionally it has been used as a source of energy only, but at an integrated biorefinery lignin can be valorised into more valuable products. There is an established practice of using lignin as an additive to concrete, and other industrial valorisation pathways include the production of vanillin, dispersants or emulsion stabilisers.

Storage and transport of residues

The storage of residues requires area capacity and monitoring and, in the case of saw dust, even coverage to safeguard losses. The quality of the residues and side-streams requires that they are handled with as little transportation as possible, which means that short distances are preferable. When a tree has been logged the wood has a moisture content of around 50% (FAO, 1990). The moisture content is different in different seasons, and varies across species. High moisture content has an impact on the heating value and on the volume. For these reasons a co-location of valorisation pathways seems the most cost-effective option. Therefore, in our analysis of the three cases we use a theoretical framework which is specially tuned to analyse local/regional path development.

4.3 Conceptual framework

In order to understand the challenges and opportunities of the forestry industry in different Norwegian regions to simultaneously diversify into new product groups and minimise waste, we draw on the literature on regional path development. In the following section we describe different types of possible path developments for the three regions included in our study.

Sydow et al. (2009) highlight that the formation of a new path involves several stages and it is during the last stages that the process first gets locked in and becomes path-dependent (p. 691). Therefore, non-predictability and the coexistence of several outcomes are typical characteristics at the start of the process, while inflexibility comes later and inefficiency is typical only of the last stage. This implies that new paths can eventually evolve into barriers for the creation of fundamentally new paths because they bind resources, such as human, economic, institutional and scientific resources. Simmie (2012) has pointed out that in this way a former competitive technology can become the basis of a declining industry (p. 758).

Building on earlier work on the path dependency of technologies and economic activities, more recent scholarly work has highlighted opportunities for developing new industrial paths (Garud, Kumaraswamy & Karnøe, 2010; Martin, 2010). Drawing on evolutionary approaches, these contributions highlight how regional path development is influenced by existing conditions, but may nevertheless take multiple forms (Coenen, Asheim, Bugge & Herstad, 2016). Declining old industrial regions have to develop regional development strategies to 'break out of locked-in paths of development by pursuing innovation, new technological pathways and industrial renewal' (Coenen, Moodysson & Martin, 2015, p. 851). Different taxonomies of possible path developments have been developed, such as the six types defined by Grillitsch and Trippl (2016), who distinguish between path extension, path upgrading, path modernisation path branching, path importation and path creation, or the taxonomy introduced by Isaksen (2015), which distinguishes between path extension, path exhaustion, path renewal and path creation. Of course, these are ideal types, which may combine in reality. Path renewal and path creation tend to require institutional change and the building of new knowledge organisations, while path exhaustion is characteristic of regional industries which are locked into activities that predominantly follow existing technological paths, limiting their opportunities for going in new directions. We use here the taxonomy introduced by Isaksen (2015).

Path exhaustion describes a development where the innovation potentials of local firms are highly reduced and these firms are not able to adapt to technological and market changes.

Path extension is defined by the continuation of an existing industrial path based on incremental product and process innovations in existing industries and well-established technologies.

Path renewal occurs when existing local industries restructure and branch into new, but technologically related industries.

Path creation is defined as the emergence of new industries based on radically new technologies and scientific knowledge, new business models or user-driven innovation.

Regional actors and their networks are central to how lock-ins and path development can be addressed. The possibilities of a firm to engage in path development are highly dependent on the firm's organisational capabilities for innovation. Such capabilities include strategies for innovation, prioritisation of innovation, innovation culture, idea management, external linkages, implementation of ideas, system rules, knowledge generation and diffusion in the organisation (Bjørkdahl & Börjesson, 2011). Firms have to decide the main focus of their innovation activities: improvements of existing products and optimisation of existing processes and value chains, which mostly require more incremental innovations, or the development of totally new types of products and processes, which require more radical innovations. Incremental innovation characterises path extension and exhaustion, while radical innovation is necessary for new path creation and also, to some extent, for path renewal. The prioritisation of incremental versus radical innovation varies significantly between industries (Pavitt, 1984), with consequences for the likelihood of different types of development. The incumbent forestry industry has generally been found to prioritise incremental innovation (Hansen & Coenen, 2017; Näyhä & Pesonen, 2014). The more radical innovations tend to be rather costly in the first stages and therefore require more refinements to improve the new products and processes, i.e. radical innovations should be complemented by incremental innovations.

The actors interact through various types of networks, including supplier networks, research and learning networks, or user-producer networks. Actors include machinery and material suppliers, firms specialising in engineering, transport, maintenance and R&D, but also business development organisations and incubators (Novotny & Nuur, 2013). The importance of different types of actors varies according to technological characteristics, and particularly in degree of technological complexity (Hansen, Klitkou, Borup, Scordato & Wessberg, 2017). The focus on the actors is motivated by our interest in processes of agency, which are essential for new path creation (Garud & Karnøe, 2001a, 2001b; Simmie, 2012).

4.4 Analysis of empirical cases

We present a comparative analysis of the valorisation of residues and side-streams that can be found and directions in which path development has evolved across the three cases. The main empirical sources for the study are interviews with several representatives of the main industry actors and their collaborators in the three cases (Treklyngen, Skogn and Borregaard) and with the national forest owner association, participation in workshops and site visits. Media analysis, analysis of relevant policy documents and descriptive statistics complement the interviews.

We have selected these three cases because they represent rather different development directions, spanning over a broad range of valorisation pathways, and based in different counties of Norway: Østfold, Buskerud and Trøndelag (see Figure 4.1).

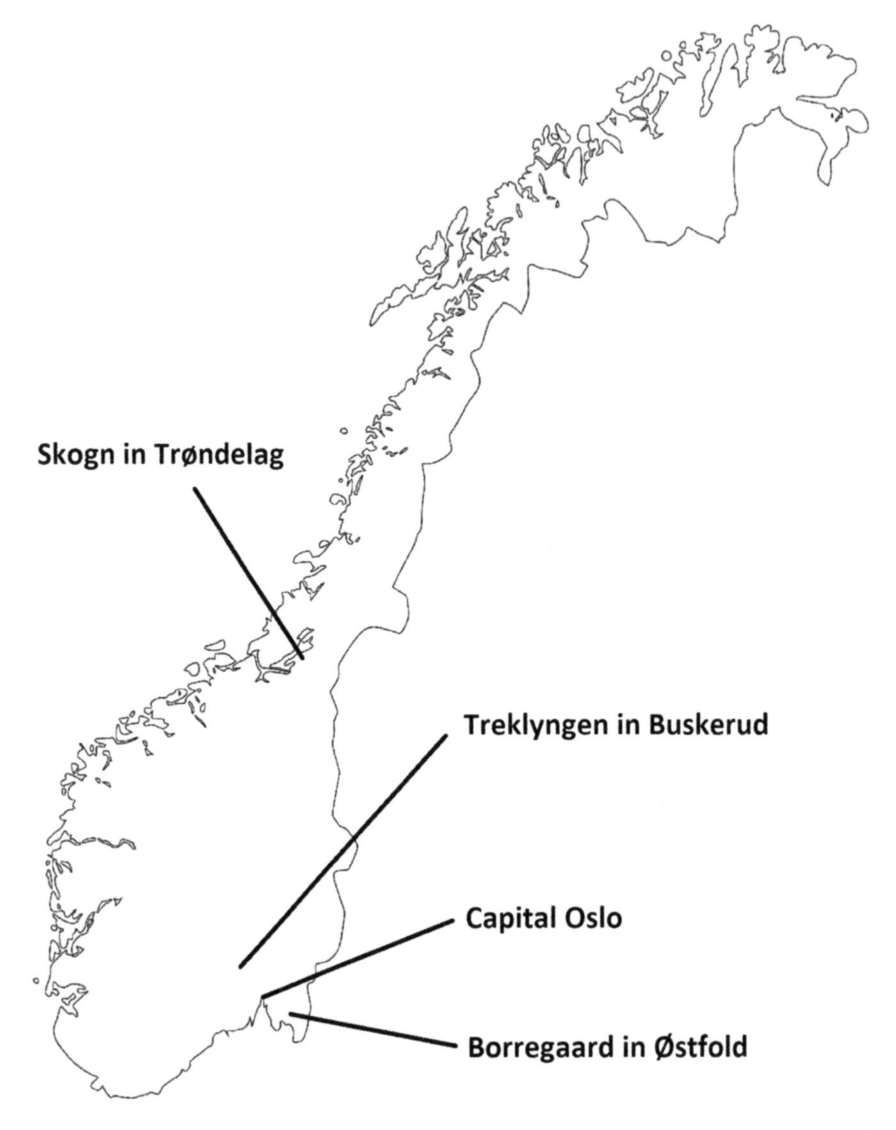

Figure 4.1 Location of the three cases: Skogn in Trøndelag, Treklyngen in Buskerud and Borregaard in south-eastern Norway.

4.4.1 Norske Skog Skogn at Fiborgtangen, Trøndelag[1]

Norske Skog has been the leading pulp and paper producer in Norway for many years. It was established in 1962 with its headquarters in Oslo, Norway, and plants in 14 countries. Norske Skog has been highly specialised in the production of newspaper paper and its first newspaper paper plant opened in

Skogn in 1966. Over the years Norske Skog put all its efforts into becoming the world's leading paper producer, acquiring many foreign pulp and paper mills globally and consequently losing a great deal of its economic power as a result of a globally diminished demand for paper. Several of the pulp and paper plants had to be closed, but the one in Skogn has remained open in Norway, with three paper machines in operation.

While the mother company has declared bankruptcy, both Norske Skog Skogn and the other Norwegian plant, Norske Skog Saugbrugs AS in Halden, which specialises in magazine paper, are still in business; indeed they are earning steady incomes and do not fear for the future because of increased demand. While global demand for newspaper paper has declined, the European production capacity has nosedived still more rapidly, leaving a potentially large market share for the paper produced in Skogn. Norske Skog Skogn owns a large area of land in Fiborgtangen where the necessary infrastructure is already in place, such as quays, railways, roads, access to clean water and renewable energy. This allows direct export by ship to the United Kingdom and the Netherlands. Norske Skog uses bark as an input for thermal energy in the production of newsprint paper and the plant recycles newspaper paper in the production of its newsprint paper. Finally, the plant produces waste water, and this is the main input for valorisation as biogas.

Norske Skog Skogn has established cooperation with Biokraft AS, a company launched in 2009 and specialising in producing biogas from residues in the Norwegian aquaculture industry. The company is mainly owned by two shareholders: Scandinavian Biogas Fuels (50%) and Trønderenergi (43%). Biokraft had earlier experience of producing bio-oil from aquaculture residues. Norske Skog Skogn and Biokraft planned to exploit rest resources from pulp and paper production together with the rest resources from salmon aquaculture (category 2) in order to produce biogas and upgrade it to liquefied biogas (LBG). Category 2 rest resources include all sick and clinically dead fish.

Biokraft bought a part of the industrial area at Fiborgtangen close to the paper plant and started building a plant in 2015, opening it in June 2018. Regarding infrastructure, it should be mentioned that Biokraft also has access to the quays, where it takes deliveries of aquaculture residuals by ship. The plant's co-location with Norske Skog Skogn allows the setting up of direct pipelines for the waste water running from the pulp and paper plant to Biokraft.

The main source of success for the biogas plant was the entrepreneurial spirit of both companies, overcoming administrative barriers at the county level concerning public procurement of LBG. A major problem for the project was that Biokraft had to buy all raw materials (the aquaculture residues) in advance and sell the LBG before the project could receive any financing. The cooperation with AGA was key to getting the funding for the plant: AGA bought all Biokraft's LBG for the next 10 years, ensuring the success of the project. AGA is a Swedish company specialising in industrial gases, including biogas, and is part of the international Linde Group. For access to the aquaculture residues Biokraft has contracts with Scanbio and other companies on the delivery of the residues.

At the national level the Biokraft project twice received funding from Enova and once received a loan from Innovation Norway. The government budget in 2014 abolished tax exemptions for LPG, which enabled LBG to become competitive in price. And in 2015 the government introduced an economic compensation scheme for public buses fuelled with biogas. These decisions enabled a long-term market for LBG in the region.

The municipality of Levanger where Norske Skog Skogn is located was very much in favour of the biogas plant and also functioned as an important enabler.

As a result of the collaboration, Biokraft will process waste water from the pulp and paper plant, which is a mixture of biological sludge and water. After processing the waste water together with the fish residues, Biokraft delivers water back to Norske Skog Skogn. A dry bio-residual will be delivered as a fertiliser to farmers nearby through a local entrepreneur. This resource will be useful for upgrading soil. The farmers already know the quality of this bio-residual. Norske Skog Skogn in turn will receive a substrate which will reduce their need for urea. The biogas is chilled down and liquefied, before being sold as LBG to AGA. Following this it is used to fuel the regional public bus transport system. A barrier to the usage of LBG is the lack of extended filling infrastructure for LBG. Norway's focus on electrical mobility has left little attention for the development of necessary filling infrastructure for LBG.

In the case of Norske Skog Skogn we see a dominance of path extension due to the focus on the incremental improvement of their traditional newspaper paper production, which remains their core product. However, we can also see elements of path renewal, since the company has engaged in collaboration with a new industry in order to produce biogas from pulp and paper residues and the waste water, and even co-create future plans for developing totally new industries within the industrial area, such as introducing land-based fish-farming and a biorefinery. Biokraft is interested in expanding its LBG producing business at Skogn.

4.4.2 Treklyngen in Hønefoss, Buskerud

In Hønefoss, the crisis of Norske Skog was the starting point for new development. In 2012, Norske Skog's pulp and paper mill at Follum, close to Hønefoss, was closed and sold for 60 million NOK to Viken Skog under the condition that the paper production had to be stopped and the equipment dismantled.

Viken Skog is a cooperative with a membership of about 11,000 forest owners, which means a very high degree of ownership fragmentation.

Viken Skog established Treklyngen as a subsidiary to establish a forest-based cluster with several firms exploiting forest resources differently to how they used to be exploited at the pulp and paper plant. Other valorisation pathways now came into focus, such as sawmills, wood-based construction materials, biorefinery and biofuel production. Viken Skog developed a vision of Treklyngen creating 1,000–2,000 new jobs, replacing the lost jobs in the forest-based industry in Southern Norway as well as over 350 jobs at Follum. Treklyngen decided that it would not repeat the mistakes previously made by Norske Skog during the failed Xynergo project, which aimed to produce biodiesel: they would exploit the whole tree and not just parts of it to avoid transport problems; they would use more mature technology than Xynergo; and they would develop an industrial symbiosis between several firms co-located at Treklyngen and share infrastructure costs regarding transport and energy. Treklyngen aimed to exploit the whole forest-based value chain from the sawmills/wood-working industry to the modern wood processing industry for processing, for instance, cellulose or biofuel, but would not be in competition with Borregaard in Sarpsborg.

The market for side-streams and residues had diminished. Over the years, wood chips guarantee instruments had been reduced gradually and at the same time low electricity prices were critical for wood-based bioenergy. With the closure of Follum it became difficult for forest owners to find a market for their pulpwood: 2.6 million m^3 of pulpwood lost their domestic market. Treklyngen planned to exploit 3–5 million m^3 timber annually in the future, about half of today's national felling volume. These plans would involve complementary businesses of different sizes, exploiting all parts of the raw material, including residues, for value creation at Follum. The forest owners at Viken Skog and the regional sawmills dependent on Viken Skog also had to struggle with a reduced market for their residuals because in 2013 the last pulp and paper plant in the region, Tofte, owned by Södra Cell, closed down. The pulp mill at Tofte had processed one quarter of all Norwegian timber. Before the closure Follum delivered 100,000 solid m^3 of wood chips a year to Tofte. After the closure the resources had to be exported to Sweden and Germany, often at low prices. Therefore, local valorisation was paramount for the development of Treklyngen. Treklyngen aims to exploit the whole log in an integrated process of three steps: (a) saw logs for construction of houses, (b) collect waste and pulpwood for producing cellulose, lignin and sugar in biorefineries, (c) collect rest streams for solid or liquid bioenergy.

Treklyngen explored different possibilities for industrial projects at Treklyngen, collaborating with: Avinor to produce bio-jetfuel, Arbaflame for a production plant for biochar, Elkem and the energy company Vardar to develop a new value chain for producing biochar and biooil, ST1, to build a bioethanol plant and an international data centre.

The production of solid wooden materials for construction purposes or for energy has been another valorisation pathway. Important projects have been Hunton Fiber, Termowood, Saga Wood and Norwegian Firewood. Several of these companies and start-ups have collaborated with the industry incubator in Hønefoss, Pan Innovation, where Treklyngen is one of the main owners. Pan Innovation functions not only as a local incubator, but as a national incubator for forest-based business development.

Hunton Fiber has specialised in producing a range of construction elements from wood fibre taken as waste chips, sawdust and off-cuts from sawmills. Hunton Fiber collaborated with Treklyngen and the Norwegian Paper and Fibre Research Institute, and would have been ideal for Treklyngen's profile. However, in 2014 the company decided to locate the new plant at another plant, near their first plant in Gjøvik.

In December 2014 Termowood signed a contract with Treklyngen to start production in 2015 at Treklyngen. Termowood developed a technology for wooden construction elements which ensures a fast and effective construction process, high quality, a good indoor climate and low costs. The business idea was based on a patent for building two-storey houses using self-supporting wooden structures with rock wool insulation in between, held together by veneer or wooden dowels. Termowood received support from Innovation Norway, SkatteFunn and Pan Innovation, but needed additional private investors. A year after the agreement with Termowood, in December 2015, Termowood decided to leave Treklyngen and start production in Hurdal, because of a new shareholder. Treklyngen has continued to collaborate with Termowood, but Termowood is not located there.

Saga Wood is a Norwegian company founded in 2015, specialising in wooden construction materials and fully utilising its experience with thermo-treated and linseed oil-impregnated wood. The company has acquired the rights to exploit Thermo 2.0 technology, developed by WTT in Denmark. Production started in autumn 2018.

Norwegian Firewood (now Varma) was established in 2015 as a start-up at Pan Innovation. The company specialises in selling wood and wooden briquettes. In 2016 the company acquired a large contract for selling wooden briquettes via a national retailer. They have developed a production capacity of 600,000–700,000 10-kg packages of wooden briquettes, exploiting sawdust from a saw mill near Treklyngen, Soknabruket.

Treklyngen does not have much manpower, but there are several highly engaged and competent people from outside the forest sector who are also trying to advance the cluster. The main advantages of Treklyngen are the location (near the capital's main airport), the size of the area and its access to infrastructure (train, road, energy). Several of the projects Treklyngen is involved in have received public funding in their early stages, but for commercialisation some of the projects, such as Arbaflame, need more investments. This has been a focus of network and lobbying activities over the last years, both within forest-based industries and with politicians.

Treklyngen has been forced to be open to many different opportunities after pulp and paper production ended. Therefore, we see here a mixture of path renewal – a restructuring and branching into new, but related industries (forest-based industries specialising in different advanced forms of bioenergy and wooden construction elements) – and path creation – aiming for biorefining and attracting an international data centre.

4.4.3 Borregaard in Sarpsborg

Borregaard is a Norwegian company, established in 1889 and located in Sarpsborg in Østfold County in south-eastern Norway. The location is favourable for industry because of access to local hydropower for the production process at Sarpsfossen, Europe's biggest waterfall, to forest resources in south-eastern Norway and to a harbour as necessary infrastructure for delivery of forestry feedstock and export of goods.

Borregaard began by specialising in pulp and paper, and at the end of the 1930s the company started producing chemicals based on timber from spruce as a raw material, exploiting the hemi-cellulose in the feedstock (Klitkou, 2013). Since the 1950s Borregaard has also used the lignin components of the feedstock to produce chemicals.

Borregaard has been working on the development of an integrated biorefinery for more than 50 years (Rødsrud, Lersch & Sjöde, 2012). Borregaard's strategy is directed towards the production of high value-added products from Norwegian forest resources. The rationale behind this strategy is to become a company which specialises in producing chemicals and not cheap commodity products such as pulp and paper or advanced biofuels. This strategy has paid off and pulp and paper plants are not trying to compete with Borregaard, because they have high labour and feedstock costs and low value-added products.

Borregaard has used lignocellulosic feedstock, sulphite spent liquor from spruce wood pulping, as feedstock for many years. The feedstock is provided by the regional forestry industry and is relatively expensive compared to other countries. Borregaard annually consumes around 400,000 tons of spruce (Johansen, 2009). The biorefinery opened in 1938. Annually, the commercial biorefinery produces 160,000 tons of speciality cellulose, 170,000 tons of speciality lignin, 20 million litres of advanced bioethanol, 1,300 tons of vanillin, 200 GWh bioenergy and 30 GWh biogas based on anaerobic digestion (Johansen, 2009, p. 4).

The company acts globally and has plants and sales offices in 17 countries. Since 2012 the company has been listed on the Oslo Stock Exchange; it has over 800 employees and a solid turnover. Today, Borregaard produces a range of products, such as performance chemicals, advanced speciality cellulose, water-soluble specialty lignin products, ingredients for food and flagrance applications and fine chemicals for the pharmaceutical market. Examples of high-performance products are vanillin and high-performance additives and ingredients for the animal feed industry, such as bypass proteins and pelleting aids produced from lignin, and MFC – Exilva – from cellulose.

Borregaard has an additional six production plants around the world which produce lignin speciality chemicals. Borregaard wants to continue being the world's leading supplier of lignin-based chemicals and to improve its position, with bio-ethanol as an important by-product. Therefore, Borregaard has bought up a high-lignin production capacity globally. The company's strategic goal is to build and operate plants in Europe or other places in the world and not become a technology supplier.

For several decades, Borregaard has been interested in optimising its different processes – debarking, pre-treatment, hydrolysis, fermentation and chemical conversion. The most recent developments have been the BALI Pilot plant, where the pre-treatment of feedstock is central, and a plant for producing microfibrillar cellulose (MFC).

The BALI pilot plant is a research plant and does not produce for the market. The plant aims to 'utilize low value biomass and convert it to various competitive products. A goal is to utilize at least 80% of the biomass to produce products, energy excluded' (Rødsrud et al., 2012, p. 52f.). The BALI Pilot is based on the biochemical conversion of lignocellulosic material and produces ethanol; lignin, specialty chemicals; single-cell protein and sugar derivatives. The BALI process is based on chemical pre-treatment, saccharification with commercial enzymes, conventional fermentation of hexoses, aerobic fermentation or chemical conversion of pentoses, and chemical modification of lignin.

Borregaard first installed a pilot plant for developing and testing the MFC technology. The plant is still in operation and is used for testing and demonstrating new applications for MFC. It has a capacity of 100 dry metric tonnes. The MFC plant started its operation in 2016 and is globally the first commercial plant for MFC. It has a capacity of 10,000 metric tonnes of 10% paste (1000 metric tonnes dry) (Borregaard Exilva, 2018).

Borregaard has its own research centre with 70 employees and collaborates extensively with knowledge organisations inside the region, such as the University of Life Sciences in Ås and Østfold Research in Fredrikstad, but also within international networks which are engaged in developing new biorefinery technology. The company has received public funding for various research projects in the field of biorefining, but also for big demonstration facilities. Important national funding agencies were the Research Council of Norway, Innovation Norway and Enova. In cooperation with other companies, Borregaard was central in the development of a national strategy for forest-based industries (Skog 22, 2014). In this project Borregaard focused on the development of wooden fibre and biorefining exploitation. The Fibre Group concluded with three main goals for the fibre-based industry: increased harvesting from Norwegian forests, increased domestic value creation and decreased dependence on exports and, finally, innovation. The short-term focus on export of timber and use of forest resources for stationary energy was assessed as risky and should in the long-term be replaced with a focus on biorefineries and production of advanced biofuels for heavy transport, aeroplanes and ships, on bio-chemicals and special lignocellulosic materials (Rødsrud, 2014).

Borregaard has been very active in European RD&D projects and has received funding from Horizon 2020, among others, under the Bio Based Industries Joint Undertaking.

Borregaard has realised many more of the potentials of path creation than the two other cases. While the other two cases more or less have plans for path creation, Borregaard has realised the transformation of a pulp and paper plant into an advanced integrated biorefinery producing a wide range of products and relying on the development and commercialisation of advanced scientific knowledge. The firm is central in international projects for advancing forest-based biorefineries.

4.5 Conclusion

Our first case study, on the recent developments of Norske Skog in the Skogn region, has shown how a successful path extension, reached through incremental improvements of a traditional production line, may have far-reaching consequences for the valorisation of organic waste. Norske Skog Skogn has been able to survive demand swings and financial turmoil because of process innovations, without moving away from paper as a core product. Incremental innovations, devoted in particular to energy optimisation within its paper plant, have enabled not only an extension of the traditional regional path, but even a path renewal by allowing a co-located production of biogas. An industrial symbiosis has emerged where the new biogas plant can employ waste water from the local paper plant as well as residues from extra-regional aquaculture activities. The symbiosis has gone so far as to include an exchange of the dry bio-residual from the biogas plant with a substrate from local farmers.

Attempts to create an industrial symbiosis in the Hønefoss area, which we have studied as a second case, have encountered several difficulties. Before the entrance of Treklyngen, the traditional path based on pulp and paper production at Hønefoss was exhausted, and Treklyngen's efforts to renew this path by branching into other forest-based sectors face an uncertain future. If successful, the current plans for exploiting the whole wood timber in an integrated process could reduce the overall amount of waste, across all the value chains involved, to a minimum. However, in spite of a strong institutional backbone, the region does not offer a firm, or group of firms, that is sufficiently successful in the traditional regional path to provide enough support for regional branching. As a consequence, some of the firms which had considered producing novel products in the area have, in the end, decided to operate in different regions, where incumbents in traditional forest-based industries are operating successfully and could offer fruitful partnerships.

We have witnessed a completely different scenario when studying the Sarpsborg area. Here the incumbent Borregaard has been able, over almost a century, to move from the production of pulp and paper to the production of a vast range of wood-related chemical products. Such path creation has been made possible by the long-term innovation strategy of the firm, directed towards the production of high value-added goods from Norwegian forest resources. An in-house research centre, active in international networks as well as collaborating extensively with knowledge organisations within the region, has indeed enabled the invention of competitive products from underutilised resources, and guaranteed an optimisation of industrial processes. Industrial symbiosis has thus been reached within the firm, by co-locating or integrating different processes pertaining to different business lines.

Summing up our findings from the three cases analysed, it seems that forest-based waste valorisation often originates with the co-location of activities belonging to different value chains. Such activities can be performed by a firm operating in different lines, or by different firms, possibly belonging to different sectors. In both cases, the presence of a strong and innovative private actor in a region raises the region's chances for waste valorisation. Innovation could be directed explicitly towards the development of high-performance products, whose production lines can be harmonised for waste-reducing purposes. On the other hand, a persistent incremental process innovation within traditional business lines could bring waste valorisation about indirectly, by shaping the right regional environment for attracting firms into the region and for creating industrial symbioses. An extension of the traditional regional path would then lead the way to a path renewal, characterised by collaborations with industries new to the region for the purpose of waste valorisation.

Note

1 Beside our main funder, the Bionær-programme at RCN, the research was partially funded by the Nordic Green Growth Research and Innovation Programme in cooperation with NordForsk, Nordic Innovation and Nordic Energy Research, Grant 83130 – Where does the green economy grow? The Geography of Nordic Sustainability Transitions (GONST). Specifically, we would like to acknowledge the GONST project for funding the empirical case work on Norske Skog Skogn and Biokraft.

References

Bauer, F., Coenen, L., Hansen, T., McCormick, K. & Palgan, Y. V. (2017). Technological innovation systems for biorefineries: A review of the literature. *Biofuels, Bioproducts and Biorefining, 11*(3), 534–548.

Bjørkdahl, J., & Börjesson, S. (2011). Organizational climate and capabilities for innovation: A study of nine forest-based Nordic manufacturing firms. *Scandinavian Journal of Forest Research, 26*(5), 488–500. doi:10.1080/02827581.2011.585997.

Borregaard Exilva. (2018). Exilva. Retrieved from www.exilva.com.

CEPI. (2013). *CEPI Sustainability Report 2013. European Paper Industry – Advancing the Bioeconomy. Summary.* Retrieved from www.cepi-sustainability.eu/uploads/short_sustainability2013.pdf.

Cheali, P., Posada, J. A., Gernaey, K. V. & Sin, G. (2016). Economic risk analysis and critical comparison of optimal biorefinery concepts. *Biofuels Bioproducts & Biorefining – Biofpr, 10*(4), 435–445. doi:10.1002/bbb.1654.

Coenen, L., Moodysson, J. & Martin, H. (2015). Path renewal in old industrial regions: Possibilities and limitations for regional innovation policy. *Regional Studies, 49*(5), 850–865. doi:10.1080/00343404.2014.979321.

Coenen, L., Asheim, B. T., Bugge, M. M. & Herstad, S. J. (2016). Advancing regional innovation systems: What does evolutionary economic geography bring to the policy table? *Environment and Planning C: Politics and Space, 35*(4), 600–620.

FAO. (1990). The potential use of wood residues for energy generation. In *Energy Conservation in the Mechanical Forest Industry*. Rome: Food and Agriculture Organization of the United Nations.

Fevolden, A. M., & Klitkou, A. (2016). A fuel too far? Technology, innovation, and transition in failed biofuel development in Norway. *Energy Research & Social Science, 23*(January), 125–135.

Garud, R., & Karnøe, P. (2001a). Path creation as a process of mindful deviation. In R. Garud & P. Karnøe (Eds.), *Path Dependence and Path Creation* (pp. 1–38). Mahwah, NJ: Lawrence Erlbaum Associates.

Garud, R., & Karnøe, P. (Eds.). (2001b). *Path dependence and path creation*. Mahwah, NJ: Lawrence Erlbaum Associates.

Garud, R., Kumaraswamy, A. & Karnøe, P. (2010). Path dependence or path creation? *Journal of Management Studies, 47*(4), 760–774.

Gregg, J. S., Bolwig, S., Hansen, T., Solér, O., Amer-Allam, S. B., Viladecans, J. P., … Fevolden, A. (2017). Value chain structures that define European cellulosic ethanol production. *Sustainability, 9*(118), 1–17.

Grillitsch, M., & Trippl, M. (2016). Innovation policies and new regional growth paths: A place-based system failure framework. *Papers in Innovation Studies, 26*, 1–23.

Grytten, O. H. (2004). The gross domestic product for Norway 1830–2003. In Ø. Eitrheim, J. a. T. Klovland & J. F. Qvigstad (Eds.), *Historical Monetary Statistics for Norway 1819–2003* (pp. 241–288). Oslo: Norges Bank.

Hansen, T., & Coenen, L. (2017). Unpacking resource mobilisation by incumbents for biorefineries: The role of micro-level factors for technological innovation system weaknesses. *Technology Analysis & Strategic Management, 29*(5), 500–513.

Hansen, T., Klitkou, A., Borup, M., Scordato, L. & Wessberg, N. (2017). Path creation in Nordic energy and road transport systems: The role of technological characteristics. *Renewable and Sustainable Energy Reviews, 70*, 551–562.

Isaksen, A. (2015). Industrial development in thin regions: Trapped in path extension? *Journal of Economic Geography, 15*(3), 585–600.

Johansen, G. L. (2009). *Creating Value from Wood: The Borregaard Biorefinery*. Paper presented at the Biofuels & Bioenergy: A Changing Climate. IEA Bioenergy Multi Task Conference, Vancouver.

Karltorp, K., & Sandén, B. A. (2012). Explaining regime destabilisation in the pulp and paper industry. *Environmental Innovation and Societal Transitions, 2*, 66–81.

Keegan, C. E. I., Morgan, T. A., Blatner, K. A. & Daniels, J. M. (2010). Trends in lumber processing in the Western United States. Part I: Board foot Scribner volume per cubic foot of timber. *Forest Products Journal, 60*(2), 133–139. doi:10.13073/0015-7473-60.2.133.

Klitkou, A. (2013). *Value Chain Analysis of Biofuels: Borregaard in Norway*. Retrieved from: www.topnest.no/attachments/article/12/Borregaard_TOPNEST_Case%20study.pdf.

Krigstin, S., Hayashi, K., Tchórzewski, J. & Wetzel, S. (2012). Current inventory and modelling of sawmill residues in Eastern Canada. *The Forestry Chronicle*, *88*(05), 626–635. doi:10.5558/tfc2012-116.

Martin, R. (2010). Roepke lecture in economic geography: Rethinking regional path dependence: Beyond lock-in to evolution. *Economic Geography*, *86*(1), 1–27.

Monte, M. C., Fuente, E., Blanco, A. & Negro, C. (2009). Waste management from pulp and paper production in the European Union. *Waste Management*, *29*(1), 293–308. doi:10.1016/j.wasman.2008.02.002.

Najpai, P. (2015). *Management of Pulp and Paper Mill Waste*. Cham, Switzerland: Springer International Publishing.

Näyhä, A., & Pesonen, H.-L. (2014). Strategic change in the forest industry towards the biorefining business. *Technological Forecasting and Social Change*, *81*, 259–271.

Novotny, M., & Nuur, C. (2013). The transformation of pulp and paper industries: The role of local networks and institutions. *International Journal of Innovation and Regional Development*, *5*(1), 41–57.

Oral, J., Sikula, J., Puchyr, R., Hajny, Z., Stehlik, P. & Bebar, L. (2005). Processing of waste from pulp and paper plant. *Journal of Cleaner Production*, *13*(5), 509–515. doi:10.1016/j.jclepro.2003.09.005.

Parikka, M. (2004). Global biomass fuel resources. *Biomass and Bioenergy*, *27*(6), 613–620. doi:10.1016/j.biombioe.2003.07.005.

Pavitt, K. (1984). Sectoral patterns of technical change: Towards a taxonomy and a theory. *Research Policy*, *13*(6), 343–373.

Rødsrud, G. (2014). *Skog22: Arbeidsgruppe Fiber*. Retrieved from www.skoginfo.no/userfiles/files/Skogforum/2014/5%20Gudbrand%20Rodsrud%20Arbeidsgruppe%20Fiber.pdf.

Rødsrud, G., Lersch, M. & Sjöde, A. (2012). History and future of world's most advanced biorefinery in operation. *Biomass & Bioenergy*, *46*, 46–59. doi:10.1016/j.biombioe.2012.03.028.

Simmie, J. (2012). Path dependence and new technological path creation in the Danish wind power industry. *European Planning Studies*, *20*(5), 753–772. doi:10.1080/09654313.2012.667924.

Skog22 – Arbeidsgruppe fiber og bioraffineri. (2014). *Skog22: Delrapport Arbeidsgruppe fiber og bioraffineri*. Retrieved from www.innovasjonnorge.no/contentassets/920a1e161a494a508f91b7a02344a47e/delrapport-skog22-arbeidsgruppe-fiber-kortversjon-endelig-versjon.pdf.

Sydow, J., Schreyögg, G. & Koch, J. (2009). Organizational path dependence: Opening the black box. *Academy of Management Review*, *34*(4), 689–709.

Talbot, B., & Astrup, R. (2014). Forest operations and ecosystems services in Norway: A review of the issues at hand and the opportunities offered through new technologies. *Journal of Green Engineering*, *4*(4).

5 Mission-oriented innovation in urban governance

Setting and solving problems in waste valorisation

Markus M. Bugge and Arne Martin Fevolden

5.1 Introduction[1]

As an ever-greater part of the world's population is living in cities, dealing with urban waste is becoming an increasingly prominent challenge for local authorities (Frantzeskaki & Kabisch, 2016). In many places, local authorities are struggling to cope with growing amounts of waste annually, as cities grow and citizens' consumption rises (Hoornweg, Bhada-Tata & Kennedy, 2013). Local authorities that lack the organizational capabilities and physical infrastructure to deal effectively with urban waste are often faced with huge environmental, economic and social problems (Hodson & Marvin, 2010; Mourad, 2016). Nevertheless, investments in organisational capabilities and physical infrastructure come with challenges and problems of their own. These investments tend to create lock-ins (see Chapter 3), which make the waste treatment system inflexible and prevent further improvements (David, 1985; Geels, 2002). The local authorities that make these investments run the risk of becoming stuck with systems that are technologically outdated and adequate only to deal with yesterday's problems. To introduce new and smarter urban waste systems, local authorities need to challenge existing policies and institutions, technologies and business models (Uyarra & Gee, 2013; Weber & Rohracher, 2012).

It is a paradox that the same investments in organisational capabilities and physical infrastructure that local authorities make to deal with problems related to waste today might prevent them from coping with tomorrow's problems. In this chapter we want to explore this paradox by focusing on how the municipality of Oslo deals with organic waste. The municipality of Oslo has made massive investments in a physical infrastructure consisting of an optical sorting plant, a biogas facility and an incineration plant. These investments have created strong incentives for the municipality to improve the sustainability of its waste treatment system by building upon existing infrastructure; for instance, by constructing a district heating infrastructure to make use of excess heat from the incineration plant and establishing a bus fleet that runs on biogas to make use of biofuels from the biogas plant. Nevertheless, the same investments also rely upon steady flows of organic waste and provide little incentives for the municipality to pursue more sustainable options for organic waste, such as reuse or prevention of food waste.

This paradox can be expressed as a dual challenge of operating in a problem-solving mode on the one hand, and of (re)defining the problems to be solved in the first place. It then becomes relevant to reflect on whether the actors who are tasked with the job of addressing and solving the challenges at hand are also the ones to decide the direction for change. This aspect has also been highlighted in the literature on mission-oriented innovation, emphasising the need for a better understanding of how the public sector can generate dynamic capabilities for addressing mission-oriented innovation (Kattel & Mazzucato, 2018), and in particular discussing who should take part in the identification and articulation of missions:

> Who decides the mission is a key issue that requires more thought.
> (Mazzucato, 2017, p. 10)

> Understanding more democratic processes through which missions are defined and targeted is tied to rethinking the notion of public value.
> (Mazzucato, 2017, p. 28)

In this chapter we will interpret the case of urban waste treatment as an example of mission-oriented innovation. How is the organisation of work in the municipality rigged to enable a balancing between problem setting and problem solving when trying to improve the sustainability of their waste treatment systems? How do they balance the needs of today with the needs of tomorrow? The objective of this chapter is thus to reflect upon how creating sustainable urban waste governance can be seen as an example of mission-oriented innovation, and how complementary forms of governance may improve the ability to develop long-term innovative and sustainable urban waste systems.

To explore these questions, the chapter will draw on theories about mission-oriented innovation and will purposely discuss the need for orchestration of broader sets of actors in order to enable a wider outlook when identifying and articulating the problems or missions to be solved. The chapter analyses innovation and sustainability in an urban waste system through the lens of valorisation pathways, and seeks to answer the following research question:

> How can the mission of sustainable urban waste treatment be understood in terms of problem setting and problem solving?

The chapter is structured as follows: following this introduction, the second section briefly outlines the theoretical framing of the chapter. In the third section, the research methods applied are presented. Section 5.4 presents the case study, and section 5.5 analyses and discusses the findings. Finally, section 5.6 concludes the chapter.

5.2 Conceptual framework

Before embarking on an analysis of the waste treatment system in the municipality of Oslo, the next sub-sections frame the case study within a mission-oriented innovation perspective. Following this, the analytical buildings blocks, consisting of the waste pyramid and notions of valorisation, will also be outlined.

5.2.1 Mission-oriented innovation

In the literature on mission-oriented innovation policies it is highlighted that long-term commitment to engaging public, private and third sector actors is key to successful implementation (Kattel & Mazzucato, 2018; Mazzucato, 2017, 2018). It has been pointed out how social movements are often central to the advocacy and development of innovative and sustainable regimes and solutions (Fagerberg, 2017), and how there is a need to include these actor groups in the selection environment when developing future strategies (Smith, Voß & Grin, 2010).

At the same time, the literature points to directionality as something vital to sustainability transitions and mission-oriented innovation (Fagerberg, 2017; Mazzucato, 2017; Schot & Steinmueller, 2018; Weber & Rohracher, 2012). The notion of directionality involves selection and priority setting, and has thus introduced and emphasised a stronger element of politics in our understanding of systems of innovation and socio-technical change (Shove & Walker, 2007; Smith, Stirling & Berkhout, 2005). This form of top-down directionality on the one hand and broad anchoring among diverse stakeholders on the other constitutes a range of actors that as of yet has scarcely been investigated. Consequently, it is acknowledged that it is important to gain a better understanding of the relationship and balance between directive and bottom-up interactions in mission-oriented innovation (Mazzucato, 2017).

As opposed to the innovation needed to solve grand societal challenges, mission-oriented innovation has traditionally been perceived as being primarily preoccupied with technological dimensions, whereas the organisational and social aspects of innovation have received less attention (Martin, 2015; Nelson, 2011). Nelson (2011) pointed out the puzzle of how a country that has managed to send a man to the moon is facing great difficulties when it comes to providing basic education and health services to overcome poverty. This is due to the intersectoral, social and complex nature of grand challenges, where there is seldom one solution that is widely agreed upon. More recently, contributions to the literature have actualised a debate on how to define and differentiate between so-called mission-oriented innovation and sustainable socio-technical transitions (Fagerberg, 2017; Kattel & Mazzucato, 2018; Mazzucato, 2017, 2018; Mowery, Nelson & Martin, 2010; Nelson, 2011). Here it is emphasised how traditional technology-oriented research and innovation policies appear deficient to address and tackle today's complex and integrated societal challenges.

Addressing this relationship, Mazzucato (2017, 2018) distinguishes between old and new types of mission-oriented projects, where the old were defined by a small and centralised group of experts, oriented towards specified technology development, and where diffusion beyond these actors was of minor importance. New mission-oriented projects, on the other hand, are characterised by broader involvement of actors in defining the direction of the mission; the missions have both technical and societal objectives, where the diffusion of solutions is paramount. The new missions also ascribe an important role to foresight analysis as part of the envisioning of potential future scenarios. Moreover, Mazzucato makes a distinction between grand challenges, missions and portfolios of projects that involve different actors and sectors in bottom-up experimentation (Mazzucato, 2018). In this sense, missions and (mission) projects can be perceived as operationalisations of the broader grand challenges. Mission-oriented innovation is seen to constitute a narrower and more clearly defined form of innovation than what is required to address grand challenges, which are more complex and multi-faceted. In parallel with the ability to set missions, it is seen as central to leave enough space for encouraging bottom-up experimentation across several types of public and private actors (Kattel & Mazzucato, 2018). Missions should also comprise a portfolio of R&D and innovation projects that allow for both success and failure, and they should have a trickle-down effect in which overall objectives should be translated into concrete policy actions. Importantly, missions should be based on a long-term agenda and draw on existing resources and policy instruments in the science and technology system (Fagerberg, 2018; Mazzucato, 2017). In order to approach pressing grand societal challenges in appropriate ways, there is a need to select missions that have enduring and democratic legitimacy, and moreover to define these missions in ways that allow sufficient breadth to motivate action across several sectors and societal actors (Kattel & Mazzucato, 2018; Mazzucato, 2017).

5.2.2 *The waste hierarchy and different treatment options for organic waste*

The various ways that local authorities can treat or deal with waste can be ranked according to a waste hierarchy or waste pyramid. In the hierarchy adopted by the EU (European Commission, 2008), the pyramid consists of five layers of progressively more sustainable options – disposal, recovery, recycle, reuse and prevention (see also Chapter 3). According to this line of thinking, less sustainable treatment options are at the base of the pyramid, while more favourable options are at the top (see Figure 5.1 below).

'Disposal' is at the bottom of the pyramid and represents the least sustainable option. Disposal implies that organic household waste is simply collected, transported and dumped at a landfill site. Although disposal can reduce

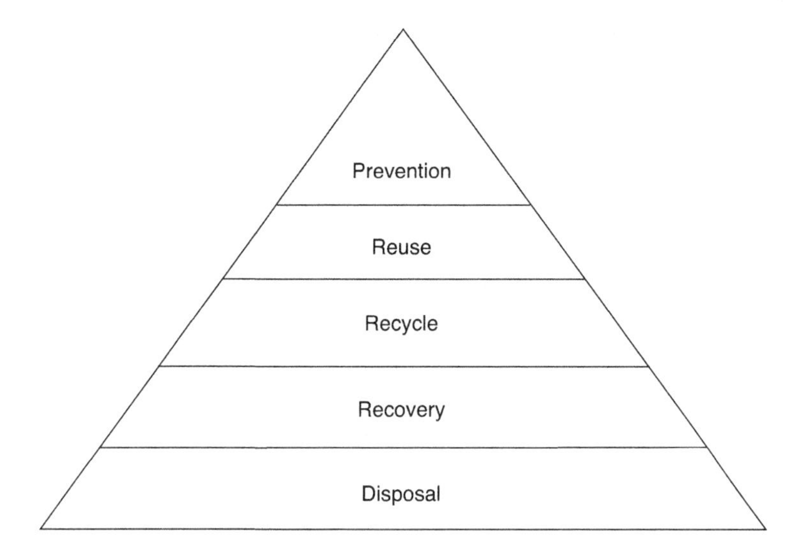

Figure 5.1 The waste pyramid.

pollution and prevent sickness within a city, landfills lay claim to land areas and can pollute ground water, air, lakes and rivers. Another implication of the use of landfills is that the waste is not sorted and precious resources cannot be recycled or recovered easily (Gee & Uyarra, 2013). 'Recovery' is a step up the pyramid and represents a more sustainable option than disposal. Recovery implies that at least some of the energy that lies within the waste can be used for some useful purpose. Recovery became a viable option in the 1960s and '70s, when urban waste was incinerated with greater and greater frequency in order to reduce the use of landfills in Europe. The incineration process contributed to air pollution and required the instalment of advanced filter systems and the use of very high temperatures. A by-product of the high temperatures used was that excess heat could be used in district heating systems, in this sense recovering some of the energy stored in the waste. Nevertheless, the ash resulting from the incineration process still has to be stored in landfills.

Recycling is another step up the pyramid and represents a more sustainable option than recovery. Recycling of organic waste often implies some sort of anaerobic process at a biogas facility where organic waste is turned into biogas (biomethane) and fertiliser. The only way to achieve the recycling of waste resources in cities is to sort the waste streams and manage them separately. For each waste stream – such as organic, paper, plastic and metal-based waste – different routes of recycling or recovery must be developed. This often implies that citizens must sort their waste before they dispose of it and that the different waste streams must be transported to and processed at different locations.

Reuse and prevention compose the top two levels of the waste hierarchy and are the most sustainable options. Reuse implies that organic resources are used again without breaking them down and reprocessing them (which is the case in a recycling process). A typical example of reuse is using leftover food as feed for animals. Prevention is the most sustainable option and forms the top of the waste pyramid. Typical examples of waste prevention are serving food on smaller plates at hotels or repacking of food items into sizes that fit the needs of the consumers and which do not generate leftovers. Reuse and waste prevention have emerged as the most important alternatives to pursue today in order to create more sustainable production and consumption patterns (Mourad, 2016).

In sum, urban organic waste can be dealt with in many different ways and these can have considerable implications for sustainability. The higher the waste treatment option is in the waste pyramid, the more sustainable the treatment option tends to be.

5.2.3 Valorisation of waste – importance of problem setting

When local authorities want to improve the sustainability of their waste treatment system, they have to engage in both 'problem solving' and 'problem setting'. When they engage in 'problem solving' but not in 'problem setting', the only possible outcome is to expand or improve their existing waste treatment system. For instance, local authorities can expand the collection and delivery of waste to its incineration plant, and thereby incinerate waste that might otherwise have been dumped on a landfill. Such activities improve sustainability, but the solutions are found on the same level in the waste hierarchy pyramid and further improvements will at some point be exhausted. To improve the waste treatment system further, local authorities need to find options outside the existing system and at a higher level in the waste pyramid. For instance, local authorities can recycle organic waste and turn it into biogas and fertiliser instead of incinerating it. We refer to this option as 'problem setting'. Of course, 'problem setting' also requires the solving of the identified problems.

When local authorities engage in 'problem setting', they are attempting to transition from one waste system to another. Transitions from one waste system to another are often very challenging, as an existing waste system will often be embedded and anchored in specific technologies, infrastructures and institutions that are not relevant to the new system (Frantzeskaki & Loorbach, 2010). The transition from landfill (disposal) to incineration (energy recovery) requires investment in incineration infrastructure to capture and exploit the energy from the waste. The transition from an incineration system (energy recovery) to a biological treatment system (recycling) requires new infrastructure, and altered behaviour from the citizens using the system, as a result of the need for sorted waste streams. In addition, there is a need for a market for the different waste streams (e.g. paper, plastics, glass, metal, textiles) and the products of biological treatment, such as biogas as a fuel and biosolids as fertiliser (Murray, 2002). The same is true for further movement up the waste pyramid to reuse and prevention. They require a 'fundamental re-think of the current practices and systems in place' (Papargyropoulou, Lozano, Steinberger, Wright & Ujang, 2014, p. 114).

Although it is possible to place different types of waste treatment at different levels of the waste pyramid, it is common for a regional waste treatment system to consist of more than one type of waste treatment. Today, the most common systems for processing organic waste are recovery and recycling-based, in the form of incineration and biological treatment systems. In this sense, problem solving and problem setting are activities that often overlap and co-exist.

5.3 Research methods and data

Our data collection is based on interviews, participation in policy and industry seminars and document analysis. We have conducted six explorative and semi-structured interviews with key stakeholders and representatives of the relevant waste management agencies and involved firms. We interviewed representatives from the following organisations and their departments (the names of the interviewees are anonymised): Avfall Norge, Østfold Research, Oslo municipality Department of Environment and Transport, Oslo municipality Waste-to-Energy Agency, Oslo municipality Agency for Waste Management and NorgesGruppen/ASKO. Most of the interviews were conducted face to face and lasted around one hour. The interviews were recorded and transcribed. We also organised two workshops on the subject, one with researchers in the field (November 2016) and the second with invited experts from the industry, public administration, NGOs and research (November 2017). In addition to the interviews and workshops, document analyses of reports and municipal strategies and media analysis have also constituted part of the data collection for the case study. Finally, field trips and participation in industry seminars and conferences have helped inform the study. The presentation of the case study is adapted from Bugge, Fevolden and Klitkou (2018).

5.4 Valorisation of urban organic waste: the case of Oslo

The governance of waste processing in Oslo is administered by three municipal departments: the Renovation Department (Renovasjonsetaten) is responsible for organising the collection and transport of municipal household waste, whereas the Energy Recovery Department (Energigjenvinningsetaten EGE) is responsible for the recycling of municipal waste (Bugge et al., 2018). Finally, the Department for Urban Environment (Bymiljøetaten) takes responsibility for the environment in the city, such as the quality of air, water and soil, and for the planning and development, management and operation of municipal urban spaces. These agencies are *coordinated* by the Vice Mayor for Environment and Transport.

These municipal departments do not operate in isolation. There are both national and international regulations that the municipality of Oslo must adhere to when developing their organic waste system. Among others, the municipality must adhere to the EU landfill ban of 2009 and the EU Waste Framework Directive of 2008, which Norwegian authorities have transposed into Norwegian law. The municipality must also adhere to The Norwegian Pollution Act, which states that municipalities have sole responsibility for the collection and processing of household waste, while private businesses are responsible for processing their own waste. According to the same Act, the municipality's handling of household waste should also be self-financed through fees and governed by waste regulations.

The municipality of Oslo has implemented a two-bin system, consisting of one bin for plastic, food and residual waste, collected one to six times a week, and one bin for paper collected one to four times a month. Citizens collect their food waste in green bags, their plastic waste in blue bags and their residual waste in neutral bags. All the bags are disposed of in the same household waste bin. Additionally, there are 910 collection points for glass, metals and textiles across the city. Moreover, the city has collections for hazardous waste, three large recycling stations, two mobile recycling stations and a regular collection of garden waste. The municipal waste processing system (Figure 5.2) includes optical sorting of waste resources from households, i.e. plastics, food waste and residual waste.

After household waste is collected in waste bins, it is delivered to a large sorting plant at Haraldrud in Oslo. At this facility, the three types of waste

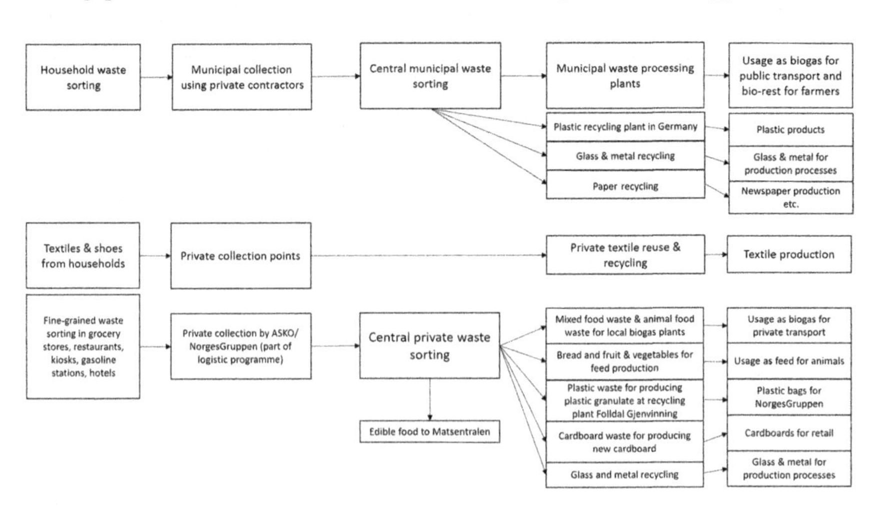

Figure 5.2 The parallel systems of waste management in Oslo (adapted from Bugge, Fevolden & Klitkou, 2018).

bags are sorted automatically by optical sensors: plastic waste goes to fine-sorting and recycling in Germany, food waste goes to the biogas plant in Nes in Romerike (outside of Oslo) and residual waste is incinerated in an energy recovery process after metal has been sorted out and removed. The ash residuals are sent to a landfill. The municipality plans to develop a pilot for carbon capture and storage at the incineration plant. The biogas plant at Nes was opened in 2013. It has the capacity to process 50,000 tonnes of food waste annually. The biogas is upgraded to liquid biogas (LBG) in a process that extracts CO_2 and reduces the volume of the biogas. The municipal biogas plant also produces bio-residuals, which are sold to neighbouring farmers as fertiliser. The produced LBG is used for the public transport system of Oslo.

In Oslo, the collection and processing of waste is divided between the public and private sectors respectively. Because of its corporate legal structure, EGE in Oslo cannot easily buy or process waste from the private sector and is instead restricted to processing household waste from the municipality of Oslo. Meanwhile, the public biogas plant at Nes in Romerike runs below capacity, and only 40% of organic household waste is treated there. Currently, the municipality has no responsibility for food waste from private businesses. This restriction makes it difficult to reduce the operating costs of the biogas plant. However, ongoing experiments with adding manure as an additional feedstock might improve the cost efficiency of the Nes biogas plant.

In parallel to the public waste collection and processing system in Oslo, the private sector has developed its own system for waste management. Over the last ten years, a large private actor, ASKO, has specialised in processing food waste from private businesses. ASKO is the wholesale and logistics business partner of one of Norway's largest grocery wholesaling groups, NorgesGruppen, which owns a number of grocery stores, restaurants, kiosks, gasoline stations and hotels. The reuse and recycling element of the logistics business includes the collection of food for redistribution by a charity organisation to reduce food loss (Matsentralen), the return of bottles and boxes, cardboard and paper recycling, plastic recycling and the reuse of different types of containers, etc. ASKO's trucks deliver food to retailers and bring their sorted waste products back for recycling on the return trip. This practice avoids the driving of empty trucks and reduces fuel costs and emissions. The collected waste streams are material-recycled: food waste as biogas, plastic waste as plastic resources, cardboard and paper for paper recycling, etc. Plastic waste is delivered to Folldal Gjenvinning, which produces a recycled interims product, a plastic granulate, which is used as a resource in NorgesGruppen's plastic bag production. The sorting of the plastic waste from private businesses is more fine-grained than the public household plastic waste, where different types of plastic are mixed, which results in a lower quality of the recycled interims product (see Table 5.1).

The introduction of the recycling system in NorgesGruppen started in grocery stores, lasting from 2004 to 2009, with the ambition of learning the

Table 5.1 Selected indicators on waste generation in Oslo, 2015

Number of inhabitants	975,744
Area in km²	266
Household waste per inhabitant in kg	336
Waste delivered to material recycling and biological treatment per inhabitant in kg	130
Share of waste delivered to material recycling, including biological treatment in %	39
Share of waste delivered to incineration in %	58
Share of waste delivered to landfills in %	3

Source: SSB Kostra.

fine-sorting of waste streams. In 2010, the first pilot including the establishment of new value chains based on the sorted waste streams was initiated. Then, after the sorting routines were well established, this was implemented by all ASKO enterprises from 2011 to 2012, and ASKO took over the logistics and transport of the sorted waste resources.

Commodity markets exist for the interim products: e.g. a secondary raw material market for items such as cardboard, plastics and metals. ASKO is able to earn revenue from these types of waste, but for food waste generating a profit is more challenging. Currently, ASKO is working to separate two types of food waste, with different objectives: bread, fruit and vegetables for feed production, and other food waste – mixed food waste and animal food waste – for biogas production. This can be done because food waste from private businesses is not mixed in the same way as food waste from households; e.g., bread that is not sold can be used as a resource for feed production. The biogas component is used in a large number of local biogas facilities. There is a conflict between the use of waste resources for energy on the one hand and for recycling, including feed production, on the other. ASKO has decided to prioritise environmental investments over economic investments and believes that this can also be legitimised economically by taking a more long-term perspective.

Besides large actors such as the municipal administration and private companies like ASKO, there are several smaller niche projects which are attempting to exploit organic waste for new purposes, such as utilising coffee gravel from coffee shops in the production of mushrooms or soap or establishing low-price lunch restaurants which serve food that would otherwise have been thrown away. This lack of coordination across parallel public and private subsectors in the processing and treatment of organic waste in Oslo shows the fragmented infrastructure of waste processing, and potentially limits critical mass and synergies across sectors and waste streams.

5.5 Analysis: urban waste valorisation as a mission

Addressing the research question presented at the outset of the chapter – *How can the mission of sustainable urban waste treatment be understood in terms of problem setting and problem solving?* – this section will discuss and reflect on the relationship between problem solving and problem setting in the governance of a mission-oriented innovation endeavour such as urban waste treatment.

5.5.1 What is the mission given – and to whom?

In Oslo, the public agency Energigjenvinningsetaten (EGE) has served as the main driver and coordinator for developing and implementing the circular system of processing and recycling the different streams of household waste generated within the municipality.

The municipality has been central in directing the system towards greater sustainability by establishing a waste sorting plant, a biogas facility and a waste incineration and district heating system. The rationale for the circular waste agenda and for building these facilities was a mandate issued by from the municipality of Oslo.

> The politicians decided in 2006 that they wanted sorting of plastic packaging and food waste from the households, and they decided to have a 50% recycling target.
>
> (Respondent from the Agency for Waste Management in the City of Oslo)

> We got the assignment in 2005–2006 to develop a system for circular waste recovery in Oslo. Then we built the sorting plant and the biological treatment plant, and the district heating company has developed and extended the district heating system.
>
> (Respondent from the Agency for Energy Recovery in the City of Oslo)

The initiatives taken to develop a circular recycling system ensured a shift of focus from energy recovery to recycling, beginning in the early 2000s. Still, after having arrived at the current recycling system, which is underpinned by heavy investment in *infrastructure* and *institutions* (e.g. the biogas plant, the sorting plant, the collection of household waste and household routines of sorting waste), there are few signals that the municipality is taking the lead with a new waste prevention system. This is not surprising, as the mandate of the Energy Recovery Department primarily targets an effective exploitation of the waste generated.

> We will sort as much as we possibly can, that is our perspective.
>
> (Respondent from the Agency for Energy Recovery in the City of Oslo)

> The biggest challenge is to get people to recycle more.
>
> (Respondent from the Agency for Waste Management in the City of Oslo)

In this sense, a waste prevention mandate or mission contrasts with the current institutional rationale or logic of the municipal Department for Energy Recovery. Thus there is reason to question whether the focus and mission to create a circular recycling system is really the best and most sustainable solution for a city such as Oslo – at least if such a system requires constant flows of waste to be economically viable, and thus constitutes disincentive to strive for waste reduction or prevention. One of our respondents confirmed that this is often the case.

> We in the waste business are constantly talking about processing waste instead of reducing waste. And to reduce must be done early on and intelligently and with the right design.
>
> (Respondent from the Agency for Energy Recovery in the City of Oslo)

One may argue that the capacity and demand for waste shown in the Oslo case constitutes a system that is oriented towards developing new value chains stemming from urban (organic) waste, and where the transition agent (EGE) has made heavy investments into the physical infrastructure enabling this production. At the same time, EGE has no financial incentives or political mandate to reduce the amount of waste in the first place. This type of sustainability mode, oriented towards recycling, is thus in conflict with demands for more circular eco-design aimed at limiting or preventing waste. In consequence, other types of actors – such as civic organisations (e.g. the student association) or private enterprises (e.g. Kutt Gourmet restaurant at the University of Oslo, and Matsentralen) – are now the ones pushing the waste prevention agenda forward as the next stage of system change in urban waste.

The Renovation Department and the Energy Recovery Department are the most central actors in any attempt to achieve higher levels of waste recycling in the municipality of Oslo. As we have seen, the Renovation Department is in charge of collecting and transporting waste from citizens, whereas the Energy Department is in charge of processing the different types of waste collected.

Our impression is that there is not much collaboration across departments in the municipality in relation to waste reduction, due to contrasting roles and mandates.

> They [the Agency for Waste Management] have a responsibility for waste reduction, and that is the opposite in relation to what we do if we only see ourselves as a producer of energy. So far, waste reduction has not been debated as much as recycling. I don't know how much they're

> working on it either, really. They have a clear historical role in ensuring that waste is safely removed from the city and handled properly. That's the most important task they have.
>
> (Respondent from the Agency for Energy Recovery in the City of Oslo)

This observed lack of coordination across the public and private sectors is an example of how existing working practices in silos do not facilitate cross-sectoral collaboration towards common goals. Such forms of silo-based organisation within and beyond the municipality represent fragmented incentive structures and a potential barrier to more radical innovation and change.

5.5.2 How are missions defined? From problem solving to problem setting

The legal regulations of waste treatment governance in Oslo restrict public actors from processing waste from the private sector and vice versa. This may constitute a somewhat rigid institutional framework and limited incentives and action space for innovation. A natural consequence of the fragmented relation across the public and private sectors in Oslo is potentially limited joint reflexivity and learning across the two domains. The tendering practices observed in Oslo associated with extensive outsourcing of municipal service provisions for collecting waste and transportation to private contractors also establish clear boundaries between the commissioner and the contractor, which may serve to hinder dialogue and mutual learning. Relatively standardised services such as waste collection and transporting have been outsourced to private contractors, whereas the development and planning of the processing of the waste streams is accomplished in-house in the municipality.

Traditional bureaucracy and silo-based working practices in the public sector typically execute power and set top-down political goals, which serves to give direction for system change. Internal direction setting in collaboration with the political level and EGE, and the creation of a circular system of biofertiliser and biogas to be used in the Oslo region, signals a traditional bureaucratic type of governance. EGE's ownership of the infrastructure and facilities also reflects a bureaucratic governance regime, which may constitute a barrier to process innovation in terms of synergies with other waste streams from the private sector.

As such, it seems appropriate to question the balance and relationship between problem setting and problem solving in the governance of urban (organic) waste in Oslo. Each municipal department has their respective mandates and there is limited coordination or joint experimentation across municipal departments. One potential limitation to such a governance mode is a weakened ability to broaden perspectives and raise ambitions in relation to sustainability in urban waste treatment. In principle, such a deficit could have been addressed at the national level. However, although Norway has a national organisation working with waste policy issues (Avfall Norge) and an interest organisation representing all municipalities (KS), there is a lack of coherence across waste systems and legislation in different municipalities, and there are several ongoing innovative projects aiming to transform existing municipal waste processing systems which unfold independently of each other. There is no dedicated national policy programme representing a coordination mechanism for joint reflexivity, learning and diffusion across municipalities in the case of waste. In consequence, the lack of coordination of experience sharing and mutual learning may paradoxically increase long-term costs and limit the effects of ongoing initiatives within the existing cost-oriented regime.

5.6 Conclusions

This chapter has sought to interpret the case of governance of urban waste as an example of mission-oriented innovation. Based on the insights gained from a case study of waste treatment in Oslo, we have discussed how governance of urban waste valorisation may be understood in terms of balancing between problem solving and problem setting.

The case study has shown how the work with waste treatment in Oslo can be interpreted as an expression of a traditional and narrow form of mission-oriented innovation policy, where the objectives have been clearly defined within the public sector domain rather than by a broad constellation of societal actors. The municipality of Oslo has been guided by political strategies aiming for a 50% recycling target. Such a target does not represent any incentive for waste prevention, but represents a technical specification that may contribute to the formation of a lock-in of the recycling stage in the waste hierarchy. The development of a circular system for the recycling of household waste constitutes a value chain that can be seen as a disincentive to support efforts to reduce waste streams in the first place. It seems as if the dynamics observed actualise a discussion of whether and how the political direction and mission given have been too specific and narrow, thus limiting the long-term action options available to the problem solvers involved. This type of (top-down) directionality contrasts with the joint and negotiated paths of directionality prescribed in the literature on transformative change (Weber & Rohracher, 2012). To avoid long-term lock-ins and to enable leaps upwards through the waste hierarchy, it appears more appropriate to operate with more open-ended and functional requirements with regard to sustainable development than to specify which sort of solutions are sought. This resembles the insight derived earlier from studies of innovative public procurement, which have concluded that functional requirements should be preferred to technical specifications in public tenders (Edquist & Zabala-Iturriagagoitiaa, 2012).

In addition to operating with functional requirements in their setting of the mission to be solved, the municipality could also have benefited from including a more diverse set of actors such as the private sector, social movements and lobby groups, into the identification and articulation of missions to be addressed and achieved. However, such working practices would call for a more networked and coordinated form of governance that contrasts with current bureaucratic, sectoral and silo-based municipal departments and working practices. It thus seems opportune to supplement current silo-based working practices with more networked governance to mobilise broader sets of societal actors into a joint reflection on possible alternative and viable ways forward towards increased sustainability in existing urban waste systems. Such an approach would be an effective response to the call for a better understanding of how the public sector can encourage more dynamic capabilities (Kattel & Mazzucato, 2018) and democratic processes in which missions are defined (Mazzucato, 2017).

Note

1 This chapter draws upon a recent paper published in *Research Policy* (Bugge, Fevolden & Klitkou, 2018). Here we take a closer look at one of the three cases presented in the original paper, and we apply another analytical framework. Instead of governance regimes, we here discuss the importance of problem setting to mission-oriented innovation in urban waste valorisation.

References

Bugge, M. M., Fevolden, A. M. & Klitkou, A. (2018). Governance for system optimization and system change: The case of urban waste. *Research Policy*, in press. doi:10.1016/j.respol.2018.10.013.

David, P. A. (1985). Clio and the economics of QWERTY. *The American Economic Review*, *75*(2), 332–337.

Edquist, C., & Zabala-Iturriagagoitiaa, J. M. (2012). Public procurement for innovation as mission-oriented innovation policy. *Research Policy*, *41*(10), 1757–1769.

European Commission. (2008). *Directive 2008/98/EC of the European Parliament and of the Council of 19 November 2008 on waste and repealing certain Directives. Official Journal of the European Union.*

Fagerberg, J. (2017). *Mission (Im)possible? The Role of Innovation (and Innovation Policy) in Supporting Structural Change and Sustainability Transitions.* TIK Working Papers on Innovation Studies. TIK Centre, University of Oslo. Retrieved from http://ideas.repec.org/s/tik/inowpp.html.

Fagerberg, J. (2018). Mobilizing innovation for sustainability transitions: A comment on transformative innovation policy. *Research Policy*, *47*(9), 1568–1576.

Frantzeskaki, N., & Kabisch, N. (2016). Designing a knowledge co-production operating space for urban environmental governance: Lessons from Rotterdam, Netherlands and Berlin, Germany. *Environmental Science & Policy*, *62*(August), 90–98.

Frantzeskaki, N., & Loorbach, D. (2010). Towards governing infrasystem transitions: Reinforcing lock-in or facilitating change? *Technological Forecasting & Social Change*, *77*(8), 1292–1301.

Gee, S., & Uyarra, E. (2013). A role for public procurement in system innovation: The transformation of the Greater Manchester (UK) waste system. *Technology Analysis & Strategic Management*, *25*(10), 1175–1188.

Geels, F. W. (2002). Technological transitions as evolutionary reconfiguration processes: A multi-level perspective and a case-study. *Research Policy*, *31*(8–9), 1257–1274.

Hodson, M., & Marvin, S. (2010). Can cities shape socio-technical transitions and how would we know if they were? *Research Policy*, *39*(4), 477–485. doi:doi.org/10.1016/j.respol.2010.01.020.

Hoornweg, D., Bhada-Tata, P. & Kennedy, C. (2013). Waste production must peak this century. *Nature*, *502*, 615–617.

Kattel, R., & Mazzucato, M. (2018). Mission-oriented innovation policy and dynamic capabilities in the public sector. *Industrial and Corporate Change*, *27*(5), 787–801.

Martin, B. R. (2015). *Twenty Challenges for Innovation Studies*. SPRU Working Paper Series. SPRU Science Policy Research Unit/University of Sussex.

Mazzucato, M. (2017). *Mission-Oriented Innovation Policy: Challenges and Opportunities* (RSA Action and Research Centre Ed.). London, UK: UCL Institute for Innovation and Public Purpose.

Mazzucato, M. (2018). Mission-oriented innovation policies: Challenges and opportunities. *Industrial and Corporate Change*, *5*(5), 803–815.

Mourad, M. (2016). Recycling, recovering and preventing 'food waste': Competing solutions for food systems sustainability in the United States and France. *Journal of Cleaner Production*, *126*(July), 461–477.

Mowery, D. C., Nelson, R. R. & Martin, B. R. (2010). Technology policy and global warming: Why new policy models are needed (or why putting new wine in old bottles won't work). *Research Policy*, *39*(8), 1011–1023.

Murray, R. (2002). *Zero Waste*. London: Greenpeace Environmental Trust.

Nelson, R. R. (2011). The moon and the ghetto revisited. *Science and Public Policy*, *38*(9), 681–690.

Papargyropoulou, E., Lozano, R., Steinberger, J. K., Wright, N. & Ujang, Z. b. (2014). The food waste hierarchy as a framework for the management of food surplus and food waste. *Journal of Cleaner Production*, *76*(August), 106–115.

Schot, J., & Steinmueller, W. E. (2018). Three frames for innovation policy: R&D, systems of innovation and transformative change. *Research Policy*, *47*(9), 1554–1567.

Shove, E., & Walker, G. (2007). CAUTION! Transitions ahead: Politics, practice, and sustainable transition management. *Environment and Planning A*, *39*(4), 763–770.

Smith, A., Stirling, A. & Berkhout, F. (2005). The governance of sustainable socio-technical transitions. *Research Policy*, *34*(10), 1491–1510.

Smith, A., Voß, J.-P. & Grin, J. (2010). Innovation studies and sustainability transitions: The allure of the multi-level perspective and its challenges. *Research Policy*, *39*(4), 435–448.

Uyarra, E., & Gee, S. (2013). Transforming urban waste into sustainable material and energy usage: The case of Greater Manchester (UK). *Journal of Cleaner Production*, *50*(July), 101–110.

Weber, K. M., & Rohracher, H. (2012). Legitimizing research, technology and innovation policies for transformative change: Combining insights from innovation systems and multi-level perspective in a comprehensive 'failures' framework. *Research Policy*, *41*(6), 1037–1047.

6 Beyond animal feed?

The valorisation of brewers' spent grain

Simon Bolwig, Michael Spjelkavik Mark, Maaike Karlijn Happel and Andreas Brekke

6.1 Introduction

The beer brewing industry has long been considered somewhat bio-circular since the major part of its organic side-stream, spent grain, returns to the biological system in the form of animal feed. In many places, spent grain has historically been given away free to farmers as livestock feed, especially for cattle. In terms of environmental sustainability, this can be a good management choice for the bio-economy, placing it between 'recycle' and 'reuse' in the waste pyramid (see Chapter 3). However, the fast deterioration of untreated spent grain requires the presence of local farmers as well as good transport infrastructure. Yet, according to Statistics Norway and Statistics Denmark since the 1980s, the number of farms has declined by 60% in Norway and Denmark, making the disposal of spent grain potentially more expensive and complicated. In addition, giving spent grain to farmers brings little or no revenues to breweries and prevents further valorisation of this resource. Globally, brewers produce about 38.6 million tons of spent grain a year (Lynch, Steffen & Arendt, 2016; Mussatto, 2014); so a change in spent grain management could therefore have significant environmental and economic impacts.

The large quantities of spent grain, along with an increasing interest in organic waste valorisation and circular bioeconomy, have spurred interest in developing new valorisation pathways as an alternative to the traditional use of spent grain as animal feed. Spent grain has a high protein content and other nutritional assets, and research projects have shown that it can be used as a feedstock in various industries, including livestock feed, food and nutrition, chemicals, pharmaceuticals and biofuels (Buffington, 2014; Mussatto, 2014; Thomas & Rahman, 2006). Yet, despite this technical potential, scholars have identified few examples of advanced uses of spent grain on an industrial scale (Aliyu & Bala, 2011; Mussatto, 2014), suggesting low levels of the deployment of research results.

Bugge, Hansen and Klitkou (2016) and Chapter 2 in this book identify different visions and perspectives on the bioeconomy: the biotechnology vision, the bio-resource vision and the bio-ecology vision. In this context, the bio-technology and bio-resource perspectives dominate the scientific literature on spent grain, which has a strong focus on biochemical and technological aspects, as is evident from the reviews (Aliyu & Bala, 2011; Mussatto, 2014; Thomas & Rahman, 2006). Socio-economic issues, such as value creation, competitive advantage, and consumer acceptance, have received less scholarly attention. And while the literature on technical options is rich and consistent, the few studies of the profitability of spent grain as an industrial feedstock report divergent results, ranging from optimistic (Mussatto, Moncada, Roberto & Cardona, 2013) to very pessimistic (Buffington, 2014).

In view of this, this chapter starts from the premise that company decisions on how to use and manage organic residues not only reflect technical possibilities but are also influenced by socio-economic, supply chain and regulatory factors. These include company-specific and industry-wide sustainability policies and initiatives, which can form part of companies' corporate social responsibility (CSR) efforts. In the brewing industry, important sustainability areas, apart from organic residues, are water consumption, waste water management, energy use and the resulting CO_2 emissions, sustainable packaging (Olajire, 2012), and responsible drinking. Integrating these sustainability areas into the core of a business can potentially increase value creation (Bocken, Short, Rana & Evans, 2013; Short, Bocken, Barlow & Chertow, 2014). In this regard, Porter (1985) argues that companies can obtain competitive advantage by pursuing a product differentiating strategy, where CSR can be a means of product differentiation (McWilliams & Siegel, 2011), through branding, for example (Roberts & Dowling, 2002), thus creating a non-imitational resource for the company (Reinhardt, 1998). In this context, Clark, Feiner and Viehs (2015) provide evidence of a positive link between CSR and a company's competitive advantage.

Against this background, this chapter aims to deepen the understanding of circularity in the brewing industry regarding organic waste. The specific aim is to investigate current management practices and options for valorisation of spent grain in brewing value chains in Denmark and Norway.

6.2 Value creation and sustainable competitive advantage

The exploration of alternative uses of spent grain encompasses potential economic gains, most obvious when a brewery incurs costs related to waste disposal but also when spent grain is given away for free. There are potential economic gains to be secured when creating a new value chain for spent grain. Yet these gains must be assessed in a cost-benefit setting. Benefits could also be related to CSR measures. How CSR can contribute to sustainable competitive advantage (McWilliams & Siegel, 2011) is elaborated on in Chapter 3.

Various industries use the principles of industrial ecology to convert waste to positive value assets, yet the literature lacks clear links to profitability and increased competitiveness (Short et al., 2014). Arguably, industrial ecology and, more narrowly, industrial symbiosis can be a foundation for business model innovation and it is important to look beyond conventional resource productivity and process innovation and find new ways of creating value (ibid.). Business models have been outlined in various ways. Whether it is the concept of marketing myopia (Levitt, 1960), value chains (Porter, 1985) or blue ocean strategy (Kim & Mauborgne, 2005), a business model essentially explains 'how a firm does business' (Short et al., 2014). Richardson (2008) sums up three aspects of business models: *value proposition, value creation and delivery, and value capture.* Business model innovation can take place in each of the three aspects, whether it depends on the product or service offered, and to whom it is offered, through which activities the value is created, or how the company handles costs and revenue (ibid.). Innovation in business models can spur sustainability (Short et al., 2014). Bocken et al. (2013) identify nine business model archetypes aiming to improve sustainability and one of them is *creating value from waste.* The strategy here is to eliminate waste and turn waste streams into inputs to other value-creating processes and production (Short et al., 2014).

Yet studies of the brewing industry emphasise the lack of change in the spent grain business model, i.e. the use of the residue as animal feed. There are many technical options for spent grain valorisation other than feed, as elaborated below, but brewers do not appear to have experimented with or implemented them (Aliyu & Bala, 2011) as a way of creating or capturing value. Even the Sierra Nevada Brewery in California, highly acclaimed for its sustainability measures, provides all its spent grain to farmers (Sierra Nevada Brewery, 2015).

Research and development (R&D) activities are playing an increasingly important role in companies' CSR policies (Baumgartner & Ebner, 2010). R&D is often integrated into the innovation and technology aspects of the economic dimension of sustainability, reducing the environmental impact of new products and business activities (ibid.). Engaging in R&D is particularly beneficial for companies that are among the first to adopt a new product, also known as 'first movers' (Robinson, Fornell & Sullivan, 1992; Srivastava & Lee, 2005). In light of this, it is interesting that the brewing industry does not have a tradition of making strong investments in R&D (Bamforth, 2000).

In view of these considerations, this chapter addresses the following research questions: What technical options exist for adding value to spent grain? What are the current status, opportunities and barriers for converting spent grain from low- or zero-value livestock feed into assets with a higher value? In this context, what R&D activities do Nordic brewers undertake regarding the management of spent grain and related areas and what motivates or hinders such activities?

6.3 Methods

This chapter is based on a case study of Nordic breweries and their handling of spent grain. The research follows a multiple case-study design (Bryman, 2015; Pettigrew & Whipp, 1991). Thirty-two breweries were selected through purposive sampling (Bryman, 2015). Seventeen were selected from Denmark from a database of 107 breweries (Ratebeer, 2017), and fifteen from Norway based on a list provided to the authors by the Norwegian Brewers' Association. The sample has breweries in a range of sizes, with a predominance of those producing up to 500,000 hectolitres of beer per year. Table 6.1 gives the key characteristics of the breweries that were interviewed for this study. We henceforth refer to the breweries by number rather than name to ensure the confidentiality of the information obtained in the interviews.

Eighteen breweries were interviewed through short telephone interviews, and an additional, in-depth interview lasting 90 to 120 minutes was undertaken with fourteen breweries following a semi-structured interview design (Bryman, 2015). The guide for the in-depth interviews was informed by a value chain approach and addressed topics such as firm characteristics, CSR strategies, spent grain value chain (structure, actors, coordination and technology), spent grain economics, alternative uses of spent grain, R&D projects, and policies and institutions. See Gereffi and Fernandez-Stark (2016) and the section on governance for waste valorisation in Chapter 3. We also reviewed academic publications and grey literature such as industry and company reports.

The interview data was analysed in two rounds. After structuring the data using the qualitative data analysis software Atlas (Atlas.ti, ver. 7.5.16), using an inductive approach, we derived five categories or factors of spent grain management from the interview data: economy, product, CSR, production and regulation. For each factor, we identified and described a number of variables, based on a review of the literature and the interview data. Table 6.2 lists these factors and variables.

Three *aspects* of spent grain management were likewise identified based on the nature of the interview statements: activities, opportunities and barriers (section 6.6), i.e. whether a statement related to an activity currently undertaken by the brewery, a future opportunity for alternative use or a barrier to realising such opportunities. The interviews were then screened again for relevant statements regarding spent grain factors and aspects. This analysis involved the quantification of the statements given in the interviews on the factors and aspects, as well as qualitative analyses. The results are presented in sections 6.5 and 6.6.

Table 6.1 Characteristics of the breweries interviewed in the case study

Brewery number	*Size (HL/Y)*	*No. of employees*[2]	*Gross revenue (€1,000)*[3]	*Year founded*	*Country*	*Interview type*[1]	*Position of interviewee(s)*	*Date of interview (month/year)*
1	116,900,000	42,062	8,422,249	1849	DK	F	Group CSR Manager	01/2015, 02/2016
2	9,700,000	2,350	852,848	1989	DK	P, F	Sourcing and Brewery Managers	01/2017, 02/2017
3	35,000	22	5,783	1902	DK	P, F	Brewmaster	12/2017, 02/ 2017
4	220,000	100	25,173	1999	DK	F	Technical and Plant Managers	03/2017
5	10,000	13	934	2012	DK	P, F	Brewmaster	11/2016, 01/2017
6	20,000	18	3,904	1998	DK	F	Brewmaster	03/2017
7	5,000	8	832	1897	DK	P	CEO	12/2016
8	800	3	93	2011	DK	P, F	Plant Manager	12/2016, 01/2017
9	15,000	7	N/A	2016	DK	P, F	Brewmaster	01/2017, 03/2017
10	740	2	N/A	2016	DK	P	Brewmaster and Owner	11/2016
11	1,500	3	N/A	2008	DK	P	Brewmaster and Owner	11/2016
12	30	3	11	1996	DK	P	Brewmaster	11/2016
13	2,000	12	348	2007	DK	P	NA	11/2016
14	500	1	87	2016	DK	P	Brewmaster and Owner	12/2016
15	3,000	7	291	2011	DK	P	Owner	12/2016
16	700	2	N/A	2013	DK	P	Brewmaster and Owner	01/2017
17	4,000	5	612	2003	DK	P	NA	12/2016
18[4]	1,410,000	1,133	533,548	1918	NO	F	Head of Development	06/2016
19	240,000	105	32,826	1904	NO	F	Brand and Product Responsible	02/2017
20	2,500	1	156	2011	NO	F	CEO and Owner	01/2017
21	20,000	38	6,366	2002	NO	F	Managing Director	03/2017
22	10,000	15	2,563	2009	NO	F	CEO	01/2017
23	3,000	17	842	2012	NO	P	CEO	02/2017
24	4,000	16	2,038	1989	NO	P	CEO	01/2017
25	3,000	7	731	2006	NO	P	CEO	01/2017

continued

Table 6.1 Continued

Brewery number	*Size (HL/Y)*	*No. of employees*[2]	*Gross revenue (€1,000)*[3]	*Year founded*	*Country*	*Interview type*[1]	*Position of interviewee(s)*	*Date of interview (month/year)*
26	5,000	8	1,653	2011	NO	P	CEO	02/2017
27	7,500	30	7,187	2009	NO	P	CTO	02/2017
28	9,000	24	2,080	2013	NO	P	Brewmaster	01/2017
29[5]	20,000	415	N/A	1918	NO	P	Marketing and Info. manager	02/2017
30	15,000	32	4,409	2003	NO	P	CEO	01/2017
31	135,000	56	22,748	2001	NO	P	CEO	02/2017
32	225,000	159	26,327	1995	NO	P, F	CEO	01/2017

Notes

1 F = Face to face, P = Phone.

2 Full-time equivalents (FTE) in 2016. Source: Own interviews.

3 The data source for breweries 1–4 and 18–22 is annual company reports, for breweries 5–17, www.datacvr.virk.dk, and for 23–32, www.proff.no.

4 This brewery is a subsidiary of brewery 1.

5 This brewery is owned by brewery 18.

Table 6.2 Factors and variables of spent grain management

Factor	*Variable*	*Description*
Economy	Direct costs	Costs that can be traced to specific cost objects such as a waste treatment fee or direct materials. Direct costs tend to be variable costs.
	Indirect costs	Costs that cannot be traced to specific cost objects. Indirect costs tend to be fixed costs or periodic costs (AccountingTools, 2017).
	Revenue	Income from the use and/or sale of spent grain.
Product		Quality and safety aspects of spent grain management that influence the quality of the main product (beer).
CSR	Water	Water use and water efficiency in the brewing process, including water content of the spent grain (Olajire, 2012).
	Climate and energy	Focuses on energy use, energy efficiency and energy conservation (Sturm et al., 2013). Includes activities to enhance biogas production.
	Waste	Activities related to the recycling, recovery or reuse of residues from brewing.
	Sustainable sourcing	Integrating sustainability into decisions regarding the production, distribution and purchasing of raw materials and products (Schneider & Wallenburg, 2012).
	Social issues	Strengthening the relationship with the local community including local farmers.
Production	Technical equipment and infrastructure	Quality or life endurance of technologies and equipment needed for brewing, storage or drying/enhancing of spent grain.
	Logistics	Logistical issues regarding the handling of spent grain. Transport and geographical distance are the main aspects.
	Size	Size and production capacity of the brewery, which can affect spent grain management.
Regulation	Regulation and policy	Influence of policy/regulations from regional, national and European authorities, including rules for subsidies and investments.
	Certification	Influence regarding standards for safety, quality or environmental compliance certification.

6.4 Technical options for spent grain use

Organic waste is defined here as 'biodegradable waste from gardens and parks, food and kitchen waste from households, restaurants, caterers, hotels and retail premises and similar waste from food processing' (Jürgensen & Confalonieri, 2016, p. 3). However, brewers' spent grain is not a waste product but the main side-stream from the beer brewing process and represents approximately 85% of all organic by-products from brewing. A large volume of spent grain is generated by the brewing process, around 20 kg per hectolitre of beer produced (Aliyu & Bala, 2011; Mussatto, Dragone & Roberto, 2006), and it is available all year round and at a low cost (Buffington, 2014).

In the wider context of sustainable production, these properties, together with its biophysical and nutritional attributes, have stimulated interest in adding value to spent grain, and numerous laboratory experiments have been performed with this objective in mind. These studies have been recently reviewed (Aliyu & Bala, 2011; Mussatto, 2014; Thomas & Rahman, 2006), revealing a great diversity in potential processes and products. Figure 6.1 shows the technical processes through which spent grain can be transformed into different product types (marked in bold), i.e. human food, animal feed,

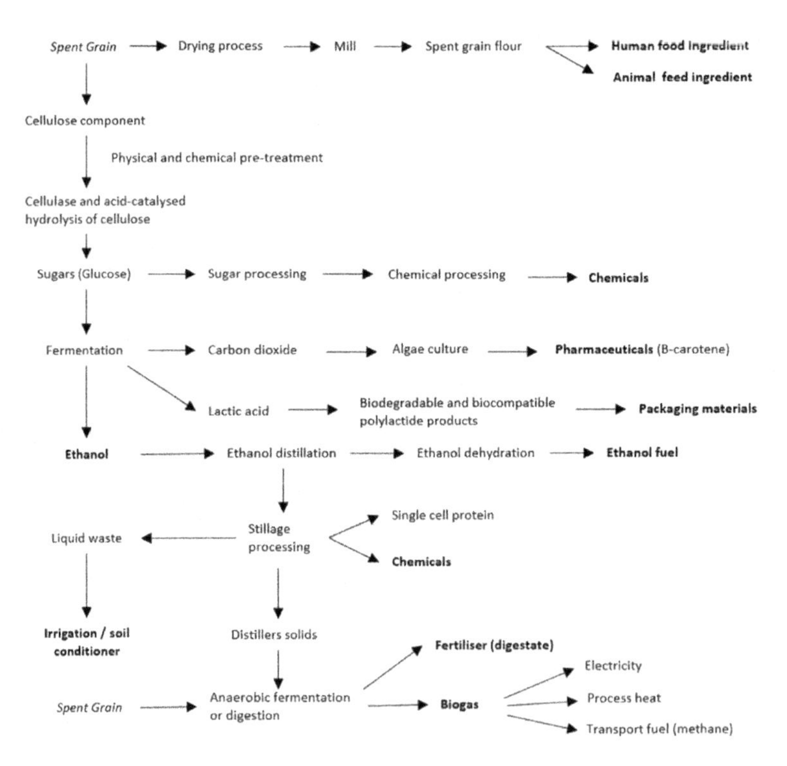

Figure 6.1 Technical processes and products based on spent grain.

Source: Adapted from Buffington (2014) and van Wyk (2001)

chemicals, pharmaceuticals, packaging materials and energy resources (biogas, ethanol), along with the by-products (e.g. fertiliser) of these processes. Below we briefly discuss alternative valorisation options, drawing on the waste pyramid presented in Chapter 3: disposal, energy recovery, recycling, reuse and waste prevention. We do not assess whether one option is more sustainable than another, as this would require detailed life cycle analyses and also depend on local-specific factors.

Disposal. A key issue in handling spent grain is its high moisture levels. Spent grain consists of 70–80% water (Lynch et al., 2016; Thomas & Rahman, 2006), meaning that transporting spent grain is costly per kg of dry matter. Second, the rich polysaccharide and protein contents of spent grain make it susceptible to fast deterioration and spoilage (Thomas & Rahman, 2006), with associated health and smell hazards. Hence disposing of spent grain as a waste requires constant effort and can be expensive for the brewery, so this option is the least preferred.

Energy recovery. Spent grain can also be used in energy production, as it can show net calorific values of 18.64 MJ per kg dry mass and is thus interesting as raw material for combustion (Keller-Reinspach, 1989). Spent grain can also be used as a substrate for biogas or second generation-ethanol production, replacing natural gas and gasoline respectively (Mussatto, 2014).

Recycling. Using spent grain in animal feed has several positive benefits, including increasing milk production by cows and improving the meat quality of livestock (Thomas & Rahman, 2006). Spent grain can also be recycled as a soil conditioner. Combining spent grain with sludge or woodchips can improve soil fertility (ibid.).

Reuse as human food. Spent grain has high contents of fibre, protein and minerals, making it potentially attractive for human consumption. Experiments have improved properties in various food products including increased levels of protein and fibres in cookies (Öztürk, Özboy, Cavidoğlu & Köksel, 2002), bread and processed meat products (Thomas & Rahman, 2006). However, consumer acceptance and the quality of the final product need more attention (Mussatto, 2014).

Reuse in chemical processes. Applying spent grain in chemical processes has also been tested. Spent grain is rich in cellulose, polysaccharides and natural antioxidants, all compounds adding value to industrial applications. Furthermore, spent grain can be used in the production of paper-based products such as paper towels, business cards and coasters (Mussatto et al., 2006; Thomas & Rahman, 2006). The most promising use of spent grain in chemical processes is as an adsorbent for organic compounds from waste gas or dye from wastewater (Mussatto, 2014). Spent grain has also proved useful in biotechnical processes (ibid.).

Waste prevention. The amount of spent grain by-product generated per volume of beer produced depends on the brewing equipment; the type and quality of the vessel that separates the wort from the spent grain is especially significant when it comes to the efficient use of malt and water and hence for reducing the amount of spent grain. In general, a mash filter, employed by many larger breweries, generates relatively lower amounts of spent grain, compared to the less efficient and cheaper mash filtration method typically used by many small or craft breweries producing in smaller batches (Lynch et al., 2016).

6.5 Overview of current spent grain management

Figure 6.2 illustrates the current management of spent grain in Denmark and Norway, identified through the case study. The top part shows the brewing process and the origin of spent grain along with the (relatively few) instances where the spent grain is used as feed or fertiliser on the brewery's own farm. The bottom part shows the firms or units that process or use the spent grain after it has been collected at the brewery, alongside the different end uses of the spent grain. The dotted arrow shows a situation where a third party organises the transport and handling of the spent grain on behalf of the end users. While the figure gives the impression of diverse management regimes and uses of spent grain, the case study revealed that in the vast majority of cases the spent grain was delivered directly to local livestock farmers, as described below.

Nearly all brewers were connected to one or more farmers who used the spent grain as livestock feed, often for cattle and sometimes for pigs. The farmers valued the nutritional quality and protein content of spent grain. These agreements were often long term and were made with one or several farmers depending on the size of the brewery and the size of farmers' herds. The farmers usually collected the spent grain at the brewery. Some breweries, especially the large ones, received payments from the farmers, but most gave

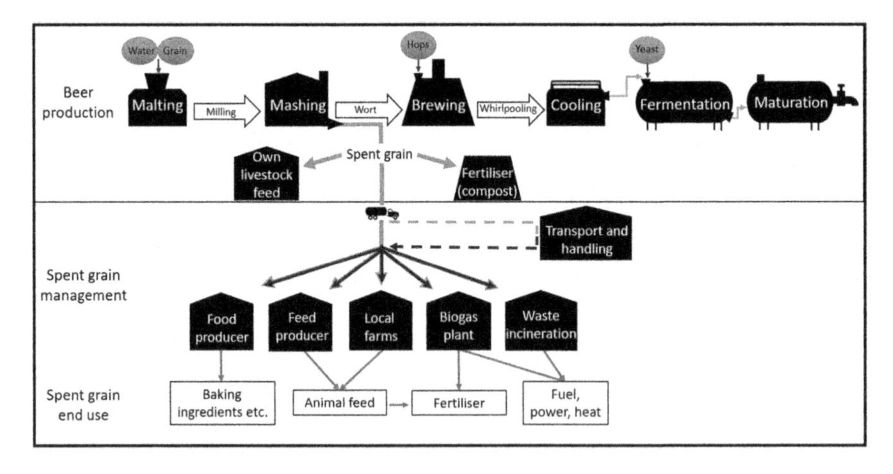

Figure 6.2 Overview of current management of spent grain.

Source: Authors' interpretation of interview statements made by brewers.

the grain away. In a few cases, an intermediary firm organised the collection of the spent grain on behalf of the farmers; in Denmark this was the case with brewery 2 among others, while the farmer buying grain from brewery 4 resold some of it to other farmers.

The price obtained for spent grain fluctuated depending on conditions in the feed market, including the price of substitute feed. The price also depended on the dry matter content, which partly depends on the brewing equipment. One large brewery (2) in Denmark received €16 (DKK 120) per ton from an intermediary, another €8 (DKK 60) per ton (3) from a farmer, while a third (4) received €134 (DKK 1,000) per ton from a local farmer acting as an intermediary. In Norway, one large brewery (32) could not obtain any payment from farmers for its spent grain, which was also the case for nearly all the small breweries (e.g. 5, 11 and 15). These significant price variations over time and space suggest a poorly functioning market for spent grain.

In a few cases, spent grain had other uses alongside or instead of simply being given to farmers. These were feed for the brewery's own livestock (5, 12), feed in high-quality wagyu beef production (30), compost on the brewery's own farm (8, 14 and 26), collection by a local biogas plant for a fee paid by the brewery (15), being sold or given occasionally to a local bakery (5, 22) and being disposed of as waste (21, 32). Two of these uses, as wagyu feed and as a baking ingredient, represent an improved use of spent grain compared to animal feed in terms of value added.

6.6 Why don't the breweries invest in alternative options?

This section analyses the interview data in more detail with a view to understanding what motivates current spent grain management, what brewers perceive as alternative management options and what hinders them in pursuing them. The last part of the section focuses on brewers' engagement in R&D projects in relation to these aspects.

Table 6.3 quantifies the interview statements of thirty-two brewers regarding the factors that influence aspects of spent grain management – i.e. current activities as well as opportunities for and barriers to alternative management options. Overall, economy (i.e. cost reduction) was clearly the dominant factor, with seventy-five statements, followed by production (47) and CSR (27). Brewers mentioned product (17) and regulation (8) less often. When considering activities only, economy is again the dominant factor, followed by production and CSR. The same pattern applies to opportunities. It is interesting to note that brewers considered production issues to be the main barrier to implementing spent grain management alternatives, followed by economy.

The interviews also produced qualitative data on the factors involved in spent grain management in terms of how they were seen to affect current activities, opportunities and barriers. The analysis below focuses on the dominant factors of economy, production and CSR.

Table 6.3 Number of interview statements by brewers in relation to factors and aspects of spent grain management

Aspect	*Activities*	*Opportunities*	*Barriers*	*Total*
Factor	$n=32$	$n=28$	$n=21$	$n=32$
Economy	39	22	14	75
Product	6	6	5	17
CSR	14	11	2	27
Production	16	16	15	47
Regulation	1	5	2	8
All	76	60	38	174

Note
The table shows the number of statements in interviews related to each factor (economy, product, CSR, production and regulation) and aspect (activity, opportunity and barrier). If more than one statement was made on the same combination of factor and aspect, then these were not counted, but when statements were made on several variables relating to a factor (e.g. direct costs and revenue in the case of economy) for a given aspect, then these statements were included in the count (therefore there are more than 32 statements on the combination of economy and activities). In total, 174 such statements were identified in 32 interviews. The sign 'n' denotes the number of breweries that gave a statement on the aspect in an interview.

6.6.1 Current activities

Economy was by far the most important factor in current spent grain management practices. Within this, cost savings were important for all breweries, while earning revenue was possible only for the large breweries (e.g. 1, 2, 3, 4 and 18). Risk management, in terms of avoiding a situation where spent grain exceeded storage capacity, generating expenses for waste disposal or even disrupting the brewing process, was also a key economic consideration. We also observed a close interplay between economy (cost savings) and production (mainly handling and storage of equipment, and location). CSR factors, in contrast, had only a minor influence on spent grain management among both large and small brewers. Where CSR was a motivation, it co-existed with or depended on other considerations such as profitability (1), compliance with public (20) or private (8) regulation, or simply convenience.

Spent grain is a potential health hazard; it quickly deteriorates, smells and fills the storage space of the breweries, and so the brewers perceived it mainly as an inconvenient by-product. None of the breweries owned technologies that allowed longer and safer on-site storage of the spent grain, such as cold storage or pressing/drying equipment. If it was not collected at regular intervals, it would force the brewery to halt production. Hence, many brewers emphasised the risks and the logistical and technical issues involved in handling spent grain. In this light, it is unsurprising that the dominant economic motivation for the current management of spent grain was reducing the costs of storage, transportation and disposal. Many brewers emphasised both indirect cost savings in terms of reduced labour time and 'hassle', as well as reduced expenditure on storage facilities, transportation and waste disposal fees (e.g. 5, 11 and 15). A case in point was brewery 1, whose strategy was to sell the spent grain at the 'highest price and with the least trouble'.

The waste disposal costs avoided due to the practice of farmers collecting the spent grain at the brewery gate are significant. In Denmark, one company specialising in the collection and treatment of food waste for subsequent use in biogas production was charging from €44 to €59 per ton plus transport costs, depending on the stability and volume of the waste, although large and stable volumes could attract a lower price (personal communication with company). In Norway, one brewery (32) paid around €101 per ton to dispose of its spent grain.

In this context, it is noteworthy that many brewers underlined the importance of having long-term agreements with reliable buyers or takers of the spent grain as a way to manage this production risk. Some brewers (e.g. 6, 32) found it difficult to get farmers to collect the spent grain year-round.

Production variables, and in particular logistical issues related to the geographical location of the brewery, also influenced spent grain management. In Norway, several small breweries were located far away from cattle farms, and the spent grain was therefore not used as feed. One brewery in northern Norway (32) had experienced reduced demand from farmers in recent years, especially during the summer, because of changes in feeding practices and fewer farm units. This meant that some of the spent grain was disposed of as waste, incurring the brewery costs in the order of €15,195 (NOK 150,000) per year. On the island of Svalbard, restrictions on waste disposal forced the brewery to ship the spent grain to the mainland, where it was disposed of as waste (21). As a solution, the brewery engaged in energy production on Svalbard replaced coal with a mixture of spent grain and demolition wood chips. Hence distance, combined with the volume of by-product, also determined whether brewers could sell the spent grain to livestock feed producers.

CSR motivated current spent grain management for some small breweries. For example, one brewery making certified biodynamic beer composted the spent grain and applied it to its own barley field to comply with the Demeter standard (8). Combatting climate change is a central part of the CSR brewery policy (1), and the company carries out CO_2 accounting across all its plants; however, spent grain management does not feature strongly in its CSR policy.

6.6.2 Opportunities

'Opportunities' refers to alternative ways of handling spent grain, which the brewers showed an interest in during interviews; some had concrete plans to adopt these in the future. Several alternatives to the current usage were mentioned. Many discussed the possibility of selling or using spent grain as a biogas substrate. Other alternatives were inputs on their own farms for composting or as animal feed, or selling to feed producers. Some brewers made general statements such as that they were 'interested in alternative ways of using the spent grain' (23, 24 and 25) without specifying the kind of alternative, suggesting limited knowledge of or interest in alternative uses.

Economy and production, followed by CSR, were the factors mentioned most often that could induce a switch in spent grain usage away from animal feed and disposal. Regarding economy, many mentioned the prospect of earning more revenue from the side-stream, often combined with technical and logistical measures. New equipment was sometimes mentioned (e.g. 32) as a key factor in changing the use of spent grain, especially equipment for dewatering the spent grain to increase its storability and reduce transport costs. Some emphasised the high quality of spent grain as an animal feed and the potential economic benefits of its nutritional value. One brewer asserted that spent grain 'contains a lot of vitamins and minerals [of benefit to humans], but these are currently not properly utilised' (18), and another that spent grain 'is quite sought after because it is good for the digestive system of animals' and this can improve the quality of meat for human consumption (19).

The most common incentives mentioned under CSR were the environmental benefits in terms of improved waste handling or recycling, sustainable sourcing and to a lesser extent climate mitigation. Sustainable sourcing was expressed in 'circularity' terms such as 'closed cycle' (10) and as an example of 'responsible thinking' (21). These incentives were often expressed as an aspect of other incentives such as improved product. One brewery believed that the reuse of spent grain in production would have simultaneous economic, environmental and product-quality benefits that could be exploited in marketing and consumer communication (22), as shown by the example of the Sierra Nevada brewery in the USA.

Finally, some brewers connected an alternative use of spent grain to certification, either regarding product and food safety, or as part of a broader environmental responsibility certification. In this regard, brewery (21) emphasised its comprehensive view on sustainability, which it planned to express and implement through Environment Lighthouse certification the following year.

6.6.3 Barriers

Barriers are the perceived obstacles to switching to an alternative use of spent grain. Economic and production factors were the barriers most often mentioned by brewers. In terms of *economics*, many expressed an unwillingness to invest time and effort in changing their management of spent grain due to the indirect costs involved in terms of staff time (e.g. 9, 10, 14 and 23). One brewery (10), for example, considered that implementing changes to install such a system was 'too difficult and too much work', especially in light of its lack of storage facilities and the limits imposed by regulations. Another brewer (14) was interested in selling the spent grain to bakeries, but time was the main constraint to developing this option, and like other alternative options it remained at the experimental level.

Brewers often mentioned such economic factors in combination with production variables, specifically an insufficiency of *equipment or storage space*, and several brewers (e.g. 18, 32) stated that they lacked the equipment to dry the spent grain and store it for longer periods. Moreover, EU and national regulations on fodder production require safe storage of spent grain, as emphasised by brewery 18. Yet it is noteworthy that many medium-sized and large breweries (6, 10, 22, 27, 28, 30 and 31) did not perceive production variables, including transport, to be a barrier to implementing new options, or as a negative aspect of current operations.

Securing the *investment finances* needed to upgrade or replace the equipment that would enable alternative uses of spent grain was also seen as a barrier. Brewery 2 noted that it had run trials to dry and burn the spent grain for energy production, but that it would need to invest in new equipment to do so on a large scale. Brewery 4 observed that its brewing equipment was old and therefore resource inefficient. In particular, the vessel that separates the wort from the spent grain is crucial for the efficient use of malt and water and hence for reducing the amount of spent grain per volume of beer produced. However, investment in such equipment had so far been outcompeted by more customer-focused investments, most recently a new bottling line.

Some respondents (7, 25) noted that the expected *low returns from investments* in alternative uses of spent grain made it difficult to access finance from within the firm, while others (e.g. 1) emphasised the importance of having a good business case, including for 'green' projects. Indeed, many mentioned the need for cost efficiencies in all parts of an operation and this was related to the strong competition and low-value nature of beer (2). In this regard, one brewery (32) observed that getting approval for a project with a payback time of more than two or three years depended on the size of the investment and how well it compared to competing projects. As mentioned above, brewery 4 had recently invested in new, expensive bottling or filling lines, requiring significant financial resources.

Firm size also influenced a firm's ability and willingness to invest. Several breweries (5, 7, 23, 24 and 28) observed that their small size reduced their ability to pursue alternative spent grain options, and brewery 24 noted 'we are too small, and the alternatives are too complicated and costly'.

Given the above considerations, the opportunities to implement greener products and techniques may well be greatest in situations where core brewing activities are undergoing significant changes, such as when production capacity is expanded (31) or the brewery is relocated to a new site (17). Regarding *CSR* as a factor for spent grain usage, one important regulatory barrier mentioned (by brewery 1) was the lack of a system whereby the brewery could receive carbon credits for the biogas produced 'off-site' by other companies from its spent grain. This limited the CSR benefits of selling spent grain to third-party biogas producers.

6.6.4 Involvement in research and development projects

Nine of the thirty-two breweries interviewed mentioned R&D projects, and thirty-nine statements relating to R&D were obtained (Table 6.4). Twenty-two statements on R&D concerned opportunities for engaging in future R&D projects, while only nine referred to current R&D activities, reflecting a generally low level of engagement in such projects. That said, the vast majority of brewers were continuously engaged in the development of new products, product variants and process optimisation without perceiving these efforts as R&D projects. Addressing sustainability issues was rarely the focus of these activities. In the words of brewer 19, 'We don't think about research because it is product based, so we say, "OK, here's something we can do better", and then we need to find a solution to make it better'.

Breweries with ongoing R&D projects were collaborating with research institutions, as well as private companies. Two Danish brewers (1, 2) were working with the Technical University of Denmark on spent grain innovations and sustainable packaging. One Norwegian brewery (30) was collaborating with the Norwegian University of Life Sciences and a food company on a project to develop cattle feed based on spent grain mixed with traditional feed ingredients and tailored to the production of high-value wagyu meat. There were also R&D collaborations in other areas of sustainable production. For example, brewer 2 was working with several firms and municipal institutions on energy efficiency and industrial symbiosis, while brewer 1 was working with the Carbon Trust to measure the carbon footprint of its value chain and develop a road map to meet its targets for reducing greenhouse gas emissions (The Carbon Trust, 2017). Lastly, some breweries were engaged in knowledge generation through meetings with other breweries or via the brewers' association in their respective countries.

Table 6.4 Number of interview statements by brewers in relation to factors and aspects of engagement in R&D projects

Aspect	*Activities*	*Opportunities*	*Barriers*	*Total*
Factor	$n=6$	$n=5$	$n=4$	$n=9$
Economy	–	3	7	10
Product	3	3	–	6
CSR	6	12	–	18
Production	–	3	–	3
Regulation	–	1	1	2
All	9	22	8	39

Note
The table shows the number of statements in interviews related to each factor (economy, product, CSR, production and regulation) and aspect (activity, opportunity or barrier). The sign 'n' denotes the total number of breweries in the size class which made a statement on the aspect in an interview. The answers are from interviews with four large and five small breweries that included R&D aspects in their answers.

Opportunities for taking up new R&D projects were a common topic addressed in interviews, where CSR was the dominant factor (Table 6.4). One brewer (1) was interested in assessing the broader environmental impact of alternative uses of spent grain in relation to energy, land use and greenhouse gas emissions. Also from a sustainability perspective, two brewers were interested in research projects on energy efficiency and renewable energy (2, 5), while a long-term objective of brewer 5 was certification in LEED (Leadership in Energy and Environmental Design), a green building rating system (US Green Building Council, 2017). Other favoured topics were water savings, sustainable packaging and ways of reducing the company's overall environmental footprint. Brewery 22 emphasised that more information and learning possibilities were critical for transitioning to more sustainable production:

> This is not a matter of subsidising investments and upgrades. This is a matter of obtaining information on feasible ways of improving production processes, i.e. consumer information in the same sense as you get when you buy domestic appliances. In other parts of the primary sector, e.g. agriculture, it is common to have knowledge and counselling centres.

Brewery 9 likewise stressed the need for increased knowledge generation and sharing within the craft brewery side of the industry.

The perceived barriers to engaging in R&D projects were mainly economic (Table 6.4). One brewer (9) found that the labour costs of preparing funding proposals were a key obstacle, while another (19) anticipated that entering into R&D projects would involve significant operational costs in terms of professional and administrative staff time. There were also perceived regulatory barriers to involvement in R&D, such as compliance with the EU Best Available Techniques (BAT) regulation (Kawa & Luczyk, 2015; Kristiansen, Johansen, Mou & Johansson, 2011; Olajire, 2012). Hence, brewery 2 anticipated the risk that their R&D projects would be viewed as either too broad or not feasible to implement under BAT rules.

6.7 Conclusion

The management of spent grain observed in the case study categorises it as part of the bio-resource vision, while alternative uses would introduce the other two visions, bio-technology and bio-ecology, as well as opening a debate around system boundaries. For example, if a brewery used spent grain for biogas production to enhance their greenhouse gas performance, it might have negative consequences for the sustainability of the wider system, i.e. for farmers, food production and land use. Our study also showed that the Nordic brewing industry has not developed a clear vision of the bioeconomy in relation to its organic residues, let alone implemented major initiatives in this area, thus confirming the common view of the industry as 'traditional in approach' (Bamforth, 2000).

Several of the alternative uses of spent grain discussed in this chapter could potentially enhance value creation and capture and perhaps even provide brewers with a sustainable competitive advantage. First, brewers could obtain a higher price for the spent grain compared to the present situation, which will have a direct financial impact. Second, brewers could promote their alternative uses of the spent grain and thus increase their branding value. Yet there are also reputational risks involved in denying farmers access to a local feed resource by diverting spent grain to other uses, and brewers could get entangled in the food-feed-fuel debate by using spent grain for energy production. Notably, given the historical lack of innovation in this sector, implemented new uses of spent grain would take time to imitate, thus creating at least a temporary competitive advantage for the first movers.

The spent grain valorisation pathways identified by this and other studies are clearly downstream and involve other value chains, meaning their realisation must include other industry actors. Brewers need to acquire new market knowledge, especially about downstream complementary assets (Roy & Sarkar, 2016). This requires internal strategies for acquiring knowledge and depends on collaboration with non-brewers who possess the required knowledge and resources, including processors, technology suppliers, researchers and financiers. Brewers' current limited engagement in collaborative R&D projects appears to be a hindrance to pursuing such a strategy.

Deploying new pathways and business models for spent grain will have consequences not only for the actors in the brewery sector and downstream sectors directly engaged in spent grain utilisation, but also for wider society when spent grain finds new uses. The geographic context of the breweries influences the possible demand for valorisation of spent grain, such as the demand for use as feed. In order to understand which valorisation pathways are 'better', one must investigate several questions: In which sustainability dimension(s) is one pathway better than another? Who is the pathway better for? What business model innovations will the pathway rely on? What kinds of breweries would it be relevant for? A further topic for future research is the development of industrial symbiosis between breweries and other industry actors with the potential to valorise the spent grain and other resources such as excess heat, water, etc. A final important topic is the interaction with intermediaries through which brewers can access better knowledge about possible new applications, markets and technologies.

Acknowledgements

This work was supported by the Research Council of Norway (grant number 244249). Mads Lykke Dømgaard created the artwork. Three anonymous reviewers commented on a manuscript which formed the basis of this chapter. Marco Capasso provided comments on an earlier version of this chapter.

References

Aliyu, S., & Bala, M. (2011). Brewer's spent grain: A review of its potentials and applications. *African Journal of Biotechnology, 10*, 324–331. doi:10.4314/ajb.v10i3.

Bamforth, C. W. (2000). Brewing and brewing research: Past, present and future. *Journal of the Science of Food and Agriculture, 80*, 1371–1378. doi:10.1002/1097-0010(200007)80:9<1371::AID-JSFA654>3.0.CO;2-K.

Baumgartner, R. J., & Ebner, D. (2010). Corporate sustainability strategies: Sustainability profiles and maturity levels. *Sustainable Development, 18*, 76–89. doi:10.1002/sd.447.

Bocken, N., Short, S., Rana, P. & Evans, S. (2013). A value mapping tool for sustainable business modelling. *Corporate Governance: The International Journal of Business in Society, 13*, 482–497. doi:10.1108/CG-06-2013-0078.

Bryman, A. (2015). *Social Research Methods*. Oxford: Oxford University Press.

Buffington, J. (2014). The economic potential of brewer's spent grain (BSG) as a biomass feedstock. *Advances in Chemical Engineering and Science*, 308–318. doi:10.4236/aces.2014.43034.

Bugge, M. M., Hansen, T. & Klitkou, A. (2016). What is the bioeconomy? A review of the literature. *Sustainability (Switzerland), 8*. doi:10.3390/su8070691.

Clark, G. L., Feiner, A. & Viehs, M. (2015). *From the Stockholder to the Stakeholder: How Sustainability Can Drive Financial Outperformance*. Retrieved from https://arabesque.com/research/From_the_stockholder_to_the_stakeholder_web.pdf.

Gereffi, G., & Fernandez-Stark, K. (2016). Global value chain analysis: A primer. *Center on Globalization, Governance & Competitiveness (CGGC)*, 1–34.

Jürgensen, M. R., & Confalonieri, A. (2016). *Technical Guidance on the Operation of Organic Waste Treatment Plants*. Vienna: International Solid Waste Association.

Kawa, A., & Luczyk, I. (2015). CSR in supply chains of brewing industry. In P. Golinska & A. Kawa (Eds.), *Technology Management for Sustainable Production and Logistics* (pp. 97–118). Berlin and Heidelberg: Springer Verlag.

Keller-Reinspach, H. W. (1989). Emissions during the combustion of spent brewer's grains. *Brauwelt, 129*, 2316–2319.

Kim, W. C., & Mauborgne, R. (2005). *Blue Ocean Strategy: How to Create Uncontested Market Space and Make the Competition Irrelevant*. Boston, MA: Harvard Business School.

Kristiansen, A. G., Johansen, L. K., Mou, C. & Johansson, C. G. (2011). *Input to TWG on BAT Candidates for Breweries*. Copenhagen, Denmark.

Levitt, T. (1960). Marketing myopia. *Harvard Business Review*, 3–13.

Lynch, K. M., Steffen, E. J. & Arendt, E. K. (2016). Brewers' spent grain: A review with an emphasis on food and health. *Journal of the Institute of Brewing, 122*, 553–568. doi:10.1002/jib.363.

McWilliams, A., & Siegel, D. S. (2011). Creating and capturing value: Strategic corporate social responsibility, resource-based theory, and sustainable competitive advantage. *Journal of Management, 37*, 1480–1495. doi:10.1177/0149206310385696.

Mussatto, S. I. (2014). Brewer's spent grain: A valuable feedstock for industrial applications. *Journal of the Science of Food and Agriculture, 94*, 1264–1275. doi:10.1002/jsfa.6486.

Mussatto, S. I., Dragone, G. & Roberto, I. C. (2006). Brewers' spent grain: Generation, characteristics and potential applications. *Journal of Cereal Science, 43*, 1–14. doi:10.1016/J.JCS.2005.06.001.

Mussatto, S. I., Moncada, J., Roberto, I. C. & Cardona, C. A. (2013). Techno-economic analysis for brewer's spent grains use on a biorefinery concept: The Brazilian case. *Bioresource Technology*, *148*, 302–310. doi:10.1016/j.biortech.2013.08.046.

Olajire, A. A. (2012). The brewing industry and environmental challenges. *Journal of Cleaner Production*, 1–21. doi:10.1016/j.jclepro.2012.03.003.

Öztürk, S., Özboy, Ö., Cavidoğlu, İ. & Köksel, H. (2002). Effects of brewer's spent grain on the quality and dietary fibre content of cookies. *Journal of the Institute of Brewing*, *108*, 23–27. doi:10.1002/j.2050-0416.2002.tb00116.x.

Pettigrew, A., & Whipp, R. (1991). *Managing Change for Competitive Success*. Oxford: Basil Blackwell.

Porter, M. E. (1985). *The Competitive Advantage: Creating and Sustaining Superior Performance*. New York: Free Press.

Ratebeer. (2017). Denmark breweries. www.ratebeer.com.

Reinhardt, F. L. (1998). Environmental product differentiation: Implications for corporate strategy. *California Management Review*, *40*, 43–73. doi:10.2307/41165964.

Richardson, J. (2008). The business model: An integrative framework for strategy execution. *Strategic Change*, *17*, 133–144. doi:10.1002/jsc.821.

Roberts, P. W., & Dowling, G. R. (2002). Corporate reputation and sustained superior financial performance. *Strategic Management Journal*, *23*, 1077–1093. doi:10.1002/smj.274.

Robinson, W. T., Fornell, C. & Sullivan, M. (1992). Are market pioneers intrinsically stronger than later entrants? *Strategic Management Journal*, *13*, 609–624. doi:10.1002/smj.4250130804.

Roy, R., & Sarkar, M. B. (2016). Knowledge, firm boundaries, and innovation: Mitigating the incumbent's curse during radical technological change. *Strategic Management Journal*, *37*, 835–854. doi:10.1002/smj.2357.

Short, S. W., Bocken, N. M. P., Barlow, C. Y. & Chertow, M. R. (2014). From refining sugar to growing tomatoes: Industrial ecology and business model evolution. *Journal of Industrial Ecology*, *18*, 603–618. doi:10.1111/jiec.12171.

Sierra Nevada Brewery. (2015). *Biennial Sustainability Report*. Retrieved from www.cdn.sierranevada.com/sites/www.sierranevada.com/files/content/sustainability/reports/SustainabilityReport2015.pdf.

Srivastava, A., & Lee, H. (2005). Predicting order and timing of new product moves: The role of top management in corporate entrepreneurship. *Journal of Business Venturing*, *20*, 459–481. doi:10.1016/J.JBUSVENT.2004.02.002.

The Carbon Trust. (2017). *Case Studies: Carlsberg Group*. Retrieved from www.carbontrust.com/our-clients/c/carlsberg-group.

Thomas, K. R., & Rahman, P. K. S. M. (2006). Brewery wastes: Strategies for sustainability: A review. *Aspects of Applied Biology*, *80*, 147–153.

US Green Building Council. (2017). *Leadership in Energy and Environmental Design (LEED)*. Retrieved from http://leed.usgbc.org/leed.html.

van Wyk, J. P. H. (2001). Biotechnology and the utilization of biowaste as a resource for bioproduct development. *Trends in Biotechnology*, *19*, 172–177. doi:10.1016/S0167-7799(01)01601-8.

7 Meat processing and animal by-products

Industrial dynamics and institutional settings

Anne Nygaard Tanner and Nhat Strøm-Andersen

7.1 Introduction

In this chapter, we examine the key mechanisms that drive the evolution of the meat and animal by-product (ABP) sector towards a circular bio-based economy. A circular bioeconomy cuts across sectors and industries and includes the production of renewable biological resources as well as the utilisation of side-streams and residues for high-value products in food, feed, bio-based materials, chemicals, cosmetics, pharmaceuticals and energy. Many of the sectors that are relevant for a circular bioeconomy are primary sectors, such as 'agriculture' and 'food, beverage and tobacco', including meat processing and animal by-product industries. In the European Union the 'agriculture' and 'food, beverage and tobacco' sectors combined are currently leading the bioeconomy in terms of turnover (estimated at 75%) and employment (80%) (Ronzon, Santini & M'Barek, 2015).

Although new technologies are a prerequisite for a bioeconomy, technological development alone is not sufficient (Pyka & Prettner, 2018). Transforming industrial sectors is a co-evolutionary process of systemic changes of interrelated elements that, besides knowledge and technology, include actors, networks and institutions (Malerba, 2002, 2005a, 2005b). It is also a highly spatially dependent process, shaped by place-specific factors such as geography and industrial structure (Coenen, Benneworth & Truffer, 2012; Hansen & Coenen, 2014). In this chapter, we focus on the dynamics of the meat processing and ABP sector and the patterns and strategies of value creation that characterise companies in this industry.

The analysis reveals a sector which is highly shaped by its regulative environment, in the sense that regulations decide input and output and define the room actors can manoeuvre within in their search for value creation. Similarly, we find that place-specific factors influence the range of opportunities available for actors in each of the nationally embedded sectoral innovation systems.

The chapter is structured as follows. The following section briefly introduces the theoretical approach the chapter draws upon. Section 7.3 introduces the empirical section with a short description of the inherent and diverse nature of ABP, followed by a presentation of the industrial characteristics in the two countries in focus for this study, namely Denmark and Norway. Finally, we present the key institutional framework that guides the behaviour of firms in this sector before analysing how meat processing and ABP companies act within these changing settings. Section 7.4 discusses and concludes the main findings in relation to the questions raised in the Introduction.

7.2 Theoretical background and approach

The chapter draws on the conceptual understanding outlined in Chapter 3 that sees innovation as key for the transformative changes towards a circular bio-economy. It builds on a systemic and evolutionary understanding of socio-economic and socio-technical systems, and their change processes. Sectors such as the meat-processing and ABP industry are formed by a set of activities that are unified by related product groups and share a common knowledge base.

Sectoral innovation systems such as the meat-processing and ABP industry are constituted by different elements such as actors (firms, universities, research institutes, NGOs, consumers, policy- and lawmakers, etc.), knowledge and learning processes, technologies, inputs and demand, networks and institutions (Malerba, 2002). The elements of the system co-evolve over time, resulting in processes of change that enable the transformation of the system. Moreover, rates and types of innovation differ greatly across sectors depending on the level of technological development, institutional settings, market opportunities and processes of selection (Malerba, 2005a).

Taking a systemic perspective implies focusing on the dynamics of sectors or industries, meaning how constituting elements co-evolve over time, rather than static comparisons of industry structures and their performance at a given point in time. In particular, the explanatory focus is on factors and mechanisms that drive these change processes; in other words, it entails understanding the laws of motion for a specific industry.

Furthermore, characteristic for the systemic perspective is its focus on interdependencies and links between related industries. Boundaries of sectoral systems are not fixed but rather change over time. In particular, for the food industry, the vertical links and coordination between different production activities or nodes in the value chain have been shown to play a significant role in the innovative behaviour of firms (Karantininis, Sauer & Furtan, 2010). Historically, the hierarchical market structure of the food industry has often been explained by a relatively weak appropriability regime (Peneder, 2010), meaning the possibility for protecting knowledge and innovations from imitators is low. However, in pace with an increasing innovative behaviour, the market structure of the industry changes, which results in a movement towards vertical integration upstream in the value chain (Karantininis et al., 2010).

Finally, in a sectoral innovation system perspective, institutions play a key role in shaping input and demand, as well as the interactions between market and non-market actors. Although institutions are most often seen as a stabilising element of the system, they may also play a role in fostering novelty and transforming sectoral systems. In particular, a disruptive change in institutional settings, such as the introduction of a radical change in regulations, fundamentally changes the belief system and established practices of an industry, and potentially leads to changes in innovative behaviour.

In the meat-processing and ABP industry, a major driving force is precisely institutional change in the form of regulations to protect the health of humans and animals. Since the 1990s, regulative changes have become the key mechanism of change affecting and shaping *input* and *demand* of the industry. The main input is defined as side-streams from meat and livestock production. Because of their human and animal health and welfare risk, ABPs are a heavily regulated raw material. The main demand is likewise shaped by market regulations such as trade barriers and sanctions, which is characteristic for agri-food products. Whether trade barriers are supported by arguments of health or environmental reasons or politically motivated as a consequence of a bilateral, diplomatic crisis, trade barriers have huge consequences for the market opportunities of the meat and ABP industry. Together, regulations shape the input and demand of the industry and thereby also influence market opportunities and innovative behaviour of firms in the industry.

The following section will elaborate and exemplify these dynamics through our empirical insights of the meat processing and ABP industries in Norway and Denmark. The analysis builds on 20 interviews: 12 interviews with actors from the industry in Denmark and eight interviews with actors from the industry in Norway. The interviews were conducted from 2015 to 2018 and aimed to cover topics such as the key dynamics of the industry, the historical development and key events of the industry, innovative practice and the development of markets and new products. In addition to the interviews, the analysis builds on secondary material such as EU directives, historical newsletters from interest organisations, articles from news media and reports and secondary literature of the sector.

7.3 The meat processing and ABP industry

The meat processing and ABP industry primarily encompass two types of firms: producers of ABPs and processors of ABPs. The production of ABPs occurs in all the nodes of the meat value chain from animal production, at slaughterhouses, and at the facilities of further meat processors (see Figure 7.1). Processing of ABPs has historically been a task for rendering companies, which are dedicated ABP processors. Rendering companies have played an important role in securing and managing the huge amounts of by-products produced from the meat industry to avoid hazardous risks. Today, dedicated ABP processors include incumbent rendering companies but also new entrants where the focus is also on upgrading specific types of ABPs (e.g., blood, hides and bones) for human consumption.

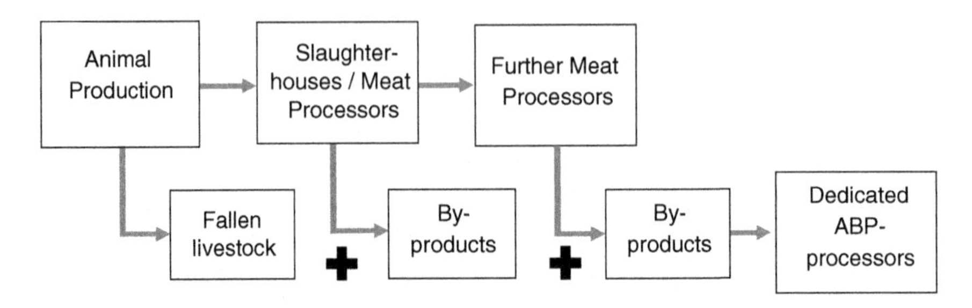

Figure 7.1 The meat processing and ABP value chain. Upper boxes are nodes (firms) in the value chain; lower boxes illustrate the flow of ABPs.

7.3.1 Potential value of ABP

ABPs are materials of animal origin that people do not consume. In the EU, over 20 million tons of ABPs are generated annually from slaughterhouses, plants producing food for human consumption, dairies and as fallen stock from farms (European Commission, 2018). ABPs have an inherently diverse nature, which offers varied possibilities for utilisation and conversion. ABPs contain miscellaneous compounds such as gelatine, protein, enzymes, fatty tissues, collagen and phosphates, which provide manifold possibilities for value-added products and applications through diverse bioprocessing technologies. For instance, animal protein can deliver a complete protein with a high biological value based on its amino acid profile (Mullen et al., 2017). Blood generation often presents a serious environmental issue because of its high pollutant capacity; however, it has exceptional nutritive value and excellent functional properties that give the potential to generate high-added-value food ingredients (Lynch, Mullen, O'Neill & García, 2017). Because of the diversity and heterogeneous nature of ABPs, technical methods that can be applied to valorise meat by-products are numerous and include, for example, ultrafiltration, extrusion, lyophilisation, isoelectric solubilisation-precipitation, solvent extraction and enzymatic hydrolysis (Aspevik et al., 2017; Galanakis, 2012; Mullen et al., 2017). The final choice of application depends on the types of the by-product and the local conditions where the raw materials are generated (Mullen et al., 2017). Local conditions relate, for instance, to the transportation distance from the slaughterhouses to the processing plant, which require methods to keep ABPs fresh for longer to ensure the best quality for further processing. Moreover, the volume of the rest raw materials at the processing plant level has implications for which processing technique is most suitable. In other words, it is not profitable to valorise small volumes of ABPs as this may lead to a negative cost-benefit analysis.

Based on strict regulations, animal by-products are classified into three categories. Category 1 (CAT1) is classified as high risk, including entire bodies and all body parts of the animals associated with TSE (transmissible spongiform encephalopathy), used for experiments and illegal treatments, infected with diseases and environmental contaminants and specified risk materials (SRMs). CAT1 should be incinerated and only approved combustion plants can receive CAT1 for treatment. Sources of energy such as electricity and biodiesel can be obtained through incineration. Category 2 (CAT2) is classified as high risk, including animals and parts of animals unfit for human consumption such as animals killed for disease control purposes, ABPs containing residues of authorised substances or contaminants exceeding the permitted levels and manure. CAT2 can be incinerated with or without prior processing, converted into organic fertilisers or soil improvers after processing, or used as fuel for combustion. Category 3 is classified as low risk, including carcases and animal parts being left from slaughterhouses, fit for human consumption, but not used due to commercial reasons. CAT3 can be processed to make pet food, or mink food, or utilised in even higher value-added applications in other industries like cosmetics, pharmaceutics or foodstuff.

The meat processing industry has utilised ABPs for centuries. However, recent literature suggests that rich and multiple opportunities exist for upgrading the utilisation of ABPs (see, for instance, Jayathilakan, Sultana, Radhakrishna & Bawa, 2012; Lin et al., 2013; Matharu, de Melo & Houghton, 2016; Mirabella, Castellani & Sala, 2014; Ravindran & Jaiswal, 2016; Toldrá, Mora & Reig, 2016). In a bioeconomy context which aims to shift upwards in the waste pyramid (cf. Chapter 3), this literature shows that ABPs have the potential to deliver on a wide range of products from high-value products to lower-value products such as fertiliser and energy. Figure 7.2 provides an overview.

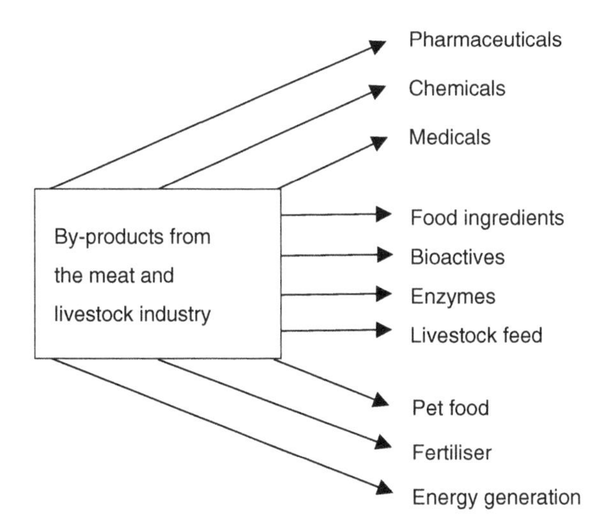

Figure 7.2 Rich and multiple opportunities to upgrade current use of meat by-products (adapted from Toldrá et al., 2016).

According to the waste pyramid, the dominating use of ABPs is characterised by the lowest part of the pyramid, namely recycling, recovery, energy and disposal. Today, the majority of by-products end up in the lowest part, recycled as pet food or feed for animals that do not enter the food chain (e.g. mink), fertiliser or as energy. The top of the pyramid – prevention and reuse – is the most preferable from an economic and environmental perspective (ECA, 2016). The main focus in this chapter is to study how companies in the meat processing and ABP industry facilitate processes of prevention, reuse and recycling of by-products to create higher added value for their resource base (e.g. for human consumption), through process and/or product innovations.

7.4 The meat processing and ABP sector

The meat processing and ABP sector has experienced a strong consolidation and internationalisation over the last 10–20 years, which has resulted in an intertwined network of companies that cuts across national borders. Companies are often connected through interest shares or in supplier–buyer relationships. This is also the case for companies in Denmark and Norway: For example, DAKA (a Denmark-based rendering company) owns 10% of the shares in Norsk Protein; Farmfood receives the majority of the Norwegian poultry ABP (CAT2 material) at their facility in Løgstør in Northern Denmark. Nevertheless, the industry structure and geography of the two countries have a huge impact on the types and volumes of ABPs available in each country, and hence for the basis of input to the industry. In the following, we give a short presentation of the key actors characterising Denmark's and Norway's meat processing and ABP sectors.

7.4.1 Denmark

The meat processing and ABP industry in Denmark is characterised by its very large animal production, especially regarding pigs (see Table 7.1 for a comparison of the size of animal production in Denmark and Norway). In particular, the Danish pig industry is very large compared to Denmark's size and counts approximately 3,300 pig farms that together produce 31.9 million pigs annually (DST table ANI9). Together, the animal production, meat processing and ABP industry comprise 45% (approximately 85,000 people) of the total employment in the food industry in Denmark (Landbrug og Fødevarer, 2017). In 2015, the turnover for the four largest co-operatives in the sector – Danish Crown, Tican, DAT Schaub and DAKA – reached US$10.7 billion (Danish Agriculture & Food Council, 2016). In 2016, Tican became a privately owned company. Key actors in the Danish meat processing and ABP industry are slaughterhouses, dedicated by-product companies, interest organisations, universities and research centres, and more recently, small- and medium-sized technology developers that have entered the industry.

Table 7.1 Animal production in Denmark and Norway (2017 figures)

	DENMARK *Animal production* *(in brackets, export of live animals)*	*NORWAY* *Animal production*
Pigs (in 1,000 ton)	1,896 (302.3)	137.2
Poultry (in 1,000 ton)	174.4 (39.2)	101.0
Cattle (in 1,000 ton)	135.2 (1.8)	85.2
Egg production (in 1,000 ton)	68	66.7*
Mink fur (1,000 units)	17,900	–
Sheep (1,000 units)	76.5 (0.9)	1,376

Sources: Statistics Denmark: ANI8, PELS1, ANI4, ANI6, ANI5, ANI9; and Statistics Norway: www.ssb.no/jord-skog-jakt-og-fiskeri/statistikker/slakt.

Note
* 2015 numbers, source: www.ssb.no/280562/produksjon-av-kjot-mjolk-egg-og-ull-sa-345.

The Danish meat industry builds on a long history based on cooperative ownership. Ever since the first cooperative slaughterhouse was formed in 1887 in Horsens, the industry has been capable of renewing itself through restructuring. In the 1970s the industry went through a national consolidation driven by technological development that justified larger facilities. In the 1980s consolidation was more market driven, where larger sizes made it easier to access new and larger markets (Tüchsen, 2014).

In the last 15–20 years, the restructuring of the sector has been characterised by European consolidation and internationalisation of markets (Hansen, Egelyng, Adler & Bar, 2015). Today the industry in Denmark is characterised by four to five large slaughterhouses (>500 employees) and around 100 smaller (1–50 employees) ones.

Danish Crown is one of the largest meat processors in the world. It is collectively owned by farmers in Denmark and is responsible for 80% of all pigs and 50% of cattle slaughtered in Denmark. TICAN was acquired by the German slaughterhouse Tönnies and is responsible for around 10% of pigs slaughtered in Denmark. Danpo is part of Scandi Standard, which is a leading producer of chicken-based food products in the Nordic region, with headquarters in Sweden. Danpo is responsible for the majority of chickens slaughtered in Denmark. Finally, Skare Beef and Himmelandskød each slaughter around 15% of cattle in Denmark. Slaughterhouses have recently intensified their interest in valorisation of by-products by engaging in the restructuring of their production facilities and establishing subsidiaries dedicated to handle ABPs. One example is Danish Crown Ingredients, founded in 2014, and Farmfood A/S, founded in 2003 (owned by Danpo, BHJ and HKScan). The vertical integration of business areas related to processing ABPs indicates that the industrial structure is changing in order to protect knowledge, innovation and new market opportunities.

In addition to slaughterhouses and meat processing companies, dedicated by-product processors include DAKA (part of the German Group SARIA) and BHJ (part of LGI Group). These are also a result of the cooperative movement and have roots that date back to the beginning of the 1900s. During the last decade, dedicated by-product companies have likewise diversified by founding subsidiaries with a focus on specific types of ABP as a strategy to increase the value of ABPs. For example, BHJ's subsidiary Essentia Protein Solutions has its main focus on food ingredients produced from Category 3 material. Essentia produces functional proteins to improve the functionality, taste and nutritional character of food products. DAKA has built similar business areas within ingredients, pet food, biodiesel, organic fertiliser, etc.

The two main interest organisations representing the industry's interests in Denmark are the Confederation of Danish Industry and the Danish Agriculture and Food Council, which also comprise SEGES, a research and innovation centre for agriculture and the food industry in Denmark. However, more importantly for the internationalised Danish industry is the European Fat Processors and Renderers Association (EFPRA) that lobby for the industry's interest at the EU level.

Finally, a low number of new technology developers have entered into partnership with some of the larger players in the industry to demonstrate and develop their technologies. These include, for instance, Lihme Protein Solutions, Dacofi and Upfront Technology.

It is difficult to estimate the exact volume of ABPs in the Danish industry. Based on our interviews we estimate >500,000 tons of ABPs are produced and handled in Denmark. Danish Crown produces 375,000 tons of ABPs per year. Besides this, Farmfood handles 140,000 tons of poultry by-products from Denmark, Sweden and Norway (including 80% of Norwegian ABPs). Interviews with actors in the industry disclose fallen volumes of ABPs, which results in increased competition on ABPs. For example, DAKA has recently (August 2018) closed their smallest processing facility because of reduced volumes of blood. The decreasing volumes are caused by new valorisation paths of ABPs (e.g. mink food and ingredients); an increasing export of live animals (e.g. piglets); and an increasing sale of products that were previously Category 3 material but are now sold to Asian markets (e.g. pig ears, gallstones).

7.4.2 Norway

The meat processing and ABP industry is the biggest sector by employment (25%) and the second largest sector by revenue (21%) in the Norwegian food and beverage industry (2016 statistics presented in Prestegard, Pettersen, Nebell, Svennerud & Brattenborg, 2017). All Norway's meat consumption is covered by Norwegian producers, with the exception of some beef that is imported duty-free from Botswana and Namibia. The sector is organised by a few large companies and a number of small- and medium-sized enterprises, with a total of 319 companies. In 2016, the sector had a revenue of US$6.1 billion and 11,477 employees (Prestegard et al., 2017). The key actors in

Norway are meat processing companies, rendering companies, the meat and poultry confederation and research centres.

Nortura is Norway's largest meat and egg producer, and is collectively owned by Norwegian farmers. The company is a major player in the food industry in Norway, which accounts for 70% market share (2014 data collection). Nortura has more than 30 slaughterhouses spread all over Norway. Nortura's total production/slaughter tonnage is 428,900 tons a year (2014 data), in which cattle account for 78,100 tons; lambs and sheep 23,600 tons; pigs 128,500 tons; eggs 94,100 tons; and chicken and turkey 104,600 tons. Its daughter company, Norilia AS, is in charge of handling by-products generated from all Nortura's slaughterhouses, which are considered to be 35% of the entire production (approximately 150,000 tons). Nortura has two other daughter companies – Norsk Hundefor AS and Norsk Dyremat AS – that produce pet food from ABPs for international and domestic markets, respectively. The rest of the meat processing industry is composed of a few medium-sized companies such as Norsk Kylling AS, Fatland AS, Grilstad AS and other small private companies.

Norsk Protein AS is the only rendering company in Norway that receives by-products from slaughterhouses and meat processing companies, specified risk materials (SMR) and dead animals. The company has five production plants (three plants for CAT3 and two plants for CAT1) receiving ABPs from all over Norway, distributed in four locations from north to south, namely Balsfjord, Mosvik, Hamar and Grødaland. There are no CAT2 plants in Norway, so CAT2 is sent to CAT1 treatment plants, and partly to Denmark. In accordance with current regulations in Norway and the EU, the company further reprocesses the CAT3 raw materials to meat and bone meal (MBM), and animal fat. Norsk Protein AS was established in the 1970s, and is currently owned by Nortura SA (67%), the Norwegian Confederation of Meat and Poultry (KLF) (23%) and DAKA Denmark AS (10%) (KLF, 2016). Norsk Protein received a total of 197,831 tons of rest raw materials from slaughterhouses in 2017 (Norsk Protein, 2017). Norsk Protein is also represented in EFPRA, the European interest organisation for rendering companies.

Other actors that play an important role in the industry network are the Norwegian Confederation of Meat and Poultry (KLF) and the Norwegian Meat and Poultry Research Centre (Animalia). KLF, founded in 1910, is an interest and industry organisation that represents the privately owned, free-standing part of the meat, egg and poultry industry in Norway. Animalia is Norway's leading research and development specialist in meat and egg production, providing knowledge and expertise through domestic animal inspections and veterinary health services, business-critical technical systems, research and development projects, e-learning and training activities, communication and other forms of knowledge-sharing. On January 1, 2018, Animalia left Nortura SA as a separate limited company. Animalia AS is now owned by Nortura SA (66%) and the Federation of Meat and Poultry Industries (KLF) (34%). Previously, the company was organised as a department in Nortura.

Because of the geographical characteristics of the country, slaughterhouses in Norway are small and scattered, which makes it difficult to collect and handle ABPs. Valorisation of ABPs involves the entire process of developing new business areas in markets and technologies, in which meat and ABPs companies have somewhat limited knowledge and experience. For example, to enter the international ingredient markets for human consumption (proteins and fats), Norwegian meat companies encounter a series of challenges related to, for instance, market entrance, distribution channels and brand reputation. The industry is aware of many new, potential technologies that can be used to process ABPs. However, it takes time to learn, acquire and select the right ones given the inherently diverse characteristics of ABPs. In addition, developing innovation is costly. Lack of risk capital funding is another issue that challenges the industry in commercialising research results. Despite these drawbacks, the ABP industry in Norway is strategically seeking higher value markets.

Summing up, the key difference between Denmark and Norway can be characterised by the word size. With almost the same population size yet Norway's land area being seven times that of Denmark, the population density is much higher in Denmark (131/km^2) than in Norway (15.5/km^2). To cover the vast area of farmland slaughterhouses in Norway are smaller and scattered across the country, whereas in Denmark slaughterhouses are fewer in number and much larger. Taking Denmark's smaller size in area into the equation, ABPs are more easily collected and transported in Denmark.

It is not only differences in the two countries' geography that influence the input and hence the possibilities for value creation, but also the industrial size and structure. Primarily, Denmark's pig production is one of the largest in Europe and the two slaughterhouses Danish Crown and Tican (Tönnies) are among the largest in Europe. Company size in terms of finances, geographical markets and volumes of meat and ABPs are important for the value creation strategy, in particular in terms of possibilities to finance new initiatives and to access the right volumes and type of ABP input. Hence, the larger volumes and easier transportation of ABP in Denmark place the Danish industry in a better position to create and utilise new market opportunities.

7.5 Regulation of the meat processing and ABP industry

To understand valorisation of ABP, it is important to acknowledge that it is a heavily regulated field. Interviewed actors in the industry unambiguously mention regulations as the key influential factor shaping the industry. EU laws and directives regulate both *inputs* to the industry in terms of types and volumes of ABP and the *demand* and market opportunities primarily through export bans and import barriers, as well as through regulations on nutritional and health claims on novel food products (see Figure 7.3). Indirectly, these

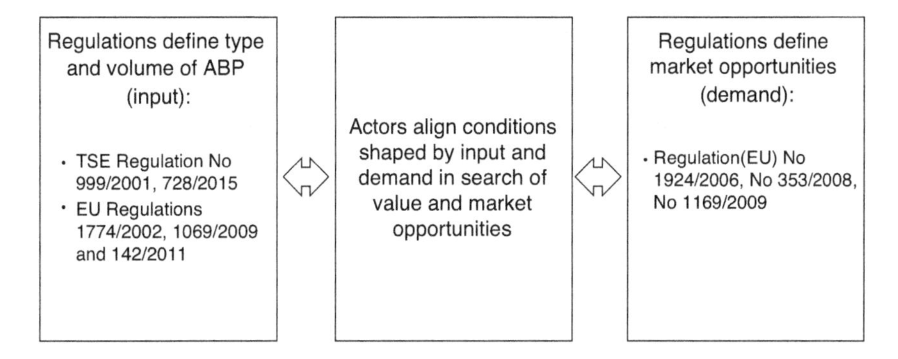

Figure 7.3 Value creation dynamics in the meat processing and ABP industry.

regulations also shape interactions between market and non-market actors as well as the practices of companies guiding innovative activities and strategies in search of new market opportunities.

7.5.1 Mad cow disease and EU-regulations

The historical background of the main regulations on the input side dates back to the outbreak of 'mad cow disease' (i.e. Bovine Spongiform Encephalopathy, BSE) in the mid-1980s, which proved to have far-reaching consequences for the industry. As a result, besides huge economic losses and the killing of millions of animals, it culminated in very comprehensive regulations on the handling and use of animal by-products, which disrupted the whole industry, and today permeate everything the industry is doing. The consequences were huge for the UK economy and the animal production industry. It is estimated that 180,000 cattle were affected and 4.4 million cows were killed during this period. By comparison, in Denmark only 15 cows with BSE have been detected along with three incidents in cows exported from Denmark.

In 1986, the first incident of BSE was diagnosed in the UK, although it is believed the disease had existed for several years prior to this. Investigations at the time showed that the spread of BSE occurred through the feed produced at rendering companies. Infected animals, either alive or fallen stock, were sent to the rendering factories and used in the production of MBM, which was used in feed to cattle and other livestock. Consequently, this cycle multiplied the spread of BSE across the UK.

The connection between BSE and the human variant, Creutzfeld-Jacobsen Disease (CJD), was not discovered until 1996. Consequently, from the late 1980s and until the early 1990s it was believed that there was no human risk from eating beef infected with BSE. However, in March 1996 when the first announcement about a possible link between BSE and CJD was made, it resulted in a total import ban on UK beef and cattle to the rest of Europe. The ban was only lifted in 2006. To date, 177 people have died because of contracting the human variant of BSE.

In 1990, the first EU regulative ban on using ruminant MBM in feed for ruminants was introduced. It was a way to inhibit the further spread of BSE by breaking the vicious cycle. In 1994 this ban was expanded to concern protein feed from all animals (including pigs, poultry, etc.) to ruminants. In 2001 the EU imposed a total ban on using any remains of all animals in feed for livestock (TSE Regulation No 999/2001). The TSE Regulation No 999/2001 was introduced throughout EU as well as in Norway from January 1, 2001 (Nærings- og fiskeridepartementet & Landbruks- og matdepartementet, 2004, 2016). The main argument was that feed for ruminants was too easily contaminated during production, storage or transportation with MBM produced as feed for pigs and poultry. Hence, it was assessed to be a risk that ruminants would be fed with ruminant proteins. Consequently, the total ban on using any animal by-products in feed for any animals was a means to meet the human health risk of consuming beef.

From one day to the next it changed the market situation for the whole industry. In Denmark, this caused a huge bottleneck in the system. DAKA, which was the main purchaser at this time, experienced their previous market for animal feed being disruptively shut down overnight. This resulted in an accumulation of 180,000 tons of MBM at DAKA with no potential purchasers. As a result, the product totally lost its value, which sent the company out in search of other markets and a process of restructuring and reorganising its business.

Also in Norway, the sudden introduction of the TSE Regulation yielded a significantly higher price than alternative solutions such as landfill, combustion or fertiliser. This new regulation changed the price of MBM accordingly from a positive value of NOK 2–3 per kg to a negative value of NOK 2–3 per kg overnight. Similarly, in Denmark, interviewees report that they had to pay the incineration and cement industry 600–700 DKK per ton. The negative prices for the rendering companies were imposed on the slaughterhouses that delivered raw materials to rendering companies.

In 2002 another set of regulations was introduced in the EU (EU Regulations 1774/2002, later replaced by EU Regulations 1069/2009 and 142/2011), which regulates the use of animal by-products throughout the entire food chain. This set of regulations introduced the categorisation of ABP in three categories, CAT1, CAT2 and CAT3, introduced in section 7.3.1.

7.5.2 Market regulations: novel food products and trade barriers

Another type of regulation is the EFSA's regulation (European Food Safety Authority) on nutrition and health claims (Commission Regulation (EU) No 1924/2006, No 353/2008, No 1169/2009). In order to protect consumers, health claims need to be justified scientifically. The EFSA regulation places high requirements on the labelling of, for example, functional food products. In practice, this means that, in order to make use of health claims on new products, the claim has to be proven by clinical trials similar to the pharmaceutical industry, which is assessed to be too costly in time and resources for companies in the food industry.

Trade with ABP and food products is highly regulated. Traditionally, food products have been included in trade barriers and sanctions supported by arguments of health or environmental reasons. Trade barriers are also often politically motivated as a consequence of bilateral, diplomatic crises, and turn out to have huge consequences for the market opportunities of the meat and ABP industry.

An example is the geo-political crisis between Russia and EU that resulted in, on August 6, 2014, a Russian decree prohibiting, for one year, imports into the territory of the Russian Federation of certain agricultural products, raw materials and foodstuffs originating from EU countries, Norway, USA, Canada and Australia (European Commission, 2017). The embargo was later extended until August 5, 2016, and then further prolonged until December 31, 2017 (European Commission, 2017).

This important event led to price fluctuations on the international raw materials market, and significantly impacted the industry on its rest raw materials base. Products on the banned list included meat of bovine animals; pork, poultry meat and edible offal in all forms (fresh, chilled or frozen); sausages and similar products of meat; meat offal or blood; and the final food products based thereon (European Commission, 2017). For example, while Nortura used to export large quantities of ABPs to Russia, after the ban it was forced to find other solutions, and to search for valorisation alternatives.

7.6 Change in innovative behaviour

This section analyses how companies navigate the highly regulated space we have outlined above. The question is how companies approach market opportunities and different market segments based on the institutional settings which regulate the type and volume of input and not least the market options. Figure 7.3 illustrates part of the regulative space which sets the overall framework of the meat processing and ABP industry.

For slaughterhouses, the introduction of the TSE regulation and the categorisation of ABP types led to increased attention being paid to the processing and collection of ABP. In the first period after the new regulation was introduced, the main activities aimed to optimise slaughtering processes and reduce losses at the slaughterhouses to increase the overall value of ABP. As a result, a significant amount of CAT1 material was upgraded to CAT3 with a much higher value. In Denmark, the Technological Institute assisted the Danish industry in optimising, sorting and collecting ABP, so the different categories of ABP materials were kept apart. If any CAT3 or CAT2 materials were in contact with CAT1 material, this would devalue the material to

CAT1. Similar process innovation took place in the Norwegian meat and ABP industry.

Rendering companies also changed their innovative practice in the years that followed the introduction of ABP regulations. They focused on organisational, process and product innovation. At the least, larger rendering companies had to restructure their production facilities to run three separate streams of material (CAT1, CAT2 and CAT3) at different plant sites. CAT1 was prepared as MBM for incineration, CAT2 had to be sterilised before being used as fertiliser and CAT3 principally found a use as pet food.

In both countries process and product innovation is still an ongoing activity characterising valorisation of ABP. For instance, Nortura collaborates with SINTEF and other research centres to optimise meat cutting and minimise waste. The process of distinctly categorising ABP and organic side-streams into different groups for optimal treatment and further processing enables more flexible and sustainable food processing. Furthermore, the company attempts to better organise process innovation by developing cooling systems at slaughterhouses and during transportation to keep the raw materials fresh.

Product innovation has been given attention where the industry seeks higher value-added applications. Norilia recently launched two innovation projects: eggshell membrane extracted and provided to the medical industry for wound treatment, and protein in various forms and applications obtained from enzymatic hydrolysis technology in a biorefinery opened in 2018. At the same time, another company, Norsk Kylling AS, has also established an enzymatic hydrolysis plant to utilise its by-products. Processing by-products to high value-added applications and products such as protein has proven to be a crucial strategy for the industry.

7.7 Regulative adjustments

As stated above, the EU regulations on ABP introduced in 2001 (amended in 2015) following the BSE scandal condition the type and volumes of *input* to the ABP industry. This also means that if this law is changed, the input foundation of the industry changes. A highly regulated field such as the meat and ABP industry provides strong incentives for the industry to engage in institutional entrepreneurship (Dorado, 2005; Leca & Boxenbaum, 2008). This occurred in 2015 when the definition of the specified risk material (SRM) in the TSE regulation was amended based on a scientific opinion published by the EFSA (2014). With this change, a large amount of bovine intestines (approximately 30 kg per cow) was moved from the SRM list (i.e. CAT1) to CAT3 material. As a consequence, the volumes of CAT3 material increased significantly and caused a significant drop in prices.

This situation threatened the European rendering industry and their interest organisation, EFPRA, started lobbying for expanding market opportunities of processed animal proteins (PAP), including lifting the export ban of PAP. Originally, the export ban of PAP was to hinder PAP from also being used in animal feed outside the borders of EU. However, the argument EFPRA put forward was that the EU had the strictest regulations in the world and whether or not EU rendering companies export PAP to be used in animal feed, PAP is used anyway in animal production outside Europe. Therefore, to assist the industry in creating a market for PAP, the EU agreed to lift the export ban on PAP. Likewise, the EU has assisted in creating new markets as a result of industry players' lobbying activities, such as allowing PAP in aqua feed.

7.8 Conclusion

This chapter has analysed the co-evolutionary development of the meat processing and ABP industry in Denmark and Norway from a sectoral innovation system perspective. We have shown how the value creation strategies of firms co-evolve with institutional change, input and demand, organisational changes and knowledge and technological development.

In particular, institutional changes in the form of regulations have a huge impact on the strategies of firms in value creation processes. First, the EU regulations on ABP introduced in 2001 after the BSE scandal condition the type and volumes of *input* to the ABP industry. Second, we demonstrate that temporary trade barriers such as import/export restrictions between countries influence the market creation possibilities, and hence the *demand* for meat- and ABP-based products. These temporary import boycotts therefore have a huge impact on price formation in the industry and influence market dynamics. As a result, actors align the conditions shaped by input and output in their search for new valorisation paths and market opportunities.

Second, we conclude that differences in the two countries' industrial structure and geography influence the input and hence the possibilities for value creation in different ways. Primarily, Denmark's pig production is one of the largest in Europe and the two slaughterhouses Danish Crown and Tican (Tönnies) are among the largest in Europe. Because of Denmark's small size in area, ABPs are more easily collected and transported to customers or processing facilities. By comparison, the slaughterhouses in Norway are smaller and scattered across the country. Company size in terms of finances, geographical markets and volumes of meat and ABP is important for the value creation strategy – in relation to possibilities to finance new initiatives and to access input as well as the technological options for efficient processing technologies. In summary, the industrial structure, size and geography of the two compared countries put the Danish industry in a better position to create and utilise new market opportunities.

Finally, in relation to our initial question of how the meat processing and ABP industry can contribute to a circular bioeconomy, we see diverging trends. During the last couple of decades, the valorisation of ABP has returned to the centre stage of the global meat industry. Slaughterhouses are increasingly interested and active in processing ABPs, which puts pressure on the supply of raw materials for meat processors and dedicated ABP processors. Another tendency is that new processors have entered the value chain and dedicated by-product processors have been forced to change focus and secure their supply of raw-materials because of increased competition.

The increased interest for valorising ABP is nevertheless faced with decreasing volumes of ABP in Denmark, partly because of a higher degree of utilisation of the animals and partly because of an increased export of piglets and live animals to Germany and Poland. The latter has consequences for the number of slaughtered animals at slaughterhouses in Denmark, which naturally drops when the export of piglets and live animals increases. The other cause, a higher degree of utilisation of the slaughtered animals, is a consequence of companies being able to sell new types of cuts to new markets (for example, pig ears, gallstones, etc. to Asian markets), which also causes a natural drop in volumes of ABP. This has, overall, led to higher competition of the remaining ABP, and prices have therefore gone up.

However, increasing prices for ABPs provide companies with incentives to utilise ABPs to a higher value. Likewise, the EU regulations on ABPs and the categorisation of by-products have caused meat processing and ABP companies to improve the utilisation of ABPs, which has led to a change in innovative behaviour. Prior to the introduction of the regulation in 2001, the industry did not innovate in relation to the use of ABPs. However, as we have argued in this chapter, the regulation caused meat processing and ABP companies to change innovative behaviour, resulting in new processes and products valorising ABPs. Put together, it is our assessment that the changes the ABP industry has faced during the last couple of decades and their interest in embracing the political agenda of circularity provide industry actors a strong incentive for valorising ABPs, adding to a circular economy.

References

Aspevik, T., Oterhals, A., Ronning, S. B., Altintzoglou, T., Wubshet, S. G., Gildberg, A., … Lindberg, D. (2017). Valorization of proteins from co- and by-products from the fish and meat industry. *Topics in Current Chemistry*, *375*(3), 1–28. doi:10.1007/s41061-017-0143-6.

Coenen, L., Benneworth, P. & Truffer, B. (2012). Toward a spatial perspective on sustainability transitions. *Research Policy*, *41*(6), 968–979. doi:10.1016/j.respol.2012.02.014.

Danish Agriculture & Food Council. (2016). *Facts and Figures: Denmark – a Food and Farming Country*. Retrieved from https://agricultureandfood.dk/prices-statistics/annual-statistics.

Dorado, S. (2005). Institutional entrepreneurship, partaking, and convening. *Organization Studies*, *26*, 385–414. doi:10.1177/0170840605050873.

ECA. (2016). *Combating Food Waste: An Opportunity for the EU to Improve the Resource-Efficiency of the Food Supply Chain*. Luxembourg: ECA, European Court of Auditors.

EFSA. (2014). Scientific Opinion on BSE risk in bovine intestines and mesentery. *EFSA Journal*, *12*, 98. doi:10.2903/j.efsa.2014.3554.

European Commission. (2017). Russian import ban on EU products. Retrieved from https://ec.europa.eu/food/safety/international_affairs/eu_russia/russian_import_ban_eu_products_en.

European Commission. (2018). What are animal by-products? Retrieved from https://ec.europa.eu/food/safety/animal-by-products_en.

Galanakis, C. M. (2012). Recovery of high added-value components from food wastes: Conventional, emerging technologies and commercialized applications. *Trends in Food Science & Technology*, *26*(2), 68–87.

Hansen, H. O., Egelyng, H., Adler, S. & Bar, E. M. S. (2015). Biprodukter fra dansk og norsk kødindustri: værdiløft står centralt i struktur- og markedsudviklingen. *Tidsskrift for Landoekonomi*, *201*, 129–139.

Hansen, T., & Coenen, L. (2014). The geography of sustainability transitions: Review, synthesis and reflections on an emergent research field. *Environmental Innovation and Societal Transitions*, 1–18. doi:10.1016/j.eist.2014.11.001.

Jayathilakan, K., Sultana, K., Radhakrishna, K. & Bawa, A. (2012). Utilization of byproducts and waste materials from meat, poultry and fish processing industries: A review. *Journal of Food Science and Technology*, *49*(3), 278–293.

Karantininis, K., Sauer, J. & Furtan, W. H. (2010). Innovation and integration in the agri-food industry. *Food Policy*, *35*, 112–120. doi:10.1016/J.FOODPOL.2009.10.003.

KLF. (2016). Daka inn i Norsk Protein. Retrieved from http://kjottbransjen.no/Aktuelt/Daka-inn-i-Norsk-Protein/(language)/nor-NO.

Landbrug og Fødevarer. (2017). *Økonomisk analyse: Fødevareklyngens nationaløkonomiske fodaftryk*, København.

Leca, B., & Boxenbaum, E. (2008). Agency and institutions: A review of institutional entrepreneurship. *Business*, *16*, ii–iii. doi:10.1111/j.1467-8683.2008.00685.x.

Lin, C. S. K., Pfaltzgraff, L. A., Herrero-Davila, L., Mubofu, E. B., Abderrahim, S., Clark, J. H., … Dickson, F. (2013). Food waste as a valuable resource for the production of chemicals, materials and fuels: Current situation and global perspective. *Energy & Environmental Science*, *6*(2), 426–464.

Lynch, S. A., Mullen, A. M., O'Neill, E. E. & García, C. Á. (2017). Harnessing the potential of blood proteins as functional ingredients: A review of the state of the art in blood processing. *Comprehensive Reviews in Food Science and Food Safety*, *16*(2), 330–344. doi:10.1111/1541-4337.12254.

Malerba, F. (2002). Sectoral systems of innovation and production. *Research Policy*, *31*(2), 247–264.

Malerba, F. (2005a). Innovation and the evolution of industries. *Journal of Evolutionary Economics*, *16*, 3–23. doi:10.1007/s00191-005-0005-1.

Malerba, F. (2005b). Sectoral systems of innovation: A framework for linking innovation to the knowledge base, structure and dynamics of sectors. *Economics of Innovation and New Technology*, *14*(1–2), 63–82. doi:10.1080/1043859042000228688.

Matharu, A. S., de Melo, E. M. & Houghton, J. A. (2016). Opportunity for high value-added chemicals from food supply chain wastes. *Bioresource Technology*, *215*, 123–130. doi:10.1016/j.biortech.2016.03.039.

Mirabella, N., Castellani, V., & Sala, S. (2014). Current options for the valorization of food manufacturing waste: A review. *Journal of Cleaner Production*, *65*, 28–41.

Mullen, A. M., Alvarez, C., Zeugolis, D. I., Henchion, M., O'Neill, E. & Drummond, L. (2017). Alternative uses for co-products: Harnessing the potential of valuable compounds from meat processing chains. *Meat Science*, *132*, 90–98. doi:10.1016/j.meatsci.2017.04.243.

Norsk Protein. (2017). Årsmelding 2017 (Annual report 2017). Retrieved from www.norskprotein.no/nyheter/årsmelding-2017.aspx?PID=86&M=NewsV2&Action=1.

Nærings- og fiskeridepartementet, & Landbruks- og matdepartementet. (2004). *Forskrift om forebygging av, kontroll med og utryddelse av overførbare spongiforme encefalopatier (TSE)*. Oslo: Lovdata.

Nærings- og fiskeridepartementet, & Landbruks- og matdepartementet. (2016). *Forskrift om animalske biprodukter som ikke er beregnet på konsum* Oslo: Lovdata.

Peneder, M. (2010). Technological regimes and the variety of innovation behaviour: Creating integrated taxonomies of firms and sectors. *Research Policy, 39*, 323–334. doi:10.1016/J.RESPOL.2010.01.010.

Prestegard, S. S., Pettersen, I., Nebell, I., Svennerud, M. & Brattenborg, N. (2017). Mat og industri 2017: Status og utvikling i norsk matindustri. Retrieved from http://matogindustri.no/matogindustri/dokument/Mat_og_industri_2017_plansjer_for_nedlasting.pdf.

Pyka, A., & Prettner, K. (2018). Economic growth, development, and innovation: The transformation towards a knowledge-based bioeconomy In I. Lewandowski (Ed.), *Bioeconomy: Shaping the Transition to a Sustainable, Biobased Economy* (1 ed., pp. 331–342). Cham: Springer International Publishing. doi:10.1007/978-3-319-68152-8_11.

Ravindran, R., & Jaiswal, A. K. (2016). Exploitation of food industry waste for high-value products. *Trends in Biotechnology, 34*(1), 58–69.

Ronzon, T., Santini, F., & M'Barek, R. (2015). *The Bioeconomy in the European Union in Numbers: Facts and Figures on Biomass, Turnover and Employment*. European Commission, Joint Research Centre, Institute for Prospective Technological Studies, Spain.

Toldrá, F., Mora, L. & Reig, M. (2016). New insights into meat by-product utilization. *Meat Science, 120*, 54–59. doi:10.1016/j.meatsci.2016.04.021.

Tüchsen, H. (2014). *Da andelen satte sig på flæsket – Den danske ledelseskanon, 6: Den danske andelssvinesektors vej til succes*. In K. K. Klausen, M. Schultz, P. Jenster, P. N. Bukh & S. Hildebrandt (Eds.), *Den danske ledelseskanon*. Copenhagen: Gyldendals Business.

8 New pathways for organic waste in land-based farming of salmon

The case of Norway and Denmark

Hilde Ness Sandvold, Jay Sterling Gregg and Dorothy Sutherland Olsen

8.1 Introduction

In this chapter, we explore the possibilities for sustainable pathways for the increased valorisation of organic waste from the aquaculture salmon industry. The chapter has a special focus on valorisation of sludge of the land-based stage of salmon production, and is based on interviews with fish producers and technology providers in Norway and Denmark.

The chapter is organised as follows: after the Introduction follows a section on the background of aquaculture, summarising important trends in general for aquaculture and explaining the developments of salmon production, including waste streams and environmental regulations. In the Findings section we explain the current utilisation of the sludge and describe challenges in the current system. The analytical section addresses the main barriers for new path development, and describes the structural elements of a development of sustainable aquaculture from the perspective of the recent literature on socio-technical transitions.

8.2 Background

8.2.1 Aquaculture trends

Global demand for animal-based food products, particularly fish, is increasing: the production of seafood from aquaculture has grown 15-fold since 1980 and has doubled since 2000 (FAO, 2016). Aquaculture now produces over half of all fish for human consumption in the world and is the world's fastest-growing food production sector (FAO, 2016). The growth in demand for seafood is largely driven by increasing wealth and urbanisation in developing regions of the world (FAO, 2016). In the developed world, concerns over sustainability issues, animal welfare, food safety and health are increasingly driving consumer behaviour with respect to seafood consumption (FAO, 2016). Nevertheless, demand is expected to increase in all areas of the world over the next decade (FAO, 2016).

Economically, salmonids are the most important fish family, comprising nearly 17% of the global seafood market, and have the largest commodity value of any group of fish, with demand steadily growing (FAO, 2016). After years of overfishing Atlantic salmon (*Salmo salar*), and habitat damage from river damming and intensive aquaculture, capture fisheries are no longer commercially viable (Parrish, Behnke, Gephard, McCormick & Reeves, 1998). Today, nearly all the world's supply of Atlantic salmon is farmed, producing 1.5–2% of the global aquaculture industry (Ernst & Young, 2018; FAO, 2016), and is the most economically important farmed salmonid (Asche, 2008; Ernst & Young, 2018). Trends and forecasts predict a large potential for growth (5% compound annual growth rate) in the Atlantic salmon industry (Ernst & Young, 2018).

Scandinavia, Norway and Denmark in particular, has pioneered innovation and technological progress in the Atlantic salmon industry, resulting in increased productivity and reduced costs (Asche, 2008; Asche & Bjørndal, 2011; Asche, Guttormsen & Nielsen, 2013; Asche, Guttormsen & Tveterås, 1999; Asche, Roll, Sandvold, Sørvig & Zhang, 2013; Kumbhakar & Tveterås, 2003; Roll, 2013; Sandvold, 2016; Tveterås, 1999; Tveterås & Battese, 2006). This innovation and expertise is also seen as a valuable commodity that can be exported and developed for other regions and fish species (Ernst & Young, 2018; Paisley et al., 2010). Atlantic salmon production in Norway grew by a factor of 10 between 1990 and 2013 (FAO, 2016) and Norway currently produces over half of the world's Atlantic salmon: 1.3 million tons of farmed salmon annually. Of this, 95% is exported, with a value of 61.5 billion NOK (6.5 billion €) in 2016 (Ernst & Young, 2018). That year, the Norwegian industry itself (which also has holdings outside of Norway) reported record revenues of 212.7 billion NOK (22 billion €), a 300% increase compared to a decade earlier (Ernst & Young, 2018). In 2017, a total of 195 licences for juvenile production and 1,015 for grow-out farming were given in Norway. Elsewhere, Denmark has a long tradition of aquaculture and is now at the forefront of land-based technological solutions, especially in the development of Recirculating Aquaculture Systems (RAS) (Nielsen, 2011, 2012). RAS has proven successful with eel and trout, and has recently been developed to produce Atlantic salmon completely on land at demonstration-scale facilities (Badiola, Mendiola & Bostock, 2012; Bergheim, Drengstig, Ulgenes & Fivelstad, 2009; Del Campo, Ibarra, Gutièrrez & Takle, 2010; Kristensen, Åtland, Rosten, Urke & Rosseland, 2009).

8.2.2 Salmon production

Currently, in production-scale firms, farmed Atlantic salmon are raised in land-based freshwater farms as smolt then transferred to sea-based cages where they stay until they are ready to be slaughtered (Asche & Bjørndal, 2011; Sandvold & Tveterås, 2014). Figure 8.1 illustrates the production process for salmon, which is divided into three main steps: the freshwater phase in hatcheries, the grow-out phase in salt water and the final processing at slaughterhouses. The hatcheries acquire fertilised eggs, which are hatched after a period in tempered freshwater. The fingerlings are kept in closed tanks until they smoltify and are ready for further growth in saltwater.[1] This usually takes place when the fish (named smolt) are 80–100 g. After vaccination, well-boats transfer the smolt to grow-out farms – floating cages at sea – and the salmon remain there until they reach a weight of 4–5 kg. Thereafter, the mature salmon are transferred to processing facilities where they are slaughtered. The whole production cycle takes three to four years.

The environmental impact of this type of production is threefold. First, organic waste can have a negative impact on the environment. While the severity of impact on the eco-system will depend on the specific local conditions of the farm, the nitrogen and the phosphorus in the waste create algae blooms and anoxic conditions in the coastal water, which can kill other aquatic life (Wu, 1995). Second, the crowded nets attract sea lice, a parasite that costs the industry up to 1.5 billion € per year (Costello, 2009). The Norwegian industry suffered a 5% loss in harvest quality (versus 2015) and a mortality rate of 19%, up from 16% in 2015, corresponding to 53 million individual fish (Ernst & Young, 2018). This has resulted in higher operational expenses in the industry and the introduction of new regulations by the Norwegian government (Ernst & Young, 2018). Chemical mitigation of sea lice is expensive and inefficient, causing additional adverse environmental impacts to coastal zones and becoming ineffectual with overuse (Burridge, Weis, Cabello, Pizarro & Bostick, 2010; Grant, 2002). Third, escaped fish can have large ecological impacts. Inter-breeding between wild and more genetically homogenous farmed salmon has been shown to reduce the life and fitness of indigenous fish populations over two generations (Thorstad et al., 2008).

Therefore, several alternatives to the traditional schedule (one year in land-based freshwater hatcheries; two years in sea-based cages in saltwater) are being considered, with the aim of shortening the period in the sea. One alternative is fully land-based production; another is moving the farm to offshore locations. These two options represent large operational, biological and technological changes, and substantial investments, with high risk. Other alternatives the industry is considering are closed underwater tanks, or floating basins in the grow-out phase. Currently, the majority of the salmon producers in Norway are using a production line with an extended land phase. This means that the smolts are not released to the sea at the standard size at 80–100 g, but rather held in the closed surroundings in the hatcheries until they reach a weight of more than 250 g.[2] In the short term, an extended land-based phase is the most realistic alternative to the traditional production schedule. Innovations and technological improvements for this alternative have become prevalent in recent years. For example, the two Danish firms are completely land-based, and Denmark is leading the innovation in this area.

This reorganisation of the production strategy and connected operational activities will lead to two major changes. First, it will require large investments for the industry in new hatcheries. These new hatcheries will all be using new water treatment technologies (RAS) (Badiola et al., 2012; Bergheim et al., 2009). Second, it means the volume of collected organic waste will increase substantially (Del Campo et al., 2010). In 2017, 330 million smolts with an average weight of 150 g were produced in Norway, generating 85,000 tons of sludge. These calculations assume a waste factor of 1.5 for the sludge (Del Campo et al., 2010). Increasing the smolt weight to 1 kg, which is now permitted, will increase the volume of stored sludge to 570,000 tons. Denmark's capacity for waste generation is much smaller, at an estimated 150 tons, but, as the innovation leader, this represents a scale where new technologies for waste handling can be developed. Farmed salmon production capacity in Denmark is currently 3,000 tons per annum. This calculation assumes 4.5 kg per finished fish.

8.2.3 Organic waste in salmon production

The waste streams from salmon production will differ in shape and volume in relation to the different phases in the production process. In Figure 8.1, the different waste streams coming from the three different phases of salmon farming are illustrated.

By-products from land-based hatcheries (first phase in Figure 8.1) consist of feed residues and fish faeces. The sludge is over 97% water but it is generally free of salt in juvenile smolt production (Badiola et al., 2012; Bergheim et al., 2009; Del Campo et al., 2010; Fivelstad, Bergheim, Hølland & Fjermedal, 2004). In contrast, waste from adult salmon in grow-out farms (second phase in Figure 8.1) will have salt content. The sludge is concentrated with a polymer and dewatered on a belt filter, which reduces the water content to roughly 80% feed (Badiola et al., 2012). Centrifuge technology can reduce the water concentration to under 70% feed (Kristensen et al.,

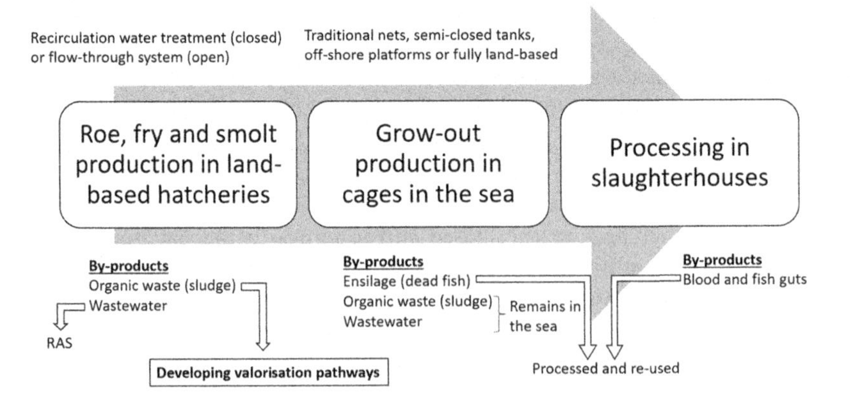

Figure 8.1 The organic waste streams along the production chain for salmon farming (by-products).

2009). Systems for drying sludge are theoretically able to reduce the water content to 20%. The sludge is rich in nitrogen, phosphorus and minerals (Del Campo et al., 2010). Because of this, there is interest in finding pathways to valorise it for re-use.

By-products from traditional grow-out farming remain in the sea (Figure 8.1). Aside from ensilage (dead fish), nothing is collected, and, with the salmon living in open nets, the organic waste is released into the local coastal ecosystem. We do not consider processing waste from the grow-out phase, because it is not collected in traditional cage systems. By-products from processing plants (blood, innards, heads, etc.) differ considerably from the waste streams from salmon production (Figure 8.1). We also do not consider waste from the slaughter of the fish because this has similar valorisation pathways to slaughterhouse waste, discussed in Chapter 7.

However, the literature on salmon production has thus far predominately focused on the grow-out phase (Asche, Guttormsen et al., 2013; Asche et al., 1999; Asche & Roll, 2013). Except for Sandvold and Tveterås (2014) and Sandvold (2016), little research has been conducted in relation to the juvenile phase. Even less attention has been given to new and sustainable applications for the increased volumes of collected organic waste in this industry. Therefore, this chapter analyses current valorisation pathways for the sludge from land-based production of juvenile salmonids. As the industry reorganises their production process in a more sustainable direction, some new challenges and opportunities appear concerning the handling of organic waste from land-based systems.

8.2.4 Environmental regulation and historical innovation in salmon production

Norwegian legislation and regulations for freshwater fish production have changed since the late 1970s, but not dramatically. As with the grow-out farms, the production of juveniles is highly regulated, and one needs a licence to legally operate in this sector. A number of requirements must be satisfied to obtain a juvenile licence, including access to a sufficient supply of fresh water, prevention of escapees, safe discharge of wastewater, as well as health, environment and safety requirements for the employees. Juvenile production has traditionally been restricted by the maximum number of units that can be produced each year, and maximum production varies by farm depending on different environmental concerns. Currently, given licences place no restriction on the number of units produced, but do place a maximum on the withdrawal of freshwater as well as a maximum on the discharge of wastewater.

Beginning in 2017, however, the Norwegian Ministry of Trade, Industry and Fisheries instituted a "traffic light" system that gives a green, yellow or red assessment to geographical areas based on losses caused by sea lice. Only in green areas may firms increase production at sea; in yellow areas, production increases are prohibited and, in red areas, firms must decrease production at sea (Ernst & Young, 2018). As such, the political landscape is evolving to address environmental concerns.

The Danish Environmental Protection Act regulates freshwater fish farms in Denmark. Regulations were introduced in 1987, and included water use and discharge, as well as waste handling. Theoretically, the regulations focused on nitrogen, phosphorus and organic matter in the effluent, although these proved problematic to reliably measure. The regulations instead focused on feed ratios, which were strict and curtailed aquaculture development, except for fisheries using new water technology, which have dramatically improved the feed ratios (Paisley et al., 2010). The 2004 Fisheries Act gives the Ministry of Food, Agriculture and Fisheries authority to regulate fisheries. The Ministry also has a goal of expanding production, but as most suitable freshwater locations have already been used, it means that RAS will be an increasingly important technology in Denmark (Paisley et al., 2010).

The increase in production volume in salmon farming in recent years has exacerbated the environmental challenges. As a response to these issues, the industry is now considering and testing different strategies for changing the production, waste handling and related technologies. RAS technology is increasingly replacing the traditional flow-through systems in juvenile production of salmon (Badiola et al., 2012; Bergheim et al., 2009). The system was developed in the 1970s in Denmark, out of small-scale laboratory equipment used to study living fish. The system was first applied to salmon smolt production as a lower-cost alternative to flow-through systems, which required more energy to heat the water. Additionally, RAS was seen as an attractive alternative to circumvent siting and regulation barriers on flow-through systems, because environmental risk factors such as escapees and polluted water spills are minimised using this technology. RAS uses 90–99% less water than flow-through systems, which also reduces energy consumption for heat. In order to recycle the water, solid waste (principally fish faeces and some feed residues) must be mechanically removed as soon as possible in order to prevent the build-up of bacteria that cause anoxic conditions. A biological filter is also used to clean the water for re-use.

Related to the organic waste, the individual Norwegian hatcheries have different restrictions in relation to their effluent. The fish farms are responsible for treating their waste in a responsible manner, yet there is no nationally required practice regarding the treatment of the outlet water. The differences in the disposal of sludge are primarily related to three different factors: (1) where they are located (north or south), (2) their age (when the hatcheries had their licences granted and (3) which technology they use; flow-through system or RAS. Restrictions tend to be lower in the northern regions of Norway than in the southern because the concentration of the farms is lower in the North and the following environmental impact will be lower. The older farms (from the 1980s) have fewer restrictions placed on the effluent, and very often still use a flow-through system. However, many of these are now closing down or being upgraded. All new farms have strict restrictions requiring waste management and emissions, and they use the RAS.

8.3 Methodology

To understand the valorisation of waste and to analyse possibilities for new and sustainable pathways for organic waste from juvenile salmon production, we have used two different sources of data. This analysis consists of both secondary and primary data.

We used secondary data in the initial phases of the study: industrial reports, government documents and the academic literature. This shaped the seven semi-structured interviews (primary data) with different fish producers and technology providers. The purpose of the interviews was to investigate waste streams in land-based farming, in order to understand the potential challenges connected to the current system from the perspective of the salmon producers. The interviews included questions about:

- How the organic waste is currently managed: processes, policy and potential challenges;
- Current and potential new value chains for organic waste: innovation, competition and barriers;
- Options for upgrading the firm, as well as incentives, trade-offs and demand factors;
- Social, economic and environmental sustainability factors that influence waste handling.

We interviewed six salmon production companies: five in Norway and one in Denmark. These firms represent the majority of the Norwegian and Danish production of smolt. Geographically, the firms are located across the whole coastline in Norway, and on the west coast of Denmark. Each of the interviewed fish production firms produce salmon all the way from hatchery to slaughter-ready fish, although our interviews focused on their land-based smolt production operations. We also interviewed a technology supplier in Denmark to understand the future trends and market outlook for this industry.

8.4 Findings

8.4.1 Current utilisation

From the data collection, we find that the sludge from land-based salmon production primarily has three different areas for use; as soil improvement and fertiliser in agricultural farming, as combustible material for heating in processing of new industrial products or as a replacement of fossil fuel after recycling, for example in transportation.

Soil improvement

Currently, the main utilisation of organic waste from land-based salmon production is soil improvement and fertiliser in agricultural production, either directly or indirectly.

The most common practice is through collaboration with local farmers, where the farmers come to the hatchery to pick up the sludge, normally by tractor and trailer. Since the sludge typically has a water content of 80–90%, the farmers need to use large tanks during transport. The organic liquid is applied directly on the soil without any kind of processing. The farmers are paid between 0.09€/kg and 0.21€/kg for the job. This collaboration is by far the most common practice to re-use the waste from smolt production.

However, some hatcheries choose to transport their wastewater to local recycling facilities. Here the sludge is dried, mixed and further processed with other kinds of organic waste (human sewage, for example). After processing, the new by-product is further sold as fertiliser as a dried biomass and also here used as soil improvement.

Replacement of fossil fuel (biogas)

When organic material goes through anaerobic decomposition, bacteria transform the waste into 60% methane and 40% carbon dioxide and other trace gases. The methane portion of the gas, often called biogas, can be used to produce new energy such as heat and electric power.

Sludge from smolt production is high in iron, because of the use of iron-chloride as a precipitate within the RA technology. This is a positive aspect of this feedstock for biogas production because it reduces the hydrogen-sulphide production as well as carbon dioxide, and is more efficiently converted to methane. The whole organic waste is generally high in fat and is therefore high in energy content; the large amount of fatty acids is nevertheless a challenge when it comes to biogas producers because they tend to inhibit methanogenic bacteria (Nges, Mbatia & Björnsson, 2012). So far, the biogas-producers have solved this by mixing in waste from agricultural farming.

Currently, biogas has two main areas for use: as a motor fuel for transport and as a heating fuel for greenhouses. In Norway, around 40 biogas plants currently exist and the sector is not yet as developed as it is in Denmark. Nevertheless, interest is increasing in Norway because it represents a sustainable energy source and the Norwegian firms see it as an interesting future possibility if the biogas facilities could be located near the fish farms, and if the requirements of mixing the waste with other agricultural residues could be met.

Biogas production has so far been driven by the agricultural sector. Yet low electricity costs have, to date, slowed down the development of biogas. At the time of the interview, the biogas production firms in Denmark were struggling financially, and although the sludge was of economic value to the biogas producers, the firm had an agreement to give it away and pay for the transport. The firm took the longer view in that they wanted to support the biogas producers as an outlet for this waste stream. The firm's representative said that they would later consider negotiating a price for the sludge, but only if the biogas producer was in a position to pay for it. The representative said it was not a high priority in terms of the fish farm's economy.

Combustible material for heating

Dried sludge can serve as a fuel in industrial production of different products, e.g. cement. In cement production, clinker is heated to 1,400–1,500°C. There is currently a collaboration between one of the salmon firms in our sample and a cement-producing firm, where the dried sludge is used to fuel the clinker ovens.

8.4.2 Challenges with current system

Waste volumes are expected to increase

One of the biggest challenges with today's waste stream management is the expected increase in volume in the years to come. The valorisation pathways are still in a nascent stage and are unlikely to be able to handle the expected increase in volume of organic waste. Because the volume is currently small, the firms have entered into simple agreements with farmers, and, in some cases, biogas producers are given the waste free or even paid to remove it. While many fish production firms have local agreements with other actors to handle the organic waste, as the scale increases, the interviewees noted that these actors may not be able to handle the increased volume of sludge, and new strategies would have to be explored.

Transport

A substantial challenge related to organic waste from salmon production is transport. If the sludge, which has a high water content, needs to be moved over large distances (which is to be expected), it is both a practical and economic issue. There is considerable distance between the smolt production facilities and the places where it can be used: typically, fish farms are located on the coast and agricultural areas are further inland. Sludge can only be economically transported short distances. The interviewees noted that the logistics are not yet in place for increased volumes of stored sludge, so transport and distribution is one of the bottlenecks in the full valorisation of fish farm waste.

Immature technology

Drying the sludge gives it a higher stability for storage, makes transport easier and cheaper and increases the fuel value. In biogas production, for example, at least 20% dry raw material is needed. However, the technology for drying is a bottleneck in re-use of the organic waste.

The experience with this technology is limited. Few of the hatcheries have invested in in-house knowledge of how to run drying facilities. Furthermore, the ones that have this are not satisfied with the technology and the labour resources required for running these systems. Current systems are difficult and costly to run. All the fish farms interviewed report that the largest motivation for this kind of investment is to reduce the costs in the end of the entire fish production process, and the drying technology has not developed to a stage where this is the case.

Another potential solution mentioned by the Danish firm was to explore algae that could break down the sludge and produce omega-3 fatty acids. Nonetheless, the volume of waste from this firm was not sufficiently large to warrant an exploration into this emerging technology. Theoretically, the waste could also be incorporated into a full aquaponics cycle, producing both fish and (hydroponic) vegetables, though this has not yet been demonstrated at scale.

Likewise, many firms recognised that there were other emerging technologies available for better procurement and transport of waste, but that this required investment in infrastructure and labour that, economically, would be more profitably directed towards other activities, such as procurement of nearby wind turbines to provide electricity to the facility. Essentially, handling of waste was not a high priority for the fish production firms in terms of their current economies and business models.

8.5 Analysis

8.5.1 Barriers for new pathways

Lack of available technology

Lack of knowledge related both to biological and technological aspects could be a barrier to realising the potential of the suggested new valorisation pathways. Many technologies are still in proof of concept or demonstration stages and it is still costly to invest in them.

Economic priorities

Fish waste does not have much value with respect to the total operating budget. The cost of producing salmon is much higher on land, and the economic incentives and willingness to invest in large-scale processing of the sludge are not yet in place. Our interviewees did not view the waste as a valuable product and it was of little interest in relation to the greater costs of running a profitable fish farm. These short-term economic priorities may also be a reason for the low levels of investment and experimentation into new ways of waste utilisation.

Resistance to go into new business areas

The firms are reluctant to diversify into new business areas. The major fish-producing firms' priorities are geared more towards producing salmon, whereas disposing waste is seen as a responsibility. Lacking expertise and seeing the waste as a small component to their bottom line, the fish firms were counting on existing recycling companies to introduce new alternatives. When asked, the firms were not interested in diversifying into other markets and did not see a financial advantage in this.

Patchwork regulation

Another barrier comes from the lack of legislation and fragmentation of policies, particularly in Norway, where the salmon farming industry is rapidly expanding. The different counties along the coastline have different guidelines for how they process and grant applications, and there are no cohesive national guidelines on how to handle the waste. Consequently, hatcheries experience different regulatory requirements regarding the treatment of effluent. On the one hand, this lack of common regulation means that salmon farms are free to use and test different solutions within the requirements. Our interviews suggest that this situation is fostering local innovation and entrepreneurial initiatives, but there is no large-scale, industry-wide solution to handle the waste in a cost-efficient manner. Therefore, on the other hand, the current regulatory environment is not conducive to developing a market for waste.

Lack of collaboration

On the local level, fish producers seem to collaborate well with external partners in order to deal with their waste. However, on the national level, there seems to be a lack of collaboration between the agricultural sector, aquaculture, producers of technology and the recycling sector. The fact that all firms and sectors specialise in their own niches results in a lack of cross-sectoral expertise. Some of the fish-producing firms we interviewed had little understanding of how other industries might benefit from their waste, and the potential it has for valorisation. Better collaboration between different industries could increase the potential of alternative uses of waste.

Co-location issues

Logistics and transport are current barriers to valorisation of waste. The fish-producing firms are typically located in the coastal districts and the processors of the waste are often located inland. Without technology for drying the waste, this results in high transport costs of biomass with high water content.

8.5.2 Socio-technical transition

As wild catch can no longer meet current demand, aquaculture will continue to expand if fish, and salmon in particular, are to continue to be supplied to the market at the current prices. Salmon-producing firms are currently in transition, in response to both a growing global demand for salmon and the calls to reduce their impact on local coastal ecologies, as well as the economic necessity of responding to the damage sea lice cause to the fish stocks. The following looks at aspects of the socio-technical transition, applying elements inspired by actor network theory (Simandan, 2018), technological innovation systems theory (Smits, 2002) and multilevel perspective theory (Geels & Schot, 2007).

Actors

Several of our interviewees claimed that consumers are becoming more interested in sustainability. The firm's image and the story of their products are therefore becoming more important. This was especially the case for the fully land-based firm in our study – market differentiation and the sustainability angle was their competitive edge in the marketplace.

Governments are also driving the transition, principally in supporting the industry, which provides jobs to rural communities and a valuable export. In Norway, we found several examples of how the industry has benefitted indirectly from regional policy. They are also responding to a growing awareness of environmental impacts through regulation, although, as yet, there is little coordination and guidance on how to handle waste.

Technology firms specialising in RAS are the main enablers of this transition, as it is now becoming more cost-efficient and easier to meet regulations regarding wastewater. Moreover, RAS allows the sludge to be more easily collected and concentrated. Further technological development is needed in centrifuge- and heat-based drying systems to make waste storage and transport more economical. Moreover, additional research is needed into new ways to process waste. Biotechnology providers could be key actors in this regard.

End-use actors (for organic waste) are not yet fully coordinated with the fish production firms. This stems from co-related factors: a lack of consistent and coordinated policy for waste handling, a lack of economic incentives, a lack of research and entrepreneurship and a lack of overlapping expertise.

Capabilities

With respect to the waste pyramid (see Chapter 3), current waste management strategies in land-based production include recycling (through transfer to wastewater recycling centres) and re-use (through transfer to agriculture, biogas production and cement industries). However, from the point of view of the fish firms, it harkens more towards disposal, since the firms are paying other actors to come and collect the waste and currently they have little interest in its fate. To move up on this pyramid will require more active coordination between actors, particularly between agriculture and aquaculture firms. Prevention is theoretically possible through integrated aquaponics systems, although there are limited capabilities and expertise in the respective industries. On-site algae production is also a possibility for waste prevention, though similar limitations in expertise exist here.

Networks

In their current state, the agricultural producers and land-based fisheries are too disparate to merge in joint production (i.e. aquaponics), yet the biogas represents a key intermediary that brings different actors together into a network. The biogas production also produces residues and there is a need to create a market for this as well (the bio-residue can be used as a soil conditioner; see Chapter 5). Thus, alternative use of the by-products, like biogas production, needs collaboration with other suppliers and users of biomass. Good collaboration requires co-location of the fish farm, the agricultural biomass producers and the biogas facility.

Industries such as cement firms utilising the waste in clinker production and energy suppliers are also emerging networks for salmon farms. However, these end-uses do not benefit from the high nutrient content of the waste; they are lower on the cascading use (see Chapter 3). Therefore, the valorisation potentials from these pathways are likely to be more limited.

A linkage that could emerge in the future could come between the RAS technology providers and key end-use actors. This would promote technology development in waste recovery: sludge dewatering, processing and handling targeted to specific ends in order to incorporate potential waste re-use and valorisation into the technology design.

Infrastructures

Because of co-location issues, infrastructure development is challenging. For example, expanding the use of organic waste from aquaculture in biogas production is a developing opportunity that potentially utilises both the nutrients and energy content of the waste. While this is performed in Denmark, it is performed in Norway to a lower degree, largely because the infrastructure is currently lacking.

Co-location of vegetable production and fish farms is not likely, as the most suitable place for aquaculture is near the coast (particularly with cage-based grow-out) and the most suitable place for agriculture is more inland. Hydroponic systems for aquaculture are costlier and would require a large capital expense for producing a low-value food source far away from existing distribution networks. The spatial embeddedness of the firms contributes to the lack of shared context.

Institutions

New business models for the re-use of organic fish waste are slow to develop. The salmon-producing companies themselves are not driving innovation in this part of the value chain. Furthermore, little research has been conducted to determine the most effective use for the sludge, either from a lifecycle assessment, or simply from an economic cost-benefit perspective. Thus, there is little to drive the industry to invent new usages for the waste.

Much of this can be explained by territorial embeddedness (Pallares-Barbera, Tulla & Vera, 2004) of the various institutions and their respective established networks. Therefore, there is little shared context for waste valorisation, and, as such, little incentive for research or entrepreneurship to link these disparate institutions.

8.6 Conclusion

8.6.1 Overcoming the barriers: key actors

There are still open questions concerning which actors and what type of institutional development will eventually emerge to handle this. Our findings show a general "wait-and-see" strategy for the business case for waste valorisation within the aquaculture industry. Disparate industries with diverse competencies, industrial expertise and institutional territorial embeddedness define the current landscape, and thus no viable market has yet emerged for sludge from juvenile salmon production. As such, there are currently no plans for large investments or any push for innovation in the near future.

Innovations at this nascent stage are accompanied with large risks and high costs, and the salmon-production firms choose to wait for innovative, new solutions before investing. Since there are no economic incentives for the firms to invest in new waste-handling technologies, it is not strategically prioritised by the fish producers. The fish-producing firms fulfil the government's restrictions, but have little incentive, currently, to explore options beyond that. National governments therefore have the potential to be key actors, especially as awareness grows of the environmental impact of this growing industry.

Other external actors also have the potential to play a key role. While the role of fish farm waste as fertiliser in agricultural systems can still be developed, further processing of the sludge through microbial conversion into value-added products could achieve future valorisation pathways. The interviews indicated some new valorisation pathways for organic waste in land-based salmon farming, such as aquaponics, niche chemicals or as an energy feedstock.

8.6.2 Impetus for future waste valorisation pathways

The development of new valorisation is therefore likely to come from three key drivers: (1) the expected scale-up of the sludge in the near future, (2) external actors discovering and seizing new business opportunities and (3) the benefits of the improved environmental reputation of the industry.

1 Globally, wild fisheries are no longer able to meet the increased demand for seafood. Consequently, the aquaculture industry in general, and the salmon sector in particular, continues to expand to meet this growing demand. In Norway, biological challenges related to the sea phase for salmon farming have forced the salmon industry to move more of its production onto land. The volumes of collected sludge will increase dramatically. Despite this, few solutions for the re-use of the waste have been successfully established. New applications will therefore be a necessity when the scale increases.
2 Organic waste from salmon farming has a high nutritional content. It is, for example, rich in nitrogen and phosphorus, which is a scarce resource globally. Processed carefully, high-value by-products can be developed from these residues. It is expected that entrepreneurs will seize this business opportunity. The high value of the raw material will therefore be a fundamental key driver for product innovation and development of new by-products in this industry.
3 Finally, by contributing to develop innovative waste-handling strategies, the salmon industry will benefit from creating a more sustainable and responsible image. Currently, pressure is being placed on the industry by environmental groups and local activists. Better utilisation of the rest products will contribute to improving the sustainability aspects. As such, fish-producing firms are becoming increasingly interested in their image and, as a result, market differentiation of their product has begun to emerge. This is particularly the case in Denmark, because it is now feasible to produce salmon completely on land. The Danish firm's model of market differentiation will become more important as the industry expands and public awareness increases about the environmental impact of this industry. The image and "selling the story" of ecologically produced fish are important components to the business model. We assume an increasing need to utilise the organic waste streams in an environmentally friendly and mutually beneficial manner for the salmon-producing companies, even though they do not want to be the leaders in the development process.

Notes

1 The smolt stage is here defined as the period where a salmon has gone through the smoltification process, where it has physically gone from being a freshwater fish to a salmon that tolerates saltwater. It attains a silver skin. There is no specific size clearly defined in the literature of the salmon at this stage. However, here we use smolt to describe a salmon with a weight between 0.1 and 250 grams, and post-smolt between 250 grams and 1 kg.
2 The size of the fish in the hatcheries has traditionally been restricted to a maximum of 250 g. However, as a pilot project from May 1, 2012, the Norwegian Ministry has given the farmers the right to grant an exemption to extend the juvenile phase in closed land-based systems until the fish reaches a size of up to 1 kg.

References

Asche, F. (2008). Farming the sea. *Marine Resource Economics*, *23*, 527–547.
Asche, F., & Bjørndal, T. (2011). *The economics of salmon aquaculture*. Chichester: Wiley-Blackwell.
Asche, F., & Roll, K. H. (2013). Determinants of inefficiency in Norwegian salmon aquaculture. *Aquaculture Economics & Management*, *17*(3), 300–321.
Asche, F., Guttormsen, A. & Nielsen, R. (2013). Future challenges for the maturing Norwegian salmon aquaculture industry: An analysis of total factor productivity change from 1996 to 2008. *Aquaculture*, *396–399*, 43–50.
Asche, F., Guttormsen, A. & Tveterås, R. (1999). Environmental problems, productivity and innovations in Norwegian salmon aquaculture. *Aquaculture Economics & Management*, *3*, 19–29.
Asche, F., Roll, K. H., Sandvold, H. N., Sørvig, A. & Zhang, D. (2013). Salmon aquaculture: Larger companies and increased production. *Aquaculture Economics & Management*, *17*(3), 322–339.
Badiola, M., Mendiola, D. & Bostock, J. (2012). Recirculating Aquaculture Systems (RAS) analysis: Main issues on management and future challenges. *Aquacultural Engineering*, *51*, 26–35.
Bergheim, A., Drengstig, A., Ulgenes, Y. & Fivelstad, S. (2009). Production of Atlantic salmon smolts in Europe: Current characteristics and future trends. *Aquacultural Engineering*, *41*(2), 46–52.
Burridge, L., Weis, J. S., Cabello, F., Pizarro, J. & Bostick, K. (2010). Chemical use in salmon aquaculture: A review of current practices and possible environmental effects. *Aquaculture*, *306*(1–4), 7–23.
Costello, M. J. (2009). The global economic cost of sea lice to the salmonid farming industry. *Journal of Fish Diseases*, *32*(1), 115–118.
Del Campo, L. M., Ibarra, P., Gutièrrez, X. & Takle, H. R. (2010). Utilization of sludge from recirculation aquaculture systems. *Nofima rapportserie*.
Ernst & Young. (2018). *The Norwegian Aquaculture Analysis 2017*. Retrieved from www.ey.com/Publication/vwLUAssets/EY_-_The_Norwegian_Aquaculture_Analysis_2017/$FILE/EY-Norwegian-Aquaculture-Analysis-2017.pdf.
FAO. (2016). *The state of world fisheries and aquaculture: Contributing to food security and nutrition for all*. Rome: FAO.
Fivelstad, S., Bergheim, A., Hølland, P. M. & Fjermedal, A. B. (2004). Water flow requirements in the intensive production of Atlantic salmon (Salmo salar L.) parr–smolt at two salinity levels. *Aquaculture*, *231*(1–4), 263–277.

Geels, F. W., & Schot, J. (2007). Typology of sociotechnical transition pathways. *Research Policy, 36*(3), 399–417.

Grant, A. N. (2002). Medicines for sea lice. *Pest Management Science (formerly Pesticide Science), 58*(6), 521–527.

Kristensen, T., Åtland, Å., Rosten, T., Urke, H. & Rosseland, B. (2009). Important influent-water quality parameters at freshwater production sites in two salmon producing countries. *Aquacultural Engineering, 41*(2), 53–59.

Kumbhakar, S., & Tveterås, R. (2003). Risk preferences, production risk and firm heterogeneity. *The Scandinavian Journal of Economics, 105*(2), 275–293.

Nges, I. A., Mbatia, B. & Björnsson, L. (2012). Improved utilization of fish waste by anaerobic digestion following omega-3 fatty acids extraction. *Journal of Environmental Management, 110*, 159–165.

Nielsen, R. (2011). Green and technical efficient growth in Danish fresh water aquaculture. *Aquaculture Economics & Management, 15*(4), 262–277.

Nielsen, R. (2012). Introducing individual transferable quotas on nitrogen in Danish freshwater aquaculture: Production and profitability gains. *Ecological Economics, 75*, 83–90.

Paisley, L. G., Ariel, E., Lyngstad, T., Jónsson, G., Vennerström, P., Hellström, A. & Østergaard, P. (2010). An overview of aquaculture in the Nordic countries. *Journal of the World Aquaculture Society, 41*(1), 1–17.

Pallares-Barbera, M., Tulla, A. F. & Vera, A. (2004). Spatial loyalty and territorial embeddedness in the multi-sector clustering of the Berguedà region in Catalonia (Spain). *Geoforum, 35*(5), 635–649.

Parrish, D. L., Behnke, R. J., Gephard, S. R., McCormick, S. D. & Reeves, G. H. (1998). Why aren't there more Atlantic salmon (Salmo salar)? *Canadian Journal of Fisheries Aquatic Sciences, 55*(S1), 281–287.

Roll, K. H. (2013). Measuring performance, development and growth when restricting flexibility. *Journal of Productivity Analysis, 39*(1), 15–25.

Sandvold, H. N. (2016). Technical inefficiency, cost frontiers and learning-by-doing in Norwegian farming of juvenile salmonids. *Aquaculture Economics & Management, 20*(4), 382–339.

Sandvold, H. N., & Tveterås, R. (2014). Innovation and productivity growth in Norwegian production of juvenile salmonids. *Aquaculture Economics and Management, 18*, 149–168.

Simandan, D. (2018). Competition, contingency, and destabilization in urban assemblages and actor-networks. *Urban Geography, 39*(5), 655–666.

Smits, R. (2002). Technological forecasting and social change: Innovation studies in the 21st century. *Technological Forecasting and Social Change, 69*(9), 861–883.

Thorstad, E. B., Fleming, I. A., McGinnity, P., Soto, D., Wennevik, V. & Whoriskey, F. (2008). Incidence and impacts of escaped farmed Atlantic salmon Salmo salar in nature. *NINA special report, 36*(6).

Tveterås, R. (1999). Production risk and productivity growth: Some findings for Norwegian salmon aquaculture. *Journal of Productivity Analysis, 12*(2), 161–179.

Tveterås, R., & Battese, G. E. (2006). Agglomeration externalities, productivity, and technical inefficiency. *Journal of Regional Science, 46*(4), 605–625.

Wu, R. (1995). The environmental impact of marine fish culture: Towards a sustainable future. *Marine Pollution Bulletin, 31*(4–12), 159–166.

9 Valorisation of whey

A tale of two Nordic dairies

Simon Bolwig, Andreas Brekke, Louise Strange and Nhat Strøm-Andersen

9.1 Introduction

Have you eaten Greek yoghurt recently? For every teaspoon of Greek yoghurt you eat, two teaspoons of surplus material are produced. This by-product, known as acid whey, has become a waste disposal issue for dairies around the world following a boom in the demand for Greek yoghurt. This development mirrors, although at a smaller scale, that of sweet whey, which is a by-product of white hard cheese production. Whey is a very strong pollutant, with a biochemical oxygen demand (BOD) 175-fold higher than the typical sewage effluent (Smithers, 2008). With increasingly strict environmental regulation of industrial wastes, disposing of whey as waste has become both difficult and costly as volumes have increased and have become more concentrated in larger production units. Together with a growing societal focus on the circular economy, these environmental pressures have forced the dairy industry to rethink the way it manages its whey side stream (ibid.).

There are valuable substances in whey that are possible to valorise, including functional proteins and peptides, lipids, vitamins, minerals and lactose (Smithers, 2008). In the case of sweet whey, dairies around the world have in recent decades developed technologies, processing capacities, products and new business models for utilising these substances. A strong driver has been the rapid growth in global markets for food ingredients, including whey-based protein powders, which are 'among the winners of several new nutrition trends and food developments' (Vik & Kvam, 2017, p. 336). Thus whey has become an important nutritional and functional ingredient for high-quality foods (NutritionInsight, 2018) and the global whey protein industry has been estimated to grow by 12–14% per year (Kjer, 2013).

This strong market trend is related to rising populations and incomes, especially in emerging economies. But it is also strongly driven by the rise of 'functional nutritionism', which refers to the increased engineering and reengineering of food in coevolution with changing corporate strategies, trends in food, diets and health, and new food and nutrition policies (Scrinis, 2013, 2016; Vik & Kvam, 2017). An indication of this trend is the large and growing market for functional food ingredients, i.e. probiotics, proteins and amino acids, phytochemical and plant extracts, prebiotics, fibres and specialty carbohydrates, omega-3 fatty acids, carotenoids, vitamins, and minerals. In 2018 these were estimated at US$68.6 billion worldwide, rising to US$94.2 billion by 2023 (PR Newswire, 2018). The development of global trade networks and infrastructures associated with globalisation has facilitated the expansion of market opportunities for whey-based products. Upstream in the5value chain, mergers and acquisitions have resulted in fewer and more specialised dairy plants with spatial concentrations of specific side streams, including whey.

Acid whey has not undergone the same development as sweet whey, due to its – so far – much lower volume and somewhat less favourable physical and taste properties (see section 9.2). Yet as this chapter will show, the market and production dynamics just mentioned may also be relevant to acid whey valorisation if issues relating to taste and processing are addressed.

In this broader context, this chapter investigates how the largest dairies in Norway and Denmark, TINE and Arla Foods, have worked to add value to whey with a focus on acid whey. Specifically, we examine key features of the dairy industry in each country, the historical development of these industries, individual firm characteristics and capabilities, and the position of each firm in national and international markets. We ask how these factors – at the level of the firm, the industry and the market – have influenced whey valorisation in the two countries. In this regard, we are especially interested in how the organisation of the firm, e.g. the separation of main product and by-product processing in different legal units, affects organisational capabilities and utilisation of whey.

This analysis draws on theories discussed further in Chapter 3: path dependence and lock-in – particularly economies of scale and scope and learning effects; and directionality through industrial practices – particularly value chain governance. Our comparative approach reveals how these drivers and mechanisms play out in different national contexts and how they influence inter-firm connections crossing these contexts.

This chapter is based on data collected through interviews with experts from a feed producer and from the TINE and Arla Foods dairies, a review of company webpages and a literature review. A total of nine interviews were carried out, of which three were with experts from Arla Foods and Arla Foods Ingredients (AFI) (senior executive R&D advisor AFI, vice president for corporate social responsibility Arla Foods, head of department separation AFI) and six with experts in TINE (research director, head of corporate social responsibility, technical director, researcher and young professional). In this chapter, the term 'whey' is used to refer to both sweet whey and acid whey, unless the type of whey is specified.

The chapter is structured as follows: the next section discusses the properties and utilisation options for acid whey. For each possible use, there is a summary of the current state of innovation, the application's commercial potential, the required technological and investment capacities and the application's place in the waste pyramid. Sections 9.3 and 9.4 present the Norwegian and Danish case studies, respectively. Here we discuss the development and key features of the dairy sectors in both countries, the main characteristics of the dominant dairy company in each country (TINE and Arla Foods) and how these two dairies utilise whey. In section 9.5 we discuss the key drivers of whey valorisation, analysing the dynamics of whey utilisation in the two companies (and countries), concerning the structure of the dairy industry, firm-level characteristics and strategies, and value chain linkages. We then discuss the implications these case studies have for the sustainability of different valorisation pathways. Section 9.6 concludes the chapter.

9.2 Properties and uses of acid whey

Never have we eaten so much Greek yoghurt worldwide, and this trend only seems to be going up. In the USA, approximately 771,000 tons of Greek yoghurt were produced in 2015, accounting for almost 40% of the yoghurt market, compared to a market share of only 1–2% in 2004. However, as the production of Greek yoghurt skyrocketed, so did production of the by-product, acid whey: for every 100 kg milk used in Greek yoghurt production, only one third ends up in the final product, while the other two thirds become acid whey (Arla Foods Ingredients, 2018b). Acid whey is a potential hazard to the aquatic environment due to its high organic matter content in the shape of lactose, resulting in a high Biological Oxygen Demand (BOD > 35,000 ppm) and Chemical Oxygen Demand (COD > 60,000 ppm) (Ramos et al., 2015; Smithers, 2015). The high BOD level means that the presence of acid whey in waters would cause a drop in biological oxygen levels, leading to the elimination of aquatic life. Hence, if other uses cannot be found, acid whey must be treated as waste water in own or municipal plants, involving significant financial costs for the dairy as well as socio-economic costs associated with waste treatment. But acid whey contains many valuable compounds, providing opportunities for companies to gain competitive advantage through value-added utilisation of acid whey, as discussed below (Guimarães, Teixeira & Domingues, 2010).

9.2.1 The properties and composition of acid whey

To understand the properties of acid whey, it is useful to understand the origin of whey. When producing cheese or yoghurts from milk, the milk is separated into a relatively solid part, which becomes the cheese or yoghurt, and a yellow liquid part known as whey (Ramos et al., 2015). Whey can be used directly for animal feed or as a biogas substrate, or processed into a number of products, especially ingredients in food and feed production; see Figure 9.1. Whey has a relatively high protein content and low fat content. There are several types of whey depending on the processing technology used for the casein removal of liquid milk, where the most common categories are

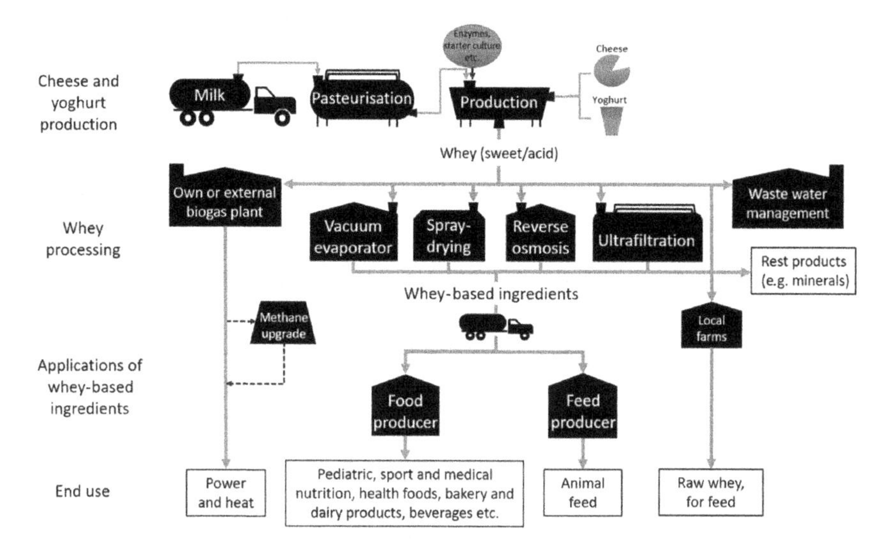

Figure 9.1 Simplified flowchart depicting the processing, application and end uses of sweet and acid whey.

Note
Whey is a by-product from the manufacturing of cheeses and yoghurts especially, as well as of calcium caseinate. Parts of the whey are processed into proteins, whey permeate (milk solids) and other compounds, which are used as ingredients in food and feed production. Whey processing generates its own side streams, including various minerals. Raw whey is used as animal feed, as a substrate in biogas production or treated as waste water.

sweet whey and acid whey (Jelen, 2011; Panesar, Kennedy, Gandhi & Bunko, 2007). Sweet whey is a by-product from white hard cheese production, and can be used to produce a range of functional and nutritional foods (see 9.3.3 and 9.4.3). Different qualities of acid whey are derived from the production of Greek yoghurt, Quark, Icelandic *skyr*, cream cheese and cottage cheese. A variant of acid whey is casein whey, which is derived from the production of calcium caseinate. Calcium caseinate is a protein produced from casein (milk protein) in skim milk and is used in coffee creamers and instant soups and as a dietary supplement by athletes. While some studies such as (Jelen, 2011) often refer to casein whey as 'acid casein whey' or simply 'acid whey', in this chapter we make a distinction between acid casein whey and acid whey derived from the production of Greek yoghurt, *skyr* and cream/cottage cheese, due to the significant differences in properties and potential uses between the two variants of acid whey (interview with AFI).

Both sweet whey and acid whey are composed mainly of water (93%) while the solid components consist of minerals (12–15%), lactose (70–72%) and whey proteins (8–10%) (Jelen, 2011). The largest difference between acid whey and sweet whey is the pH, which lies within 6.0–6.5 for sweet whey (Jelen, 2011) and 3.6–4.5 for acid whey (Gami, Godwin, Czymmek, Ganoe

& Ketterings, 2016), while acid casein whey has a pH of 4.5–5.1 (Jelen, 2011). Overall, compared to sweet whey, acid whey has less protein, is more acidic and has a more distinct (sour) taste. Table 9.1 shows the breakdown of component composition in acid whey derived from yoghurt and cream/ cottage cheese production. Below, we focus on the properties and utilisations of acid whey and to some extent acid casein whey, which have been less documented than those of sweet whey (de Wit, 2001; Jelen, 2011).

The challenges of utilising acid whey occur in the processing procedure. The most common way to process whey into a product suitable for industrial use is to dry it through evaporation in multistage vacuum evaporators followed by spray-drying. However, spray-drying acid whey with conventional technology is not feasible due to the high content of lactic acid, which makes the whey powder more likely to absorb moisture, resulting in an increased stickiness of the powder (Chandrapala et al., 2016). Moreover, a low pH makes the proteins less stable and it is more difficult, for instance, to remove water from acid whey than from sweet whey. Because of the low pH and the proximity to the isoelectric point, the protein will readily precipitate, which may make it difficult to recover. Proteins are thus more readily available and easily isolated from sweet whey than from acid whey. Moreover, sweet whey can easily be heat-treated and used in new products (e.g. creamy cheese). If acid whey is heated, it will not become an acidic gel, but unites and acquires a slightly granular consistency.

9.2.2 Utilisation of acid whey

Over the past decades, research and technological innovation have transformed the utilisation of sweet whey from waste (or feed) to a resource for

Table 9.1 pH and nutrient composition of acid whey from the production of Greek yoghurt, cottage cheese and cream cheese

	Unit	*Min*	*Max*	*Average*
pH		3.55	4.48	4.11
Solids	%	2.49	6.53	5.16
Total nitrogen	mg/100 mL	22.4	258.3	85.1
Ammonia-N	mg/100 mL	0.0	16.1	2.8
Organic-N	mg/100 mL	18.9	258.3	82.3
Phosphorous (P_2O_5)	mg/100 mL	120.5	194.0	169.1
Potassium (K_2O)	mg/100 mL	142.5	212.5	192.8
Calcium	mg/100 mL	90.7	136.8	121.6
Magnesium	mg/100 mL	7.1	11.3	9.9
Sodium	mg/100 mL	31.3	44.1	39.3
Sulfur	mg/100 mL	5.0	17.0	7.3
Zinc	mg/100 mL	0.4	0.5	0.4
Chloride	mg/100 mL	79.0	189.0	108.0

Source: Gami, Godwin, Czymmek, Ganoe and Ketterings, 2016.

value-added products (Smithers, 2015). Can such a transformation also occur for acid whey? Below we introduce the most common valorisation pathways for acid whey (including casein whey), drawing on a review of research articles and company websites.

Animal feed

For decades, using acid whey and sweet whey for animal feed has been one of the most common uses of whey due to their nutritional properties. Acid whey is especially suitable for piglets because of the acidity and nutritional composition. In piglet diets acid whey is mainly a source of energy (milk sugar) rather than a protein source.

Many Greek yoghurt and cheese manufacturers still pay their milk suppliers to take the acid whey, which they then mix with silage for animal feed (Smithers, 2015). Although acid whey is a suitable feed, the transport and storage of liquid whey is costly and barely profitable. Some yoghurt makers such as Chobani in the US sell it to local farmers for animal feed supplement (Erickson, 2017b). Chobani has invested in reverse-osmosis filtration technology, which separates the water from the whey using high-pressure systems, resulting in a more compact by-product that is easier and cheaper to transport to farms (Erickson, 2017b).

Biogas

Another common application of acid whey is as a substrate for biogas production, where the biogas is used for electricity production. No estimate of the financial gain for this utilisation could be found. The first environmental gain of using acid whey for biogas production is the avoided risk of spillage of acid whey into the environment. The second gain occurs if the electricity produced replaces electricity made from non-renewable sources and thereby reduces greenhouse gas emissions. Biogas can also be upgraded into pure methane and fed into the natural gas grid, where it exists. In Switzerland, several cheese plants have biogas digesters installed onsite (Jelen, 2011). Alphina Foods (NY) is an example of a company that transports acid whey to a nearby farm, where it is used in biogas production. Some biogas plants, for instance in the UK, run entirely on whey.

Bakery

Some research articles examine the use of the acid whey which stems from the production of the Indian cottage cheese known as paneer, as a bread ingredient (Divya & Rao, 2010; Paul, Kulkarni & Rao, 2016). They report that the inclusion of a moderate amount of acid whey did not have any negative influence on the bread quality, while higher amounts increased the dough proofing time. Research design and results reveal the focus to be on solving the waste problem connected to acid whey rather than finding a truly valuable output for the resource stream. The authors do, however, observe that acid whey contains valuable nutrients suitable for human diets.

Whey beverages

A whey beverage refers to a drink where whey is the main component in liquid form. There are several examples of both soft drinks and alcoholic beverages containing casein whey. Acid whey derived from Greek yoghurt and similar products is not suitable for whey beverages due in part to a too high mineral content. The Swiss brand Rivella is the only one with a reasonable market share and lasting success. Interestingly, Rivella does not advertise the use of acid casein whey in their products, and studies have shown that several international marketing attempts to promote whey beverages have failed (Jelen, 2011). The most typical beverage combination of casein whey is with citrus fruit juices due to the high content of lactic acid. The literature also contains examples of alcoholic beverages such as beer, wine and champagne, which are produced with acid whey. The start-up Alchowhey, created by scientists at the Technical University of Denmark, has developed a DIY patented bacteria solution that converts lactose into ethanol suitable for spirits. The technology is applicable at both large and small scale, which makes it relevant for small dairies which struggle to valorise their surplus whey onsite. Although whey beverages have had very little commercial success due to processing challenges and the unusual flavour of raw whey, a more refined technology and an increasing awareness of the nutritional benefits of whey proteins may encourage the further development of whey beverages (Jelen, 2011).

Nutritional products

Acid whey contains several nutritional components that can be isolated. Research projects have been conducted to isolate valuable compounds, such as lactose, proteins and vitamin B_{12}, while several articles deal with the isolation of substances such as β-lactoglobulin and α-lactalbumin. The latter two are valuable proteins, which are present in significantly smaller concentrations in acid whey in comparison to sweet whey (interview with AFI). Such single components can be used as food additives as well as for medical purposes, where earnings can be even greater. However, most of these processes require complex technologies, which many dairies are reluctant to include in their production lines, and few articles contain references to the commercial recovery of single compounds and substances from acid whey. This can be due to two reasons. First, this process requires new investments and a reorganisation of production, which is unlikely to be supported by dairy managers. Second, the isolation of useful components will result in even less valuable residual material, which will be less suitable for animal feed and require further processing. Casein whey is commonly used in medical nutrition, while acid whey derived from Greek yoghurt is not suitable for this purpose due to its high mineral content. For example, MyProtein® (myprotein.com) used to provide casein whey protein powder used for fitness purposes, but the powder is now out of stock.

Nutrilac®

A recent innovative solution which diminishes the generation of acid whey waste while also maximising product capacity is the protein-based product solution, Nutrilac®, developed by Arla Foods Ingredients (AFI). Nutrilac® does not consist of acid whey but is derived from milk and can be added directly to the acid whey together with water, transforming the acid whey into a low-cost raw material, which can be added to products such as stirred yogurt, beverages, soups, jar cheese, dips, dressings, processed cheeses and cream cheeses (Arla Foods Ingredients, 2018a; Erickson, 2017a). For example, Nutrilac® can be added to acid whey, water and cream, resulting in a high-quality dip. Nutrilac® has won several awards including the 'Beverage Innovation Award' in 2013, the 'IFT Food Expo Innovation Award' in 2014 and the 'IFT Best Dairy Innovation' in 2015. According to AFI, Nutrilac® requires few or no adjustments to existing production lines such as yoghurt or cream cheese while offering a sustainable solution that adds value by expanding companies' product portfolio (Arla Foods Ingredients, 2016).

9.2.3 Innovation, commercialisation and technological requirements

The previous section has outlined the various ways in which acid whey can be transformed into or provide components for value-added products, and at the same time avoid waste disposal or spillage. Yet it also pointed to a generally low level of commercialisation of these conversion technologies, products and components, aside from low-value uses as biogas and animal feed in raw form. In view of this, Table 9.2 provides an overview of the current state of acid whey valorisation for each main use category in terms of: research and innovation, commercial potential, technology and investment requirements, place in the waste pyramid, and use examples.

9.3 The dairy sector in Norway

Historically, the dairy sector has been the backbone of agriculture in Norway and the processing of milk has been dominated by co-operatives owned by milk farmers. Most dairy products found in the cool counter at supermarkets in Norway are labelled TINE while a smaller portion comes from the brand Q-Meieriene. Foreign dairies such as Danone and Arla supply an even smaller portion. The Norwegian dairy sector has experienced a strong structural development, going from many small, almost independent dairies to a few, large and highly efficient ones. A similar transformation has taken place

Table 9.2 Overview of acid whey valorisation

Use	*Research and innovation*	*Level of commercial potential*	*Technology and investment requirements*	*Place in the waste pyramid*	*Use examples*
Animal feed	Acid whey is commonly used for animal feed. Important nutrients are maintained during processing and drying using modern technology. This makes the product highly palatable and digestible for calves, goat kids and lambs. Acid whey is especially suitable for piglets due to its acidity and nutrients. It is mainly a source of energy (milk sugar) rather than protein.	High level of commercial potential due to the wide range of different physical parameters available in whey products.	Feed producers need high-quality processing equipment, storage space, silos, transportation and warehousing.	Reuse/upcycling of nutrients in the whey.	The US yoghurt producer Chobani has adopted advanced filtering systems in its Idaho and NY plants and provides whey as supplemental feed to local farmers.

Biogas	Anaerobic digestion treatments are well developed for biogas production.	Biogas can be used locally for electricity or process heat, or can be upgraded to and sold as bio-methane. Installing the digester at the location of whey production saves transport costs but requires more volume.	Requires no changes in production processes. Investment in a biogas plant can be costly for small dairies; state support has in some cases helped overcome this barrier. Whey can also be sold to an external biogas plant.	Recovery (energy) and recycling (digestate applied on fields).	The world's largest biogas plant, located in the UK, runs on whey. In the US, Alphina Foods (NY) uses a bio digester on a nearby farm, resulting in 100% processed acid whey.
Bakery products	Using raw acid whey material in bakery products has not reached a solution. Bakery products can be upgraded by using acid whey together with Nutrilac® (see below).	There is little nutritional advantage to be gained from using acid whey as a substitute for water in baking. Moreover, the demand for bakery products containing acid whey is still low.	Does not require major changes to production systems.	Reuse.	No examples could be found.
Whey beverages	The food industry and scientists are increasingly assessing the use of acid whey in beverages due to its favourable nutritional composition. Innovations within alcoholic beverages are also gaining ground.	Acid whey beverages have not yet been successfully adopted by the market, so commercial potential is still low. Solutions like Nutrilac® may increase the market potential, as may the production of alcohol from whey.	Requires handling of acid whey and product development in the production of whey beverages.	Recycling.	Rivella is the only successful acid whey beverage on the market. Alcowhey is using a patent bacteria solution to make ethanol from whey. Arla Foods has also created acid whey beverage possibilities.

continued

Table 9.2 Continued

Use	*Research and innovation*	*Level of commercial potential*	*Technology and investment requirements*	*Place in the waste pyramid*	*Use examples*
Nutrition	The food industry is now focusing on the health aspects of acid whey with regard to nutritional opportunities in food products. Extracting functional components of sweet whey has become a major industrial activity, but for acid whey it is still mainly in a research state.	Using the many nutritional properties of acid whey has high market potential, when considering the many current uses in food ingredients of similar properties of sweet whey.	Requires investments in complex technology for separation and for the production of specific components.	Reuse/upcycling of nutrients in the whey.	MyProtein® fabricates acid whey protein powder for fitness purposes.
Nutrilac®	Nutrilac® technology has created new possibilities for the use of acid whey, according to AFI.	Nutrilac® can increase the commercial potential of the uses of acid whey in a variety of by-product developments.	No technological investment is needed. Companies will have to purchase Nutrilac® from AFI.	Enables reuse and recycling.	Arla Foods is producing value-added products using Nutrilac®, and Nutrilac® is sold to other food producers.

upstream in the value chain. The number of milk farms fell drastically from 148,000 in 1959 to 16,000 in 2009, further declining to 8,800 in 2015 (Almås & Brobakk, 2012; Statistics Norway, 2016).

TINE is obliged to collect all milk from farmers in Norway, irrespective of the distance of the farm to the nearest dairy. This requirement is connected to a general national policy of decentralised development in a vast, sparsely populated country.

In Norway the agricultural sector is highly protected and TINE does not have the 'political mandate' to source from other countries (which for instance led to what is known as the 'butter crisis' in 2011 (Aftenposten, 2011)). In addition to high transportation costs, Norway faces tariffs when exporting food to the EU. Trade between Norway and the EU is regulated through the 1992 Agreement on the European Economic Area (EEA Agreement) and tariffs and quotas for agricultural products are negotiated under this Agreement, most recently in April 2017 (European Commission, 2017). The agricultural trade balance favours the EU (ibid.) and EU exports to Norway are increasing, with the product category 'fresh milk and cream, buttermilk and yoghurt' rising from €14 to 19 million between 2013 and 2017. Norwegian exports to the EU of products in the same category remained stable at €3 million from 2013 to 2017 (European Commission, 2018). According to Vik and Kvam (2017), EU import tariffs mean that the EU is not an important market for Norwegian whey protein concentrate.

9.3.1 TINE in Norway

TINE is by far the largest dairy cooperative operating in Norway. It traces its history back to 1856 when Rausjødalen Meieri – the first dairy cooperative in Norway and Northern Europe – was founded with 40 members. During the second half of the 19th century, the organisation expanded as still more dairies were established, and the Norwegian Dairy Association was founded in 1881. The year 1900 saw the highest number of dairies in operation with 780 sites in total. A few cheese brands were interesting to the international market, and in the 1920s, an organisation was established to handle the export of dairy products from Norway. The name TINE was introduced in 1992 and the company structure has changed a few times since (TINE, 2018a).

Today, the TINE cooperative has 10,500 members (owners) and 8,500 cooperative farms (TINE, 2018b). Farmers are guaranteed milk sales to dairies through the cooperative and dairies are guaranteed a supply. Milk production is the most regulated activity in the Norwegian agriculture sector. A central agreement on the milk price ensures an equal price to all farmers, irrespective of the use of the milk or the farm's geographical location.

From the year 1900 up to the 1980s there was a gradual decrease in the number of dairies from 700 to about 200. Instead of having a dairy at or near each farm, regional dairies were established. They were located to efficiently collect milk from the surrounding farms and supplied almost the entire product portfolio to the regional market. The decrease in the number of dairies between 1900 and 1990 was accompanied by an increase in the size of regional dairies. Since the introduction of the TINE brand in 1992, restructuring has been associated not only with larger dairies but also with a higher degree of specialisation. Today, TINE has 31 dairies (TINE, 2018c) and no ambition to produce the entire product portfolio at each dairy. Instead, the company operates a complicated logistical system where intermediate products are transported between dairies and end products are transported from dairies to regional and central warehouses.

Yet TINE is more than a processor of milk. As mentioned, transport and storage logistics are important parts of the company. There are also strong connections upstream in the value chain back to the farms. This is emphasised in TINE's contribution to R&D within animal health and milk production especially, which focuses on a strong relationship with the Norwegian University of Life Sciences and with other Norwegian universities and research centres.

9.3.2 Utilisation of acid whey in Norway

The 'discovery' of large streams of by-products such as acid whey has been a consequence of the restructuring and specialisation of Norwegian dairies since the early 1990s. When each dairy still produced a great variety of products, the amount of each type of by-product was small and easy to handle through wastewater treatment. The moment cottage cheese and Greek yoghurt were produced at a single facility, at the same time as the demand for these products grew tremendously, the volume and spatial concentration of acid whey also increased and became difficult to handle.

TINE is one of the main suppliers of Greek yoghurt and cottage cheese in Norway and is therefore also a major producer of acid whey. Because of the rising volumes of acid whey, adding value to this by-product has become a new focus area for TINE. TINE produces several thousand tons of acid whey each year, which come mainly from one of its 31 dairies, located in southern Norway. It has been a challenge for TINE to handle the increasing volumes of acid whey because it cannot be treated or used in the same way as sweet whey. Today, acid whey is mainly used in low-value applications, namely directly as animal feed (to pigs) and biogas production. Acid whey can be used in the production of piglet feed, and can obtain a rather high price per kg dry matter, but no Norwegian dairies sell acid whey to feed producers (interview with feed producer). This is probably because it is not deemed profitable given the investment costs and control of processing needed to make a raw material suitable for the feed industry.

TINE delivers free acid whey directly to farmers, who pay half of the transport costs. Another dairy company in Norway also delivers acid whey to farms, with a more stable chemical composition and more dry matter, which the farmers pay for. A reason for this difference is that TINE makes both cottage cheese and Greek yoghurt, which tends to give more variation in acid whey properties, as it is derived from two different processes, while the other manufacturer only produces Greek yoghurt (interview with TINE and feed producer).

9.3.3 Utilisation of sweet whey in Norway

Although TINE treats acid whey almost like it has no value, the same is not true for sweet whey. Sweet whey is made into whey protein concentrate (80% total whey protein content) and its coproduct whey permeate (milk solid) at two new facilities located in Jæren and Verdal. Whey permeate contains 85% lactose and only 3% protein, and can be used as an ingredient in bakery products, confectionary, dairy products, sauces, drinks and similar food products. Whey raw materials are supplied by TINE from its production of yellow cheese. The facilities went into operation in 2013 and were built (Jæren) or upgraded (Verdal) with technical assistance from AFI, with whom TINE has had a partnership based around whey since 2008 (Arla Foods Ingredients, 2018a). This partnership also stipulates that AFI is the sole distributor in export markets of all the whey protein concentrate and whey permeate produced by TINE. TINE exports around 2,800 tons of whey protein concentrate and 21,000 tons of whey permeate each year through AFI (Vik & Kvam, 2017). Prior to 2008, TINE relied on different traders for whey powder exports (ibid.). The partnership gives TINE access to AFI's global sales channels and in turn helps AFI to increase its global market share. AFI has furthermore invested in infrastructure, logistics and quality systems to facilitate the export of Norwegian whey-based substances (Vik & Kvam, 2017).

9.4 The dairy sector in Denmark

Danish dairies have received worldwide recognition for their processing skills and product quality, and their products are present across the world. The Danish dairy sector comprises 28 companies and 54 production plants, processing around 4.9 billion kg milk (Danish Dairy Board, 2018a). Cooperatives of milk producers are the dominant ownership model and 97% of the milk produced in Denmark is supplied to cooperative dairy companies. Over the last century the number of dairy companies fell dramatically to the present level (see below).

Arla Foods is by far the biggest dairy company and cooperative in Denmark and Sweden. In 2017, Arla Foods processed 87% of the Danish milk pool (Danish Agriculture & Food Council, 2018) and 66% of the Swedish milk pool. In Denmark, the remaining dairies are both cooperatively and privately owned companies. They typically specialise in specific product areas within cheese, butter and liquid milk production (Danish Agriculture & Food Council, 2018). A large part of their production is exported by specialised exporters.

As in Norway, the Danish dairy sector has gone through tremendous structural changes over the last decade, with production now taking place on a small number of large farms. According to Statista (2018), the number of Danish dairy farms decreased by almost 50% (from 6,253 to 3,293 farms) in the period 2005–2016. In 2010, approx. 4,100 dairy farmers each had an average of 127 cows and a milk quota of 1,142 tonnes (Danish Agriculture & Food Council, 2018), which made the Danish dairy sector among the largest and most modern in Europe.

Danish dairies are strongly export oriented and two-thirds of the total Danish milk pool go into export products, placing Denmark among the world's top five exporting nations (Danish Dairy Board, 2018b). The historical focus of Danish agriculture on export markets (butter and bacon to the UK) has enabled the specialisation and growth of the dairy industry, generating increasing volumes of side streams. The value of all Danish dairy exports equals €1.8 billion annually (Danish Agriculture & Food Council, 2018), representing more than 20% of all Danish agricultural exports (Danish Agriculture & Food Council, 2018). The largest export market for Danish dairy products is the EU. Exported dairy products are mostly cheese, preserved milk products and butter. The domestic market is mainly covered by domestic dairy production (Danish Agriculture & Food Council, 2018). The market share of imported milk is modest.

9.4.1 Arla Foods

Arla Foods is part of the Arla Group, which has its headquarters in Denmark. It is Europe's largest dairy group and the fourth largest in the world in terms of milk intake (Hansen, 2018). It is owned by 11,262 milk producers across seven European countries, most of them in Sweden (2,780), Denmark (2,675), Germany (2,327) and the UK (2,395) (Arla, 2017). In total, Arla processes 13,937 million kg of milk, of which 35% comes from Denmark, 23% from the UK and 13% from Sweden and Germany respectively (ibid.).

Arla's journey began back in 1882 when the first Danish dairy cooperative was established, while farmers in Sweden set up their first cooperative in 1881, calling it Arla Mejeriforening (Arla Foods Ingredients, 2018a). The dairy cooperative movement quickly took off and in 1890, there were 900 cooperative dairies in Denmark, and 10 years later the number had reached 1,000. The cooperatives grew in size and scope throughout the 20th century and in 1970, four large cooperatives and three individual dairy farmers formed Mejeriselskabet Danmark (MD). The creation of MD required the construction of a more professional organisation to include modern strategic and functional management. MD was continuously developed through mergers and growth in the following two decades, resulting in the establishment of the more internationally oriented MD Foods in 1988. In 2000, MD Foods and Swedish Arla merged and changed their organisation's name to Arla Foods. At the time, the two companies processed 90% of the Danish and 66% of the Swedish milk pool respectively.

Arla Foods Ingredients: AFI

In Denmark, the discovery of the potentials of whey protein began in 1974 when the first pilot production of whey proteins was established in the cellar of a milk powder plant owned by the dairy cooperative in Holstebro (today, Arla Hoco). Whey, which was hitherto regarded as waste, proved to be, as Arla declares, 'full of wonders', and in 1980 Danmark Protein opened the world's largest factory for the production of whey protein concentrate and lactose (milk sugar). In 1988, Danmark Protein established a subsidiary in Japan, while MD Foods established MD Foods Ingredients in Germany in 1990 (Arla Foods Ingredients, 2018a). In 1994, MD Foods then acquired Danmark Protein, which was merged with MD Foods Ingredients. A similar development was seen in Sweden. With the merger of the mother companies in 2000, MD Foods Ingredients became AFI.

Today AFI is an independent subsidiary of Arla Foods and is responsible for all Arla Food's whey protein products destined for the global B2B market for functional and nutritional ingredients. The head office is located next to Arla Foods in Viby, Denmark, and it has subsidiaries across the globe (see below). The company has 1,370 employees. AFI has experienced high growth rates over the last few years. In 2017, the revenue of AFI was €655 million, and €766 million including its joint ventures (Arla, 2017). This represents a revenue increase of 19.6% compared to 2016. AFI has also seen a relative growth in both the scope and quality of its product portfolio; the share of the volume of 'value' products compared to 'standard' products grew from 68% to 74% between 2012 and 2017 (ibid.).

Maintaining or increasing market share in a fast-growing global market for whey-based products requires large investments. AFI invested €220 million in value-adding activities between 2012 and 2017 (Arla, 2017). A large share has been directed towards expanding and refining its domestic production capacity. This includes a €57 million investment in 2012 that nearly doubled AFI's whey processing capacity in Denmark, and a new lactose factory in 2013 worth €120 million. In 2016, a new production facility for whey protein hydrolysate, used in infant, sports and clinical nutrition, went into operation, tripling AFI's existing capacity in this production category.

Today AFI processes 6–7 million tons of sweet whey every year on its Danish facility in Nr. Vium. This includes 400,000 tons of casein whey, derived from the production of calcium caseinate at Arla Hoco (interview with AFI). To utilise the full capacity of the plant, AFI also imports sweet whey from Sweden, the UK and Germany, and it buys sweet whey from other Danish dairy companies. The products produced from sweet whey are described in section 9.4.3.

AFI invests significant resources in research and development (R&D). Its R&D division employs 70 scientists and technicians, accounting for 10% of all AFI employees, and collaborates closely with universities. The R&D covers 'tailoring of molecular functionality, advanced separation technologies to isolate specific components, heat treatment and pasteurization technology to improve functionality and shelf-life. R&D further develop powder technologies to form safe and functional powders for dry blend and optimal rehydration' (Arla Foods Ingredients, 2018c).

In 2009 AFI established a large high-tech ingredients application centre in Aarhus, Denmark to substitute the more modest application facilities at the R&D pilot plant in Nr. Vium. Another, smaller centre was established in Buenos Aires in 2002. The innovation centres are used for both whey protein research as well as for product trials. A key purpose is to assist AFI's customers in utilising the whey-based ingredients sold by AFI in their product development.

The Danish dairy ingredients subsector has shown an even stronger export market orientation than conventional dairy products, responding to the rapid growth in demand worldwide. Since 1990, MD Foods Ingredients/AFI has established overseas subsidiaries in Europe, the US, Asia (Japan, China, Singapore and South Korea) and South America (Argentina, Brazil and Mexico). The internationalisation strategy includes foreign direct investments in whey processing plants in Argentina (from 2002 to 2018) and Germany (1993, 2002 and 2011) facilitated by joint ventures and acquisitions (Arla Foods Ingredients, 2018a). A key purpose of these investments is to increase access to whey raw material, which is critical to maintaining AFI's market share (Kjer, 2013). In the other countries mentioned, direct investments have mainly been in the form of sales offices. In 2008, a partnership was created with the TINE dairy in Norway (see 9.3).

9.4.2 Utilisation of acid whey in Denmark

Arla Foods generates limited amounts of acid whey from the production of *skyr*. This whey contains mainly lactose and calcium and very little protein. It is therefore not further processed but until today only used as a biogas substrate, as markets need to be found for potentially innovative solutions (interview with AFI). As mentioned, AFI has developed a whey protein product called Nutrilac® that can upgrade acid whey functionality so it can be used in the production of a range of products (see 9.2.2). AFI sells Nutrilac® to customers including producers of dairy products.

9.4.3 Utilisation of sweet whey in Denmark

Up to the 1980s, whey from cheese production in Denmark was mainly used for animal feed, but today whey-based products have become the main commodities, which may even create a higher value than the cheese production itself. Arla Foods Ingredients is the only company in Denmark that processes sweet whey on an industrial scale. It also owns processing facilities in Germany and Argentina and it markets whey products produced in Norway. AFI products are sold in more than 90 countries (Arla, 2017). They include a range of functional and nutritional ingredients (proteins) to support industries within paediatric nutrition, sport nutrition, health foods, medical nutrition, bakery products, dairy and affordable food (Arla, 2017). The proteins produced include 15–20 'pure proteins' such as whey protein concentrate, whey protein isolate, Lacprodan and Capolac Nutrilac, among others. From these proteins, other and new 'mixed' proteins are constantly being developed so that the total number of proteins produced and sold at any given time ranges between 80 and 150 (Hansen, 2018). Likewise, AFI produces whey permeate, a coproduct of whey protein concentrate, on its plants in Denmark and Argentina and sells the whey permeate produced by TINE (Arla Foods Ingredients, 2018d) (see 9.3.3). A substantial part of the permeate production is further valorised into different qualities of lactose mainly used for infant formula production (interview with AFI).

9.5 Discussion

9.5.1 Drivers of whey valorisation

In just a few decades, the production of whey-based food ingredients has become an important part of Arla Foods's overall strategy. The development of these side streams has been significant for the company's strong position within value-added activities today (Lange & Lindedam, 2016) and the aim of AFI is to become 'the global supplier of value added whey' (Arla, 2017, p. 34).

Arla Foods has employed several strategies to strengthen its position in a very dynamic global market for whey-based products. A key factor has been the establishment of the subsidiary AFI dedicated to whey processing, as well as joint ventures between AFI and foreign ingredients companies. This has provided the needed organisational framework for their large investments in raw material sourcing, production capacity, R&D and marketing of value-added whey over the last two decades. In general, subsidiaries bring internationalisation opportunities, and they also enable the creation of more direct customer contact as well as consumer control, resulting in increased governance features for the subsidiary in the respective location (Biisgård, 2006). AFI thereby enables the creation of a more tailored product portfolio within their main markets (Danish Dairy Board, 2006). Moreover, AFI has created three business units for specific products, thereby enhancing its capacity for product differentiation and specialisation: nutrition, functional milk protein, and permeate and lactose (Andersen, 2014). In support of their business units, AFI has a large R&D division and runs two customer-focused application centres. All this has contributed to Arla Foods becoming a 'lead firm' (Gibbon, Bair & Ponte, 2008) in the global value chain for whey proteins, sourcing whey not only from its own member producers but also from other dairies in Denmark and Norway (see below).

Nutrilac® may be an efficient way of dealing with acid whey today, at least for high-value end products. Yet it is presently only available in limited volumes and the challenge for AFI is therefore to identify markets with a moderate demand for the innovative solutions (interview with AFI). Thus it remains to be seen whether Nutrilac® will become a dominant technology in acid whey valorisation and if other processing solutions are developed and commercialised alongside this one.

As noted in Chapter 3, lock-in mechanisms not only create barriers to innovation in the bioeconomy, but can under certain conditions promote and reinforce innovation and sustainable business development. In this regard, as a result of a series of mergers and acquisitions, foreign direct investments and technological specialisation over the past decades, Arla Foods has realised positive lock-in mechanisms in whey-based activities. These are economies of scale (in whey sourcing, processing and marketing), economies of scope (regarding production and marketing of different whey products) and long-term learning effects (regarding whey sourcing, processing technology and markets). A key factor has been the establishment of the subsidiary AFI with an R&D department dedicated to whey product development. Important economic and political contexts have been the fast-growing global ingredients market, good access to regional (Northern Europe) whey resources and a basis in an export-oriented agricultural sector (see Table 9.3). All this represents a strong 'directionality through industrial practices' (Chapter 3) in whey valorisation, which was originally motivated by a growing waste problem.

Currently, TINE has had no explicit strategy for adding value to its side streams and is still establishing a strategy for acid whey in cooperation with a Norwegian research institute (interview with TINE). Most of TINE's R&D focuses on end products, health, packaging and animal health. The main focus within TINE has been on how to manufacture end products most efficiently, while the handling of by-products has received less attention. The department called 'Ingredients' does not have a special focus on surplus resources, but rather on the needs of the food industry. This means that they do not search for alternative uses of surplus resources from the dairies, but respond to demand from a specific market. There is an emphasis on keeping production and products stable, and on not interacting with existing production infrastructure in the utilisation of side streams.

For side streams TINE has emphasised reducing the costs of management rather than increasing the value added. For acid whey, this choice is related to the rather small volume of acid whey generated (even with the increase in Greek yoghurt and cottage cheese production), which makes valorisation less attractive in a cost-benefit analysis. It is also related to TINE's current lack of an international market channel for acid whey-based products (interview with TINE). In the case of sweet whey, which is much more abundant, TINE's partnership with AFI has enabled the co-production of two bulk commodities – whey protein concentrate and whey permeate. But TINE depends on AFI for the marketing of these products and does not engage in more refined processing or product development, and it also depends on AFI for building or upgrading its processing facilities. From a value chain governance perspective (Gereffi, Humphrey & Sturgeon, 2005), this places TINE in an

Table 9.3 Summary of dairy industry dynamics and whey valorisation pathways

Firm and country	*Characteristics and dynamics*	*Economic context*	*Political context*	*Whey valorisation*
Dairy industry in industrial countries	Fewer and more specialised dairy plants, resulting in spatial concentration of specific side streams.	Fast-growing global market for functional and nutritional food ingredients.	Stricter environmental regulations: need to process whey or pay for waste disposal.	Ranges from simple energy recovery to advanced processing into high-value food ingredients.
TINE (Norway)	Large cooperative dairy in national context. No strong focus on by-products. TINE Ingredients is a department, not a separate company. Own, large R&D centre. Relatively small plants.	Strong value chain relations with milk suppliers, regulated by law. Geographically dispersed farms.	Strong regulation of dairy industry; protection from foreign products. Not a member of the EU.	Producer of few sweet whey-based ingredients. No processing of acid whey.
Arla Foods (headquarters in Denmark)	Large cooperative dairy in national and global context. Has grown through mergers and acquisitions, joint ventures and investments in whey processing capacity. High R&D and marketing capabilities related to food ingredients. Has set up a separate ingredients company, which handles and sells by-products. Relatively large plants.	Export-oriented agricultural sector. Liberalised national market. Large farms located close to each other and to the dairies.	Easy access to EU markets. Export-friendly policies.	Producer of multiple sweet whey-based ingredients. No processing of acid whey. Developed Nutrilac®, which eases the use of acid whey as an ingredient in high-value products.

asymmetric and 'captive' supplier relationship with AFI (Vik & Kvam, 2017), where the latter's control of research-based knowledge of processes and products as well as customer relations makes it costly and risky for TINE to switch to another buyer (ibid.). The partnership with AFI has, however, also meant a technological upgrading of TINE in terms of an improved ability to produce bulk commodities from whey (ibid.). It remains to be seen whether the partnership can help TINE to further upgrade its whey-based business in terms of learning about more advanced products and processing techniques, or develop deeper market knowledge, and so strengthen its position in the value chain. In this regard, Vik and Kvam (2017) observe that AFI does not share knowledge about customers and prices with TINE to avoid losing control over TINE as a supplier. Finally, any upgrading strategy chosen by TINE needs to consider how to access new sources of whey raw material to achieve scale economies, given the geographical context and the domestic orientation of milk sourcing and sale of dairy products (see Table 9.3).

9.5.2 The sustainability of different valorisation pathways

Arla Foods has strategised on the valorisation of acid whey in a clever way. Instead of bringing large amounts of acid whey to a central processing facility, AFI sells a product to enable its customers to turn acid whey into a useful product. This is, however, not the same as saying this strategy is the most sustainable one for handling acid whey. For any product or product system to be deemed sustainable, it should at least have a better economic, social and environmental performance than other ways to achieve the same function. The term 'product system' does not only refer to the acid whey itself. For instance, when acid whey is used for human food production instead of feed production, one must also consider that other feed ingredients are needed to replace acid whey.

There are likely economic gains both for AFI and the dairy companies who buy Nutrilac® to valorise their acid whey. Livestock farmers, who lose a cheap feed ingredient, will have to replace this source with a probably costlier alternative, such as imported feed. This will of course create extra income for the farmers producing the feed alternative and for other actors in this value chain. This will again increase the demand for feed products with consequences for other value chains dependent on such products. Similarly, if acid whey is originally used for biogas, alternative substrates need to be found. Hence there are complex distributional and economic knock-on effects on the wider food (and energy) system of changing valorisation pathways for side streams such as acid whey, which can be important but at the same time difficult to measure.

A hypothesis in this regard is that when valorisation becomes more technically complex and costly, there is a risk that the distribution of wealth becomes more unequal. Repercussions in the environmental dimensions will follow similar paths to those in the social and economic dimensions, although the magnitude and the direction of impacts may be very different. As such, evaluating the valorisation of by-products cannot be performed in isolation. A usage which is conventionally seen as higher order, such as human food or medicines, may from a holistic perspective be less sustainable than solutions with lower levels of sophistication. While by-products should not be wasted, the development of sustainable valorisation pathways requires sophisticated assessment tools, such as life cycle assessment.

9.6 Conclusion

There are several potential valorisation pathways for acid whey, including as a component in animal feed, nutritional products, bakery products and beverages. Yet the dominant use today is as animal feed or as a biogas substrate in unprocessed form. Nutrilac® is a new whey protein, which enables the use of acid whey in a range of high-value products. In contrast, virtually all sweet whey is upcycled to value-added products, notably a variety of whey proteins and whey permeates used in the manufacturing of nutritional and functional foods. It has become a scarce resource.

Whey is the central side stream in the dairy sector and has been utilised differently in Denmark and Norway. Arla Foods, based in Denmark, has a stronger focus on utilising by-products, enabled by international market connections, R&D capabilities, a series of strategic domestic and international investments and the size of the regional dairy industry. A key factor has been the establishment of a subsidiary dedicated to whey processing, product development and marketing. Through these and other means, Arla Foods has gained a significant share of the market for whey-based ingredients, taking a leading position in the global value chain for these products. In Norway, the agriculture sector has mainly supplied the domestic market; in this context, TINE has focused on developing and manufacturing a variety of fresh dairy products for Norwegian consumers, rather than adding more value to its side streams. Hence, TINE lacks the advanced knowledge of whey products and processing techniques, the market connections and possibly the milk resource base, which are a necessary part of being a lead firm in the ingredients value chain. Instead it plays the role of 'captive' supplier of generic whey products to Arla Foods' ingredients subsidiary, AFI.

Even though AFI has found an efficient way of dealing with acid whey, the solution might also create a lock-in that hinders potential new solutions. Ideas that are generated in individual and possibly across industrial sectors may guide TINE in utilising their acid whey in innovative ways, although it lacks the systematic processes to search for these solutions.

Using whey as a raw material in production processes can not only improve a company's sustainable competitive advantage; it also aids the transition towards a sustainable bioeconomy. Innovation in waste systems results in lower disparate rates while increasing the ability to upcycle whey, adding value to by-products in the dairy value chain. When considering alternative valorisation options for by-products such as whey, it is important to take a holistic and whole-system perspective on sustainability. The most technologically advanced options are not necessarily the 'best' ones in terms of environmental, economic and social impacts and there will likely be trade-offs between these three dimensions of sustainability.

Acknowledgements

This work was supported by the Research Council of Norway (grant number 244249). Mads Lykke Dømgaard created the art work. Antje Klitkou and Marco Capasso provided comments on an earlier version of this chapter.

References

Aftenposten. (2011). Nå er det smørkrise. Retrieved from www.aftenposten.no/norge/i/y3oVr/Na-er-det-smorkrise.

Almås, R., & Brobakk, J. (2012). Norwegian dairy industry: A case of super-regulated co-operativism. *Research in Rural Sociology and Development, 18*, 169–189. doi:10.1108/S1057-1922(2012)0000018010.

Andersen, H. (2014). *Arla Foods Ingredients: Business case on innovation and export of bio economy products: A journey from side streams to main products*. Paper presented at the Bioeconomy – Potentials for New Growth, Copenhagen.

Arla. (2017). *Consolidated annual report 2017: Driving innovation in dairy*. Retrieved from www.arla.com/company/investor/annual-reports.

Arla Foods Ingredients. (2016). Press release: Dairies can say goodbye to by-products with Nutrilac® HiYield. Retrieved from www.arlafoodsingredients.com/about/press-centre/2016/pressrelease/dairies-can-say-goodbye-to-by-products-with-nutrilac-r-hiyield-1398233.

Arla Foods Ingredients. (2018a). History. Retrieved from www.arlafoodsingredients.com/about/company/history.

Arla Foods Ingredients. (2018b). Make acid whey an asset to your dairy business. Retrieved from www.arlafoodsingredients.com/49cc24/globalassets/restricted/2018/dairy/white-paper-make-acid-whey-an-asset-to-your-dairy-business_web.pdf/?DownloaderId=97792ba972294ddc827ecff1b2e0c41c&SourceId=7ea207ff-ecd1-445c-a3d2-109b0dc3b02a.

Arla Foods Ingredients. (2018c). Research and development. Retrieved from www.arlafoodsingredients.com/about/research-and-development.

Arla Foods Ingredients. (2018d). Whey permeate: Taste and texture enhancement at a cost benefit. Retrieved from www.arlafoodsingredients.com/globalassets/global/products/permeate-and-lactose/arla_whey_permeate_0815_web.pdf.

Biisgård, P. (2006). Interview with Peder Tuborgh. *Mejeri* (pp. 6–7), Mejeriforeningen.

Chandrapala, J., Chen, G. Q., Kezia, K., Bowman, E. G., Vasiljevic, T. & Kentish, S. E. (2016). Removal of lactate from acid whey using nanofiltration. *Journal of Food Engineering, 177*, 59–64. doi:10.1016/j.jfoodeng.2015.12.019.

Danish Agriculture & Food Council. (2018). Danish dairy industry. Retrieved from https://agricultureandfood.dk/danish-agriculture-and-food/danish-dairy-industry.

Danish Dairy Board. (2006). Vore nordiske naboer. In *Nyhedsmagasin om mejeribranchen*.

Danish Dairy Board. (2018a). Dairy companies. Retrieved from https://danishdairyboard.dk/danish-dairies.

Danish Dairy Board. (2018b). History of the Danish dairy industry. Retrieved from https://danishdairyboard.dk/history.

de Wit, J. N. (2001). *Lecturer's Handbook on whey and whey products*. Brussels: European Whey Products Association.

Divya, N., & Rao, K. J. (2010). Studies on utilization of Indian cottage cheese whey in wheat bread manufacture. *Journal of Food Processing and Preservation, 34*, 975–992. doi:10.1111/j.1745-4549.2009.00432.x.

Erickson, B. E. (2017a). Acid whey: Is the waste product an untapped goldmine? Retrieved from https://cen.acs.org/articles/95/i6/Acid-whey-waste-product-untapped.html.

Erickson, B. E. (2017b). How many chemicals are in use today? *Chemical Engineering News, 95*(9), 23–24.

European Commission. (2017). EU and Norway conclude negotiations to enhance trade of agricultural products. Retrieved from https://ec.europa.eu/info/news/eu-and-norway-conclude-negotiations-enhance-trade-agricultural-products-2017-apr-07_en.

European Commission. (2018). Agri-food trade statistical factsheet: European Union – Norway. Retrieved from https://ec.europa.eu/agriculture/sites/agriculture/files/trade-analysis/statistics/outside-eu/countries/agrifood-norway_en.pdf.

Gami, S. K., Godwin, G. S., Czymmek, K. J., Ganoe, K. H. & Ketterings, Q. M. (2016). *Acid whey pH and nutrient content*. Cornell Agronomy Factsheet #96. Ithaca, NY: Cornell University Press. Accessed 10 September 2017. Retrieved from http://nmsp.cals.cornell.edu/publi cations/factsheets/factsheet96.pdf.

Gereffi, G., Humphrey, J. & Sturgeon, T. (2005). The governance of global value chains. *Review of International Political Economy, 12*(1), 78–104. doi:10.1080/09692290500049805.

Gibbon, P., Bair, J., & Ponte, S. (2008). Governing global value chains: An introduction. *Economy Society, 37*(3), 315–338. doi:10.1080/03085140802172656.

Guimarães, P. M. R., Teixeira, J. A. & Domingues, L. (2010). Fermentation of lactose to bio-ethanol by yeasts as part of integrated solutions for the valorisation of cheese whey. *Biotechnology Advances, 28*, 375–384. doi:10.1016/j.biotechadv.2010.02.002.

Hansen, E.-K. K. (2018). [Arla Foods Ingredients. Personal communication dated 02.11.2018].

Jelen, P. (2011). Whey processing: Utilization and products. *Encyclopedia of Dairy Sciences: Second Edition*, 731–737. doi:10.1016/B978-0-12-374407-4.00495-7.

Kjer, U. (2013). Valle i høj kurs. Retrieved from https://mejeri.lf.dk/tema/temaer/2013/populaer-ingrediens/valle-i-hoej-kurs#.W-V0tdVKguV.

Lange, L., & Lindedam, J. (2016). *The fundamentals of bioeconomy: The biobased society*. Retrieved from http://orbit.dtu.dk/files/140638164/Lange_L_Lindedam_J_2016_The_Fundamentals_Of_Bioeconomy_The_Biobased_Society.pdf.

NutritionInsight. (2018). 'Big protein, small bottle': Arla Foods Ingredients to introduce whey shot concept. Retrieved from www.nutritioninsight.com/news/big-protein-small-bottle-arla-foods-ingredients-to-introduce-whey-shot-concept.html.

Panesar, P. S., Kennedy, J. F., Gandhi, D. N. & Bunko, K. (2007). Bioutilisation of whey for lactic acid production. *Food Chemistry, 105*(1), 1–14. doi:10.1016/j.foodchem.2007.03.035.

Paul, S., Kulkarni, S. & Rao, K. J. (2016). Effect of Indian cottage cheese (paneer)-whey on rheological and proofing characteristics of multigrain bread dough. *Journal of Texture Studies*, *47*, 142–151. doi:10.1111/jtxs.12168.

PR Newswire. (2018). Functional food ingredients: Global market outlook to 2023 – Development of different techniques to create high-value natural carotenoids. Retrieved from www.prnewswire.com/news-releases/functional-food-ingredients-global-market-outlook-to-2023-development-of-different-techniques-to-create-high-value-natural-carotenoids-300683874.html.

Ramos, O. L., Pereira, R. N., Rodrigues, R. M., Teixeira, J. A., Vicente, A. A. & Malcata, F. X. (2015). Whey and whey powders: Production and uses. *Encyclopedia of Food and Health*, 498–505. doi:10.1016/B978-0-12-384947-2.00747-9.

Scrinis, G. (2013). *Nutritionism: The science and politics of dietary advice*. New York: Columbia University Press.

Scrinis, G. (2016). Reformulation, fortification and functionalization: Big food corporations' nutritional engineering and marketing strategies. *Journal of Peasant Studies*, *43*(1), 17–37. doi:10.1080/03066150.2015.1101455.

Smithers, G. W. (2008). Whey and whey proteins: From 'gutter-to-gold'. *International Dairy Journal*, *18*(7), 695–704. doi:10.1016/j.idairyj.2008.03.008.

Smithers, G. W. (2015). Whey-ing up the options: Yesterday, today and tomorrow. *International Dairy Journal*, *48*, 2–14. doi:10.1016/j.idairyj.2015.01.011.

Statista. (2018). Number of dairy farms in Denmark in selected years from 2005 to 2016. Retrieved from www.statista.com/statistics/608994/number-of-dairy-farms-in-denmark.

Statistics Norway. (2016). Structure of agriculture, 2015, preliminary figures. Retrieved from www.ssb.no/en/jord-skog-jakt-og-fiskeri/statistikker/stjord/aar/2016-01-07.

TINE. (2018a). About TINE. Retrieved from www.tine.no/english/about-tine/about-tine.

TINE. (2018b). Om TINE. Retrieved from www.tine.no/om-tine.

TINE. (2018c). The TINE cooperative. Retrieved from www.tine.no/english/about-tine/the-tine-cooperative.

Vik, J., & Kvam, G. T. (2017). Trading growth: A study of the governance of Norwegian whey protein concentrate exports. *Proceedings in Food System Dynamics*, *8*, 145–154.

Part III

Cross-sectoral perspectives

10 What knowledge does the bioeconomy build upon?

Linn Meidell Dybdahl and Eric James Iversen

10.1 Introduction

Modern economies have long been characterised as 'knowledge-based economies' (e.g. David & Foray, 2002), whereby more advanced economies are distinguished by their ability to generate, disseminate and use new scientific and technological knowledge. Going forward, however, the ability of all our economies to successfully address society's daunting grand-challenges is recognised as something which is not solely about how to increase innovation in firms, universities and research institutes; it is increasingly seen as being related to improving the efficiency of innovation systems when leveraging existing investments from different parts of the economy (Bessant & Venables, 2010).

This chapter focuses on the underlying 'knowledge base' in the formative bioeconomy which extends across the established boundaries of different sectors and encompasses a range of scientific and engineering disciplines. It involves learning that takes place within organisations, but it also involves learning processes at a higher level of aggregation, including those that take place across different fields of science (agricultural science, engineering, biomedicine) and across different sectors of the economy (primary sectors, manufacturing, energy and research sectors). Although, we know something about the research agenda in the bioeconomy, less is known about the 'knowledge bases' that the bioeconomy builds on and, not least, how they are organised.

The contribution of this chapter is to provide an empirical look at how the knowledge production process is organised in the formative bioeconomy and, moreover, at which knowledge bases are involved in this important area. We have chosen two levels of empirical analysis which address the following related questions:

a *How are the links and interactions (e.g. of researchers) organised in the knowledge-base?* This question explores what we refer to as the 'organisational capital' dimension of the knowledge creation process. Organisational capital refers here to the way in which scientific and technological production are organised across organisations such as universities, research laboratories and private enterprises. Examples from the literature include the important roles that venture capital or collaborative centres housed at universities play in certain contexts. We focus on a comprehensive set of research and innovation projects financed by a central funding instrument of the bio-based economy in Norway, namely the Bionær programme organised by the Research Council of Norway (see also below).

b *What disciplines and capabilities make up the knowledge-base?* This question addresses the 'human capital' dimension of bioeconomic research more directly, in terms of the knowledge that researchers and others have accumulated in their educations and their professional careers. At this second level we utilise the CVs of researchers who are participating in Bionær projects. Following earlier work in other contexts, we use CVs by analysing the fields and disciplines that the researchers represent as well as other aspects of their current positions and their educational backgrounds.

The underlying argument of our approach is that publicly funded research and innovation programmes are important instruments, especially for a formative meta-sector such as the bioeconomy. Public policy interventions expressly seek to promote the creation, dissemination and accumulation of new knowledge in this context. One result is that the projects they support bring together one of the leading edges of the research community. We proceed on the assumption that by using the comprehensive information from this central funding instrument, we can learn more about the types of knowledge that are involved and how they interrelate. This is seen as an important endeavour since there appears to be a lack of consensus around what types of research areas the bioeconomy is based on (see Chapter 2 and Bugge, Hansen & Klitkou, 2016) and since it could help direct future public policies.

The chapter is organised as follows: the next section discusses the role of knowledge starting from the evolutionary economics and extending to the science and technology studies (STS) literature which informs our empirical approach. We go on to present our approach, introducing some generic aspects about CVs as an analytical lens. This lays the basis for our presentation of what this approach tells us about the knowledge base and how it is organised. We will then conclude with suggestions for future research, emphasising CVs as a promising data source that should be explored further.

10.2 Background

The creation, diffusion and use of knowledge are of course fundamental in advanced economies. Their importance has long been recognised, particularly in the heterodox literature by authors such as Freeman (1995), Nelson (1993) and Lundvall (1992). Improving the frameworks that promote knowledge processes has been a central focus of a range of literature such as systems literature as well as affiliated approaches such as the Triple Helix (e.g. Etzkowitz & Leydesdorff, 1995). Below, we focus on Sectoral Systems of Innovation (Malerba, 2004) as the most relevant approach with which to frame our study of the formative bioeconomy.

In this light, the circular bioeconomy can be considered a meta-discourse that engages a range of interests in academic spheres and political spheres (Pülzl, Kleinschmit & Arts, 2014). The growing academic work on the bioeconomy is echoed by policy discourse around the world that has repeatedly underlined the necessity of building knowledge for the future bioeconomy (Staffas, Gustavsson & McCormick, 2013). For instance, the European Commission's strategy for the bioeconomy (2012) calls for investments in research, innovation and skills as central policy interventions. In the US, the National Bioeconomy Blueprint has outlined support of research and development (R&D) investments, as well as updating training programmes to secure the right competences needed in the bioeconomic workforce (The White House, 2012), while in Norway, the government's bioeconomy strategy emphasises that knowledge building and investments in research and innovation are important aspects of developing a modern bioeconomy (Departementene, 2016).

10.2.1 Knowledge and the bioeconomy

In the formative 'bioeconomy', it is particularly worth emphasising the importance of knowledge and learning and the role that public policy can play to promote it. Our empirical look at how knowledge production is organised and what knowledge bases are involved in the bioeconomy starts from a longstanding evolutionary tradition in economics. The case for the importance of knowledge to innovation, industrial change and, in turn, the changing sectoral composition of the economy has been consistently and convincingly made in the evolutionary economics literature that grew out of Nelson and Winter (1982).

Innovation systems are understood to emerge from the complex interaction between a broad range of actors that create and share knowledge, involving both the creation of new knowledge and/or the combination of elements of knowledge in new ways (Lundvall, 1992). In general, systems of innovation are seen as being 'constituted by elements and relationships that interact in the production, diffusion and use of new and economically useful knowledge' (Lundvall, 2017, p. 86). Edquist (2005) argues that the systems of innovation approach focuses on three kinds of knowledge/learning: (1) Research and development, which is conducted by universities, research institutes and companies; (2) Competence building – creates human capital through various forms of training and education; (3) Innovation – the knowledge-related asset controlled by companies.

Studies of biotechnology and information technology have shown that relationships between companies and actors such as universities and research centres can be a source of innovation and change (Nelson, 1993). The variety of connections among actors influences the dynamics in the innovation system. New knowledge can result in novel links with other innovation systems, stimulate the entry of new actors and institutions and alter the system boundaries (Malerba & Adams, 2015). The heterogeneous area of nanotechnology is another example where new knowledge and techniques help to promote innovation in a range of existing industrial contexts. Knowledge transfer across cross-sectorial connections can lead to transformation processes in sectoral systems (Malerba, 2005). This suggests a cross-sectoral perspective which we argue is particularly germane to the so-called bioeconomy. We therefore invoke the Sectoral Systems of Innovation (SSI) perspective, which serves to 'focus on systemic features in relation to knowledge and boundaries, heterogeneity of actors and networks, institutions and transformation' (Malerba, 2005, p. 398).

The systems approach is useful for designing policies which support the stimulation of a sector. The development of a meta-sector like the bioeconomy is guided by knowledge-based processes which help to direct 'the patterns of firms' learning, competencies, behaviours, and organisation of innovative and production activities in the sector' (Malerba, 2004, p. 23). Governments play a key role in the absence of existing markets. They are seen as critical in promoting learning and innovation by promoting research and innovation across the boundaries of economic sectors of universities and other higher education institutions (HEIs), public research organisations (PROs) and the full range of relevant private and non-private entities. These can help existing knowledge systems to reorganise themselves in ways that can promote the creation and sharing of new knowledge within the sector. In addition, this type of dynamic may beget new sectoral institutions and organisations (such as research centres or new educational fields), creating more knowledge variety, which again can influence the evolution of a sector (Nelson & Rosenberg, 1993).

A current focus of the innovation studies landscape is on improving the coordination of existing and emerging knowledge of different types in order to address what are known as 'societal challenges', i.e. challenges that primarily involve social payoffs rather than individual payoffs and which involve systemic change in which public policy is expected to play a more central and coordinative role. The literature has more recently recognised the new role that knowledge can play in addressing the societal challenges of the 21st century (Bessant & Venables, 2010) as well as the roles that public policy can play in this process.

What individual actors know and how they learn is thus a key component of any innovation-oriented system. This includes both new and existing types of knowledge, processes of creation as well as coordination, and theoretical as well as more practical types of knowledge. Components of knowledge are one dimension of this picture, in terms of what economic actors know. In addition, the way that the knowledge processes are organised across existing knowledge bases and learning contexts also helps to define the direction of innovation. The knowledge that researchers and other agents have (human capital) at any given point is integral to the emergence of innovative new fields, and the way the knowledge is organised (organisational capital) is instrumental in shaping the trajectory that innovation takes in contexts such as the bioeconomy.

10.3 Approach

We consider the bioeconomy to be a formative meta-sector as it cuts across several sectors and industries. Furthermore, as Chapter 2 points out, the notion of the bioeconomy is multifaceted and includes three visions. For some, the bioeconomy is about biotechnology and the promises of breakthroughs in this area (the bio-technology vision). Some see it as being about sustainability and ecological processes that, for example, will support biodiversity and prevent monocultures and soil degradation (bio-ecology vision). For others, the bioeconomy is about advances in resource-based sciences more generally and how different fields can be coordinated through new research to improve how organic resources are used (the bio-resource vision).

The bio-resource vision raises a number of questions for us, including (i) whether the underlying knowledge is firmly based on a specific field of science or whether it draws on a wider range of knowledge from different areas, and (ii) whether development is linked to specific lead entities such as universities or whether knowledge creation is more distributed. We argue that it is important to get a clearer idea of the disciplines, sources and the organisation of new knowledge in the bio-based economy. A better understanding of what knowledge the bioeconomy builds upon can be useful in several ways: it can help to clarify the boundaries of this economy; it can help consolidate the population of entities that see their own missions in terms of the bioeconomy; it can help identify knowledge strengths and gaps; and it can help inform future public policy interventions, etc.

This chapter undertakes a systematic empirical analysis of the sources and organisation of knowledge production in the bioeconomy. It has a specific focus; namely publicly financed projects in the area of research and innovation activity for food and bio-based industries. Norway is among the OECD countries that have earmarked public funding to promote research and innovation of the bioeconomy. The argument is that publicly funded research and innovation programmes are important instruments to promote the creation, dissemination and accumulation of new knowledge in the area of societal grand challenges (see e.g. Mowery, Nelson & Martin, 2010).

We use information from one of Norway's key programmes in this area, the Bionær Programme, to learn more about the knowledge system of the bioeconomy. As mentioned in the introduction, we will explore two dimensions of knowledge creation and accumulation. The first involves what sorts of actors are involved and how their work is organised. This level, which we refer to as the 'organisational capital' dimension of the knowledge creation process, is recognised as being important in formative fields (see Bozeman and others in the discussion of different U-I partnerships to promote specific research agendas). We are particularly interested in the profile of entities that are involved in research and innovation activities on this front, in terms of the spread between sectors (HEI, PROs and the private sector) and the international dimension. The second level explores 'the human capital' dimension of the bioeconomy by delving into the CVs of participating researchers to understand the educational backgrounds and the positions that are involved in this research and innovative activity: is it from one scientific area or several?; is it domestic or foreign?, etc.

10.3.1 CVs as an analytic lens

Although the use of CV data is not new, it constitutes an innovative (and labour-intensive) approach which deserves special comment. CVs are rich data sources of longitudinal information about a person's career (Bozeman, Dietz & Gaughan, 2001). Researchers include in their CVs information about their educational backgrounds, their current positions and their publications. Gläser (2001, p. 698) argues that research careers are 'theoretically and practically important because they link individuals with institutions as well as social structures with knowledge production'. CVs of researchers include information about who they have collaborated with (identified as either co-authors or research collectives). Consequently, it is possible to use CVs to map researchers' networks in addition to their scholarly disciplines, affiliations and various work experience.

The use of CVs in research has become more and more prevalent since the 1990s, although its growth is still hampered by the availability of CVs and, moreover, the lack of tools to automate their analysis (Geuna et al., 2015). Cañibano and Bozeman (2009) point out that the use of this unique data source is primarily found in the research evaluation sphere where its use has shifted over time from a focus on output (in terms of publication) based on specific inputs (e.g. to evaluate the success of education and research policies) to include a greater focus on capacities (i.e. the ability to develop relevant competences). They indicate that CV-studies have generally focused on one of three topics: career trajectories, mobility and mapping of collective capacity (Cañibano & Bozeman, 2009).

In the literature, notable themes include mobility and research performance (Cañibano, Otamendi & Andújar, 2008), commercial activity (e.g. Dietz & Bozeman, 2005; Lin & Bozeman, 2006), collaboration and productivity (Lee & Bozeman, 2005) and career transitions (e.g. Mangematin, 2000). A relevant approach is suggested by Lepori and Probst (2009), who used CVs to understand the structure and dynamics of a scientific field which is characterised by conceptual, theoretical and methodological pluralism. They argue that CVs offer an easier and quicker way to look at such a community than a survey could, for example. A more recent study of entrepreneurship scholars which aimed to understand the field's knowledge base also used CVs as a data source (Landström, Harirchi & Åström, 2012). We follow the mapping focus to explore the human capital dimension in terms of how researcher capacity and competences are arrayed in the formative 'bioeconomy'.

10.3.2 Data

We utilise data provided by the Bionær (Sustainable Innovation in Food and Bio-based Industries) programme to study the organisational and human capital dimensions of the bioeconomy. The Bionær programme coordinates funding allocations from a range of ministries into research and innovation activity for food and bio-based industries. The programme aims to trigger research and innovation for enhanced value creation in Norwegian bio-based industries. The objective is to increase knowledge and expertise in order to promote sustainable industries and foster policy development and innovation in bio-based companies and bio-resource management. The requirements for being accepted into the programme are interdisciplinarity and international research collaboration, as well as having a market-oriented focus, and incorporating the concepts of sustainability and circularity (RCN, 2013). The outcome should be both strategic basic research and industry-oriented research.

The Bionær project portfolio provides a unique – if imperfect – empirical approach to research and innovation in the bioeconomy. Several strengths that recommend this programme as a lens are:

a Topicality: it focuses on research and innovation activities in the bio-based industries in general. This definition is sufficiently topical; it focuses on an array of bio-based projects including a category of projects that explicitly focus on the 'bioeconomy'.
b Duration: it has existed for over a decade.
c Extent: the funding frame is substantial, with 100 million NOK in 2018 earmarking 'bioeconomy' projects alone (in conjunction with other programmes). The projects therefore tend to be long-term and involve larger numbers of partners.
d Quality: the quality of the projects in terms of research and innovative degree is approved by a panel of international experts.

There are certain characteristics of the Bionær programme that are relevant to mention:

a The programme does not account for all innovation and research activity in the area. It does not include activities that are carried out internally in companies or in universities that are not funded by the programme. For example, universities and firms may fund their own R&D work, which does not benefit from this programme (see also Chapter 11). It is also worth mentioning that there are other complementing funding programmes. As an example, RCN coordinates the BIA-programme and SkatteFUNN, which are more generic research and innovation instruments directed towards industry actors. To deepen our understanding of the competences involved in the bioeconomy, some of these programmes were considered for inclusion in the dataset of this study, but RCN encountered challenges in extracting the relevant projects from their database (due to issues of categorisation). In addition, projects involving mainly marine bioeconomy are largely organised in separate programmes.

b Although it focuses on research and innovation activity for food and bio-based industries, some individual projects may not be seen as directly relevant to the bioeconomy, depending on one's definition.

c Disclaimer: this chapter grows out of a project that is itself receiving significant funding from Bionær.

10.3.2.1 Project data

The Bionær programme funded 333 projects in the period 2005–2016. In this period the programme focused on agricultural, forest and bio-based value chains, and also included most of the seafood value chain. We obtained project and CV information from the programme itself. In this chapter, we focus on the 136 research and competence-oriented projects that were still active in 2016. The 136 targeted projects involved between one and 20 team members (lead, collaborator, associate) each and had an average team size of 5.3 members. They lasted an average of 3.7 years and involved a total funding amount of an average of 9.2 million NOK.

The projects which were active in the period 2007–2016 can be broken down into two main types. The first type is *Research Projects* (60 or 44% of the total), and as the name suggests, these tended to be explorative projects driven by research enquiry. The second type of projects tended to involve industrial partners more directly. This category includes so-called *Innovation Projects* (66 or 49%) as well as other collaborative projects with a focus on competencies and the needs of the industry (the remaining 7%).

10.3.2.2 Researcher data

A total of 611 individual participants from a total of 498 entities were identified as being directly involved in one or more of the projects. The entities represented the higher education institution (HEI) sector, the public research organisations (PRO) sector, the private enterprise sector and the government sector. From the CVs, we extracted information about what type of positions the project participants held, their field of expertise and their education levels and profiles. We also included other characteristics to inform specific questions: for example, educational degrees and experience from foreign institutions are potentially useful when investigating sources of knowledge spill-over.

In this study, access to CV information at the researcher level is constrained by two main considerations. The first is a formal constraint. A number of researchers were no longer engaged at the partner organisations at the time of our study (2017–2018) and were therefore not given the chance to opt out of the study. This led to the exclusion of 27 CVs. The second constraint is more formalistic. Not all variables (e.g. year of birth, field of science, degrees) were included in all the individual CVs. This made processing of the information difficult despite the reliance on manual processing. This constraint led to further exclusion of other CVs.

Table 10.1 indicates that roughly 570 CVs provided at least patchy information for the variables we were interested in, such as education levels, field of science (fos), age, year of graduation, etc. Most of the work involves around 430 researchers for whom we had sufficiently extensive information (either good or complete).

10.4 Empirical findings

In the following we will present our findings concerning organisational and human capital in the emergent bioeconomy. This empirical section starts by focusing on publicly funded projects in Norway. A point of departure is the literature which debates how fruitful mission-based funding can be in addressing societal grand challenges (see above). We focus therefore specifically on projects designed to promote research and innovation under the Bionær programme.

Table 10.1 The number of individual CVs available based on the earliest project participation of the researcher

Earliest project start	*Quality of processed CV*					
	Complete	*Good*	*Patchy*	*Poor*	*Insufficient information*	*Total*
2007	2	1	0	0	0	3
2010	10	2	2	2	1	17
2011	44	5	19	7	5	80
2012	11	1	3	0	0	15
2013	65	22	18	6	1	112
2014	84	13	39	12	2	150
2015	75	12	30	4	0	121
2016	74	10	25	2	2	113
Total	365	66	136	33	11	611

Source: Bionær Programme: for active projects in the period 2007–2016.

10.4.1 Organisations and organisational capital

The 136 projects involved a range of organisations, from private companies and universities, to research organisations and a range of public and quasi-public organisations such as interest organisations. A total of 498 entities from around the world contributed to the projects. The nature of the Bionær programme promotes collaborations with Norwegian actors in general: a number of the programme areas particularly promote collaborations with private entities. The breakdown of project participation reflects this (see Figure 10.1). The majority (364 or 73%) were based in Norway, with a further 21% (105) from the rest of Europe; primarily the other Nordic countries (36 or 7%). The remaining 6% (29 or 5.8%) came from the US or other countries. This suggests that knowledge in the bioeconomy is indeed global but that it is organised and anchored nationally or regionally.

10.4.1.1 Norwegian partner entities

Roughly 360 of the entities that participated in the Bionær programme during the period of study were based in Norway. The following Figure 10.2 groups the activities of these Norwegian entities into aggregates of primary NACE rev2 classification. Following from the figure above, the bar-diagram can be divided into two broad sectors. The private sector, which accounts for slightly more than half of the entities, is arrayed on the lower part of the

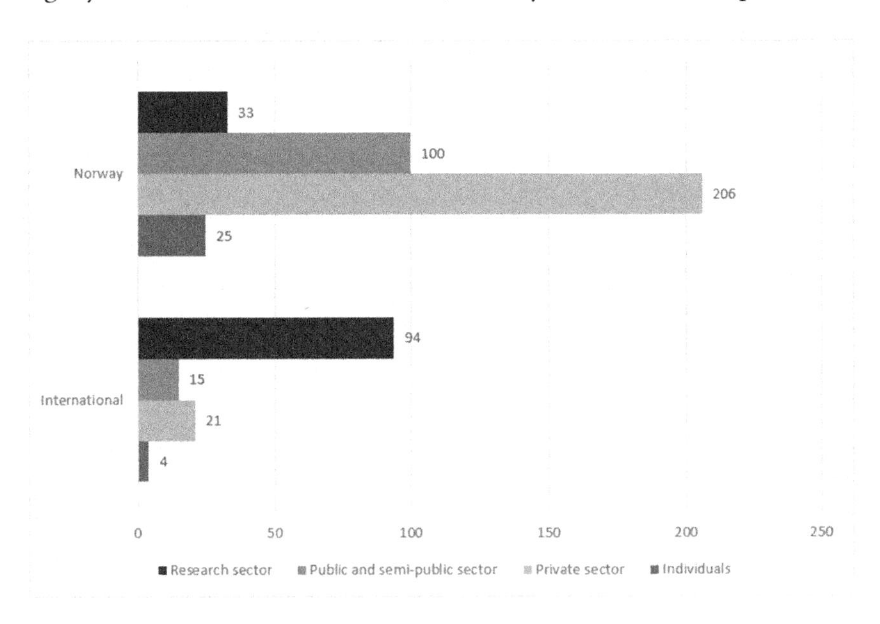

Figure 10.1 Types of organisations by region: gross breakdown ($n = 498$): projects active in the period 2010–2016.

Source: Compiled by NIFU based on raw data from the Bionær programme, 2017.

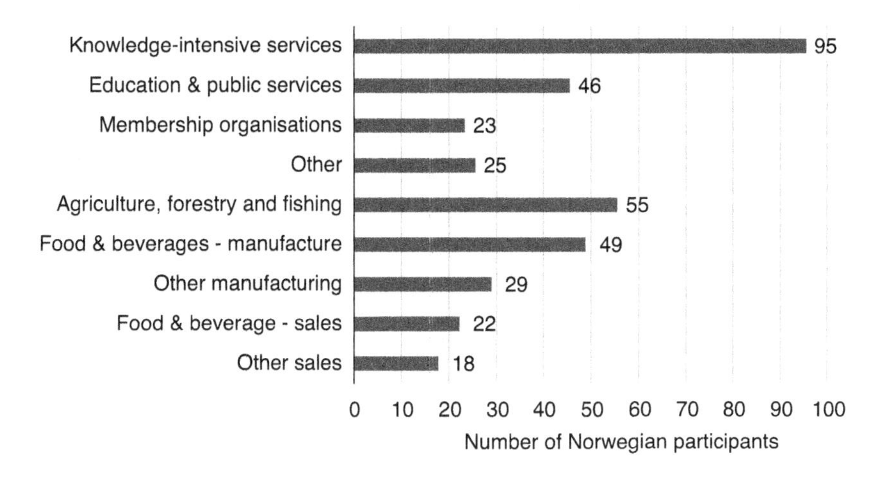

Figure 10.2 Norwegian participants by NACE activity ($n = 362$), 2007–2016.
Source: Compiled by NIFU based on information from RCN's Bionær programme.

figure. Primary industries (agriculture, forestry and fisheries) feature prominently here, followed by the manufacture and sale of goods (food, beverages, lumber, etc.) from these industries. In addition, there is a smaller share of companies involved in the manufacturing and sale of other products.

The upper part of the diagram consists primarily of the non-private sector entities, including higher education institutions, public research organisations (the 'knowledge-intensive services'), as well as an array of government and quasi-governmental organisations ranging from municipal authorities to interest organisations for involved industries. In reality, the division is not so stark between the sectors: a number of entities involved in knowledge-intensive services are in fact private while a number of entities in the primary sector are not purely private (they include publicly owned/controlled companies).

Public research programmes provide a vehicle for bringing together different types of expertise and knowledge to explore/exploit the research and innovation possibilities that are emerging in this field. The bioeconomy is not only about 'biotech' firms. An important point is that the bioeconomy involves an interrelationship between different types of organisations in different sectors. We emphasise here that the bioeconomy builds on competencies which are located across a range of entities from the HEI, the PRO, the governmental area and the private sector.

Chapter 11 looks in more detail at a broader register of Norwegian entities that are involved in research and innovative activities in the 'bioeconomy' in the country. The 360 entities included in this study are central to that register.

10.5 Human capital and researchers

We turn now to the question of the type of fields science researchers in the bioeconomy represent. As discussed above, the CVs provide detailed information about the disciplines the researchers represent, their affiliation (university, research institute, private sector) and their seniority (professors, post-docs, PhD students). This offers us a vantage point on the human capital that goes into RD&I projects and so allows us to better understand the knowledge bases that the bioeconomy builds upon. This section will review the sectors of the economy, the fields of science and the types of research bioeconomy research stems from.

10.5.1 Researchers

The 611 researchers who participated in funded Bionær projects in the period were predominantly affiliated with research organisations. Almost 80% stemmed from public research institutes (295) and from universities (188). The following Table 10.2 provides an overview of the general characteristics of the project participants by sector of employment. Two thirds of project participants were male, although the share is higher in the HEI sector (77%) and lower for PROs (61%).

Three quarters of the population held degrees from Norwegian universities. Of the 420 researchers for whom we have valid data, over 80% had PhDs. One hundred researchers (of the 420) held degrees from abroad, where the majority of degrees were from Sweden (19%), Denmark (14%), the UK (12%), the USA (12%), Germany (9%) and Finland (6%).

The average year of birth was 1964, which means that the average age of the researchers was 54 years. However, the average age of the whole bioeconomic research population as a whole is likely to be lower. One aspect that will affect the CVs represented in the applications is the strategic or tactical choices made during the application phase. Although the Bionær programme plan states that the participation of young researchers is valued, it is not unlikely that more experienced researchers will be considered a positive asset for funding probability. As a result, there might be a biased representation of the experience level among the persons in the application teams (limitations are discussed further in the concluding section).

The breakdown does, however, reveal some interesting aspects of ongoing research, development and innovation (RD&I) activities that focus on the bioeconomy. It indicates that this economy involves a broad range of sectors. Research institutes and universities lead the effort, but they work together and with private enterprises as well as with the government sector. In addition, over 25% of the 500 organisations that were involved are located abroad. The share of PhD holders was higher than usual for the sector. In part this reflects the point made above. What is perhaps more interesting is that a large proportion of those PhD holders took their

Table 10.2 General breakdown of researchers by sector of employment, gender, mean age and academic degree

Sector	*Number*	*Percent of valid data* (n = 421)			*Mean (year)*		
	Total (valid)	*Male*	*Norwegian degree*	*PhD*	*Degree*	*Birth*	*Position*
HEI	188 (118)	77	57	92	1997	1964	2006
PRO	295 (241)	61	84	87	2000	1966	2007
Private	69 (44)	71	82	43	1996	1965	2006
Government and other	15 (7)	67	70	45	1992	1962	2008
N/A	44 (7)	56	57	71	2003	1968	2013
Total	***611 (421)***	***67***	***76***	***83***	***1999***	***1965***	***2007***

Source: compiled by NIFU based on information from RCN's Bionær programme.

doctorate degrees in another country, indicating a level of spill-over from other innovation systems.

10.5.2 Field of science

Formal education provides an important indication of where knowledge of the formative field comes from. We used the CV information to categorise the fields of science in which the project participant held his/her highest degree. The fields of science were translated and manually categorised into a standardised (ISCED) schema which distinguishes between a number of main fields, namely agricultural sciences, engineering sciences, health sciences, physical and life sciences, social sciences and other categories such as humanities or applied service fields (e.g. accountancy).

The following Table 10.3 indicates that the broad area of agricultural sciences was the most represented field. Of the 384 individuals for whom we have data, the greatest number hailed from the broadly aggregated fields of agricultural sciences (123), followed by the social sciences (79), physical and life sciences (63), and engineering, manufacturing and construction (61). Most of the researchers with foreign doctorates were in the fields of agriculture (including forestry and fisheries), physical and life sciences, and social sciences, business or law.

Table 10.3 Researcher field (and subfield), n = 384

Fields	*Subfields*	*Number of researchers*
Agriculture	Agriculture, forestry and fishery	109
	Veterinary	14
Engineering, manufacturing and construction	Engineering and engineering trades	47
	Manufacturing and processing	10
	Engineering, manufacturing and construction	4
Health and welfare	Health	21
Humanities and arts	Humanities and arts	16
Physical and life sciences	Life sciences	50
	Engineering and engineering trades	1
	Physical sciences	6
	Computing	3
	Mathematics and statistics	3
Services	Services	21
Social sciences, business and law	Business and administration	43
	Social and behavioural science	22
	Law	14
Grand total	Total	384

The table confirms that RD&I work in the bioeconomy is not isolated to a single field of science. It shows that the bioeconomy is based on an array of fields, ranging from the agricultural sciences and the life sciences, engineering and the physical sciences, to a variety of social sciences and professional studies. To further get a sense of the contributing knowledge, the table disaggregates these broad fields into more specific subfields. For example, veterinary sciences can be distinguished from other parts of the broader field of agriculture, while life sciences can be separated from physical sciences. This helps us to appreciate the relative importance of life sciences, engineering and business administration in particular.

10.5.3 Sectors and seniority

As indicated above, many of the project participants are currently affiliated with the PRO sector, principally SINTEF, NOFIMA and the former Norwegian Forest and Landscape Institute, now merged with Bioforsk and the Norwegian Agricultural Economics Research Institute into the Norwegian Institute of Bioeconomy Research. The importance of this sector extends through many of the fields and subfields. The HEI sector accounts for the second largest group, largely from the Norwegian University of Life Sciences, the Norwegian University of Science and Technology (NTNU) and the University of Oslo. We note that a number of the major institutes were merged or otherwise reorganised during the reference period, and some research institutes became part of the HEI. In addition, we see that bioeconomy research is not led solely by universities or even the HEI and the PRO sector; the private and government sectors are also very much involved.

The final table (10.4) illustrates how project participants are distributed at different levels of seniority. The first category of researchers (R1 and R2, according to the EU schema) consists of PhD students and post-docs, and general researchers, while R3 consists of associate professors and researchers. Lead researchers include professors and research professors, while administrative leaders include heads of departments and other directors. In addition, a range of other positions such as R&D coordinator or operating managers were placed in the non-classified category.

Table 10.4 Researchers by sector of affiliation and level of seniority

	R1–R2 researchers	*R3 established researcher*	*R4 lead researcher*	*Administrative leader*	*Not classified*	*Grand total*
HEI	29	56	126	8	2	221
PRO	105	83	14	31	21	254
Private				20	48	68
Government and other		2	1	2	10	15
Total count	134	141	141	61	81	558

The categorisation system is not perfect. For instance, the HEI and the PRO sectors employ different rankings which, in the latter case, may depend on the given institute. In addition, the rankings are skewed towards the more advanced due to how CVs are included in project proposals. The table does, however, provide a good indication of which parts of the organisations are involved in this activity. Again, we see a balance between the most senior positions (200 professors, lead researchers and administrative leaders) and the younger researchers (275 from the low- and middle-rank positions). In the PRO sector especially, we also find a range of ancillary positions in the unclassified category that do not correspond to generic positions but are more involved in the running of projects, etc.

10.6 Concluding discussion

The chapter provided an empirical look at how knowledge production is organised in the bioeconomy and which knowledge bases are involved. Its purpose has been to improve our understanding of the 'knowledge base' which will hopefully comprise the knowledge necessary to identify and exploit new and sustainable value propositions in organic waste streams. We noted a general need to solidify what is known about this dimension of the formative meta-sector. One dimension involves public policy, which, as we observed, is dedicated to increasing the allocation of resources to this area. In this light, it is useful for public policymakers to know what fields of science are involved in innovation and how they are organised, as this will help to appreciate the strengths and challenges present in this changing context.

To do so, the chapter has drawn on literature about the role of knowledge in emerging areas. Our starting point was the tradition of sectoral systems of innovation. The relevant literature pioneered analysis of the role of knowledge in order to understand the integral role of innovation in industrial change and, in turn, in the changing sectoral composition of the economy. A second strand of literature that we followed in our empirical strategy is from the STI literature. Here we have used the work of Bozeman and colleagues, who integrated the concepts of human capital and organisational capital into the STI tradition and, in doing so, have pioneered the use of CVs as data in their studies.

This strand of the literature has especially inspired our empirical work. Not least, this is due to its focus on human capital and the way that it is organised, and to the fact that this approach has previously explored the role of scientific and technological knowledge in emerging meta-sectors such as biotech (Corolleur, Carrere & Mangematin, 2004) and nanotech (e.g. Bozeman, Larédo & Mangematin, 2007). These are themes that lend themselves well to our study on the emerging bioeconomy.

The chapter has furthermore followed their pioneering work by using CVs to study human capital and how it is organised. Taking our cue from this literature, we differentiated between two dimensions of knowledge creation and accumulation in the bioeconomy:

1 *How are the links and interactions (e.g. of researchers) organised in the knowledge base?*
The basic dimension, which we have dubbed 'organisational capital', looked at how the collective knowledge is being brought together to create new forms of scientific and technological knowledge in the bio-economy. Here we focused on the role of project data from a large-scale research and innovation programme in Norway to study the links between different agents in different sectors. Several interesting observations emerged from this exercise.

   a The 500 entities that contributed to the 136 projects were used to study the distribution of what sorts of agents are involved in Norwegian RD&I projects in the bioeconomy. The chapter indicates that around three quarters are domestic, while a further 20% are from other Nordic countries or elsewhere in Europe. This suggests that knowledge in the bioeconomy is indeed global but that it is organised and anchored nationally or regionally.
   b We also found that the projects were based on collaborations between different sectors of the economy: the PRO sector, the HEI sector, the private enterprise sector and the government sector. Although the inclusion of different sectors may in part be shaped by the Bionær programme requirements, the material illustrates that there is an active division of labour between the different sectors.
   c The chapter focused on the Norwegian participating entities. It showed that, next to the HEI and PRO sectors, the private sector involvement largely featured the primary industries (agriculture, forestry and fisheries) and the manufacture and sale of goods (food, beverages, lumber, etc.) from these industries.

2 *What knowledge and capabilities make up the knowledge base?*
We then took stock of the 'human capital' that is embodied in the individual contributor to the researcher project (the 'researcher'). Here the CVs of project participants were used to gauge inputs to RD&I projects, in terms of the knowledge that researchers and others had accumulated through their education and their professional careers. This labour-intensive exercise revealed a number of aspects about the knowledge base that the bioeconomy is building upon. The chapter indicated:

   a that the different sectors contribute with different types of knowledge to the bioeconomy RD&I;
   b that participants in the Bionær programme tend to be male, although the balance differs between sectors;

c that their highest degree tends to be from a Norwegian institution, although the share of foreign PhDs is markedly higher in the HEI sector;
d that they tend to hold PhDs, where the degrees of PRO researchers tended to be slightly more recent than those of participants from the HEI sector.

Coordinated, integrated R&D efforts are important to the Norwegian bioeconomy agenda. The projects integrate a range of knowledge across different science fields. It is worth recapping the role that different fields of science play in the bioeconomy, as this is an important contribution of the chapter. We find among the CVs for which we have good data that the highest share represents the agricultural sciences, broadly construed. This is a confirmation of what one might expect in the bioeconomy field. It is interesting that a range of other fields complement this core area. Prominent among these is the field of life sciences (combined with medicine and health). A second major component is the participation from engineering and the physical sciences, while a third, made up of social sciences, humanities and professional degrees (business administration, law), is also important in these cross-sectoral collaborations. We argue that this broad involvement of various disciplines and capabilities is especially important in the development of a circular and sustainable bio-based economy. If the evolving bioeconomy is to contribute towards solving some of the 21st century's complex societal challenges, its knowledge base must be inter- and transdisciplinary.

10.6.1 Limitations

Before we discuss the possibilities for future work, some of the limitations associated with using this data should be mentioned. The chapter has previously stated a number of recognised problems associated with utilising CVs. In addition, we review the more specific limitations our work encountered:

1 The Bionær programme is a major public policy intervention to promote RD&I in the bioeconomy in Norway. However, it clearly does not represent the full scope of all work being done here. First, we excluded information from unsuccessful applications. Second, the selection was skewed more towards the HEI and the PRO sectors and offers comparatively little insight into what is happening in private enterprises (see Chapter 11 for more information on the contribution of the private sector).
2 The data covers a period of time during which researchers may develop in ways that are important to the analysis. We focused on the first project a researcher participated in and may have excluded updated information (e.g. PhD year). In addition, a number of entities changed sectors during the period (from research institute to a university).

3 Not all project participants were included in the granted applications. We noted that project CVs tended to include those contributors with the longest track-records and favoured PhD holders over non-PhD holders, due to requirements of the funding agency. The teams were also subject to change, especially in the longer projects. Researchers changed jobs or retired, which left the actual composition of the team quite different to its composition at the time of application. A small number of CVs were also not available in our analysis due to formal reasons.
4 CVs are notoriously labour intensive to work with, although techniques are improving. There are recognised challenges associated with non-standardised formats of the documents themselves (Cañibano & Bozeman, 2009; Dietz, Chompalov, Bozeman, Lane & Park, 2000). In addition, there are several types of translation involved that may lead to non-standardised categorisations. For example, career descriptions vary according to institution, country and disciplinary context. The type of unification problem this creates can make the integration of sectors, seniority and fields of science a challenge.

10.6.2 Future paths of research

This study builds on a composite set of linked data, involving basic project data, information about participant entities and the CVs of individual researchers. This approach has proved to be time-consuming and has involved formal hurdles as well as the practical challenges of compiling the datasets, especially the CV data. However, data-extraction tools are continuously improving (see Geuna et al., 2015). This is taking analysis in a direction where the challenges of extracting and coding data from CVs will be reduced. This can help to make CV studies an important source of information that can help shape the national bioeconomy agenda going forward.

In this context, our chapter represents an explorative starting point which can open the way for other studies. We see several avenues available to explore. The first broad avenue is to more fully exploit the information from the combined dataset. A unique aspect of the dataset that we developed here is that it combines information about the project participant ('researcher'), information about the affiliated enterprise or institute and information about the collaborative project. This combination affords a number of potential vistas for exploration, including studies of the subsequent direction of collaboration and careers of involved researchers, or of the publication or patenting profiles of their affiliated organisations. There is further scope to explore how research is organised in a formative meta-sector like the bioeconomy. This line of study can help indicate potential links between research sectors and the private enterprise sector and other stakeholders. This would then have implications, for example, in terms of identifying which configurations work well in which contexts.

A second broad avenue involves using more specific information from the CVs. So far, we have primarily looked at current affiliation and latest degree, but CVs contain considerably more information about the careers of the researcher, their publication records and other projects that the researcher has been involved in. This information can be used, in general, to follow up existing studies that have used CVs to focus on scientific careers and on research evaluation (see Cañibano et al., 2018 for discussion). There are multiple possibilities for pursuing existing or emerging lines of enquiry. One important topic involves sectoral mobility. Here there is a need to better understand how researchers contribute to the integration of intersectoral research, especially those that link the private enterprise sector to the research sectors and other stakeholders. Another important topic involves career trajectories, especially those of recent PhDs. One current question is what happens to PhD holders after graduation and during their early careers. A more general question is whether and how they contribute to emerging RD&I agendas such as the bioeconomy. Finally, we note that CV-based studies can be used to map human capital and to understand future needs for the appropriate training of tomorrow's workforce. Such knowledge can support educational institutions and policy makers in their planning of educational programmes and of interventions to support industry development.

References

Bessant, J. R., & Venables, T. (2010). *Creating wealth from knowledge: meeting the innovation challenge*. Cheltenham, UK: Edward Elgar Publishing.

Bozeman, B., Dietz, J. S. & Gaughan, M. (2001). Scientific and technical human capital: an alternative model for research evaluation. *International Journal of Technology Management, 22*(7–8), 716–740.

Bozeman, B., Larédo, P. & Mangematin, V. (2007). Understanding the emergence and deployment of 'nano' S&T. *Research Policy, 36*(6), 807–812.

Bugge, M. M., Hansen, T. & Klitkou, A. (2016). What is the bioeconomy? A review of the literature. *Sustainability, 8*(691), 1–22. doi:10.3390/su8070691.

Cañibano, C., & Bozeman, B. (2009). Curriculum vitae method in science policy and research evaluation: the state-of-the-art. *Research Evaluation, 18*(2), 86–94.

Cañibano, C., Otamendi, J. & Andújar, I. (2008). Measuring and assessing researcher mobility from CV analysis: the case of the Ramón y Cajal programme in Spain. *Research Evaluation, 17*(1), 17–31.

Cañibano, C., Woolley, R., Iversen, E. J., Hinze, S., Hornbostel, S. & Tesch, J. (2018). A conceptual framework for studying science research careers. *The Journal of Technology Transfer*, 1–29. doi:10.1007/s10961-018-9659-3.

Corolleur, C. D. F., Carrere, M. & Mangematin, V. (2004). Turning scientific and technological human capital into economic capital: the experience of biotech start-ups in France. *Research Policy, 33*(4), 631–642.

David, P. A., & Foray, D. (2002). An introduction to the economy of the knowledge society. *International Social Science Journal, 54*(171), 9–23.

Departementene. (2016). *Kjente ressurser – uante muligheter*. Oslo: Nærings- og fiskeridepartementet. Retrieved from www.regjeringen.no/contentassets/32160cf2 11df4d3c8f3ab794f885d5be/nfd_biookonomi_strategi_uu.pdf.

Dietz, J. S., & Bozeman, B. (2005). Academic careers, patents, and productivity: industry experience as scientific and technical human capital. *Research Policy*, *34*(3), 349–367.

Dietz, J. S., Chompalov, I., Bozeman, B., Lane, E. O. N. & Park, J. (2000). Using the curriculum vita to study the career paths of scientists and engineers: an exploratory assessment. *Scientometrics*, *49*(3), 419–442. doi:10.1023/a:1010537606969.

Edquist, C. (2005). Systems of innovation: perspectives and challenges. In Jan Fagerberg, David C. Mowery & Richard R. Nelson (Eds.), *The Oxford handbook of innovation*. Oxford, UK: Oxford University Press.

Etzkowitz, H., & Leydesdorff, L. (1995). The triple helix: university–industry–government relations: a laboratory for knowledge based economic development. *EASST Review*, *14*(1), 14–19.

European Commission. (2012). *Innovating for sustainable growth: a bioeconomy for Europe*. Luxembourg: European Commission.

Freeman, C. (1995). The 'National System of Innovation' in historical perspective. *Cambridge Journal of Economics*, *19*(1), 5–24.

Geuna, A., Kataishi, R., Toselli, M., Guzmán, E., Lawson, C., Fernandez-Zubieta, A. & Barros, B. (2015). SiSOB data extraction and codification: a tool to analyze scientific careers. *Research Policy*, *44*(9), 1645–1658.

Gläser, J. (2001). Macrostructures, careers and knowledge production: a neoinstitutionalist approach. *International Journal of Technology Management*, *22*(7–8), 698–715. doi:10.1504/ijtm.2001.002987.

Landström, H., Harirchi, G. & Åström, F. (2012). Entrepreneurship: exploring the knowledge base. *Research Policy*, *41*(7), 1154–1181. doi:10.1016/j.respol.2012.03.009.

Lee, S., & Bozeman, B. (2005). The impact of research collaboration on scientific productivity. *Social Studies of Science*, *35*(5), 673–702.

Lepori, B., & Probst, C. (2009). Using curricula vitae for mapping scientific fields: a small-scale experience for Swiss communication sciences. *Research Evaluation*, *18*(2), 125–134.

Lin, M.-W., & Bozeman, B. (2006). Researchers' industry experience and productivity in university–industry research centers: a 'scientific and technical human capital' explanation. *The Journal of Technology Transfer*, *31*(2), 269–290. doi:10.1007/s10961-005-6111-2.

Lundvall, B.-Å. (1992). *National systems of innovation: toward a theory of innovation and interactive learning*. London, UK: Pinter.

Lundvall, B.-Å. (2017). *The learning economy and the economics of hope*. London, UK: Anthem Press.

Malerba, F. (2004). *Sectoral systems of innovation: concepts, issues and analyses of six major sectors in Europe*. Cambridge: Cambridge University Press.

Malerba, F. (2005). Sectoral systems: how and why innovation differs across sectors. In J. Fagerberg, D. C. Mowery & R. R. Nelson (Eds.), *The Oxford handbook of innovation*. Oxford, UK: Oxford University Press.

Malerba, F., & Adams, P. (2015). Sectoral systems of innovation. In M. Dodgson, D. M. Gann & N. Phillips (Eds.), *The Oxford handbook of innovation management*. Oxford, UK: Oxford University Press.

Mangematin, V. (2000). PhD job market: professional trajectories and incentives during the PhD. *Research Policy*, *29*(6), 741–756.

Mowery, D. C., Nelson, R. R. & Martin, B. R. (2010). Technology policy and global warming: why new policy models are needed (or why putting new wine in old bottles won't work). *Research Policy*, *39*(8), 1011–1023. doi:10.1016/j.respol.2010.05.008.

Nelson, R. R. (1993). National systems of innovation: a comparative study. In R. R. Nelson (Ed.), *National innovation systems: a comparative analysis*. Oxford, UK: Oxford University Press.

Nelson, R. R., & Rosenberg, N. (1993). Technical innovation and national systems. In R. R. Nelson (Ed.), *National innovation systems: a comparative analysis* (pp. 3–22). New York, Oxford, UK: Oxford University Press.

Nelson, R. R., & Winter, S. G. (1982). *An evolutionary theory of economic change*. Cambridge, MA: The Belknap Press of Harvard University Press.

Pülzl, H., Kleinschmit, D. & Arts, B. (2014). Bioeconomy: an emerging meta-discourse affecting forest discourses? *Scandinavian Journal of Forest Research*, *29*(4), 386–393. doi:10.1080/02827581.2014.920044.

RCN. (2013). *Work programme 2012–2021: research programme on sustainable innovation in food and bio-based industries – BIONAER*. The Research Council of Norway Retrieved from www.forskningsradet.no/prognett-bionaer/Programme_description/1253971968649.

Staffas, L., Gustavsson, M. & McCormick, K. (2013). Strategies and policies for the bioeconomy and bio-based economy: an analysis of official national approaches. *Sustainability*, *5*(6), 2751–2769.

The White House. (2012). *National bioeconomy blueprint*. Washington, DC: The White House. Retrieved from https://obamawhitehouse.archives.gov/blog/2012/04/26/national-bioeconomy-blueprint-released.

11 Actors and innovators in the circular bioeconomy

An integrated empirical approach to studying organic waste stream innovators

Eric Iversen, Marco Capasso and Kristoffer Rørstad

11.1 Introduction

This chapter addresses the need for a reliable way to identify actors in the 'bioeconomy' and to take stock of the innovative activities they engage in, especially in terms of how these activities involve organic waste paths and the circular economy. The focus is on what the actors do, not how they are initially categorised. Economic actors engage in activities that they either directly associate with the bioeconomy or that can be associated with the bioeconomy via scientific activity. We have screened a comprehensive range of available data sources based on both more objective measures (e.g. patents or projects linked to the bioeconomy) and on more subjective links (e.g. affiliation with relevant interest organisations, or survey responses). This identification procedure yields a population of actors whose contribution to the bioeconomy can be linked to one or more measures, allowing us to say something about the population itself as well as the activities that the actors are involved in that contribute to the bioeconomy.

The chapter starts by introducing the basic challenges that emergent and/or sector-bridging industries face, before laying out an identification procedure to help stabilise that population and their activities. We then explain the approach we use to identify actors according to clear criteria based on a range of available data sources. The chapter has three empirical sections:

i R&D baseline: The first section establishes a baseline for research in the bioeconomy, starting with the most standard measure possible, official national research and development (R&D) statistics. It builds on a customised study carried out in Norway in 2016 using 2015 data (Rørstad & Sundnes, 2017).

ii Population of bioeconomy firms: The second section reports on the empirical strategy used to flesh out the who and what of the emerging circular bio-economy. The Norwegian Inventory of Bioeconomy Entities (Iversen, 2018), which systematically integrates R&D, patenting and project data, is used to identify 'economic actors' involved in the circular bioeconomy, highlighting the six subsectors considered in the previous book chapters. 'Actors' are primarily private-sector entities ('firms') but may include research organisations in the governmental sector such as research institutes, universities, etc.

iii Survey of bioeconomy innovation: Finally, the chapter presents results from a novel firm-level questionnaire that focused on innovation in organic waste activities in Norway in 2017. The population-frame for this mapping exercise, which was carried out at the TIK centre (University of Oslo), came from the NIoBE dataset. The survey also provides a quantitative backdrop to the cases previously presented in this book within specific subsectors.

Together these steps provide a consistent and comprehensive view of the actors and their activities and how they contribute to the emerging circular bioeconomy in Norway.

11.2 Background[1]

The inaccuracy of industrial classifications for emerging fields and sectors: Addressing the question of who is involved in the bioeconomy acts as a stumbling block for empirical and policy-relevant research. Recognised industrial classifications, such as the NACE taxonomy (Statistical Classification of Economic Activities in the European Community), are not reliable guidelines in this context. This is because the bioeconomy is emergent and not yet fully stabilised as a recognised and distinct field; instead, it continues to take shape as existing sectors utilise both new and existing technologies, inputs and ways of (inter)working to explore emergent possibilities. A deductive approach to defining the bioeconomy based on the existing foundations of established industrial classification does not get us very far and may even lead us down the wrong road.

Industrial classifications are also unreliable here for another reason. The bioeconomy is largely a meta-sector that extends across more narrowly defined industries. In many cases, the products of activities here are not new: outputs, such as energy, may simply be substitutes for established activities that use other inputs. This raises its own set of problems in creating reliable metrics (see discussions elsewhere, e.g. OECD, 2018).

Current approaches: There are ongoing activities to capture the bioeconomy in figures. These efforts are especially current in jurisdictions where policy intervention is targeting this nascent sector or area. This includes the OECD countries in general and Europe in particular. To match the focus of European policy-makers, the EU statistical agency (Eurostat) has recently attempted to coordinate data-collection efforts across Europe. These have a focus on primary biomass production and on waste resources.

Establishing the industrial sectors related to economic activity that can be categorised under the bioeconomy is more difficult. One EU project (BIC

Consortium, 2018), for example, tries to estimate the boundary of existing sectors and presents the turnover linked to these estimates. In addition, there are a number of other national (LUKE Natural Resources Institute Finland, 2018) and sectoral efforts, such as for the cellulose industry (CEPI, 2018), and cross-sectoral efforts available as data sources for bioenergy and biofuels (EurObserv'ER, 2018).

Metrics for the bioeconomy are largely based on estimates of how biomass production (agriculture, forestry, fisheries) is processed/refined in discrete sectors (food and beverages, paper) to produce organic-based products. The first problem is that this relationship is not one-to-one. Estimates are used to allocate subpopulations of established categories (like chemicals and plastics) to the 'bioeconomy', based on various estimates.

The most formalised efforts focusing on the bioeconomy involve measures of biomass and 'organic residuals' or 'side-streams'. New rules have been introduced to more accurately account for the generation and treatment of real resources. In Europe, the more accurate measures of organic waste are in keeping with the revision of statistics in line with Eurostat WStatR. However, this data is not linked directly to the firm level (yet). In the case of Norway, the revision of sector-based estimates (before 2011) to improve data collection exposed estimation errors of up to 100%. This suggests a need for a stronger micro-level foundation for accounting in this area.

The problem extends further, notably to our ability to map not only economic activities that produce organic residuals or 'waste', but also those that process them: it is difficult to properly size up the bioeconomy. However, efforts to link economic activity to biomass, such as those undertaken by EuroObserv'ER, should be encouraged. Being unable to frame the bioeconomy reliably and accurately in metrics has important consequences. We highlight the difficulties in properly framing and focusing on the role that innovation plays in the circular 'bioeconomy'.

11.3 Empirical sections

There is a range of ways to design a procedure that can identify the target population in these circumstances. As sizing up emerging technologies, industries and sectors is not a new problem, the chapter references the sectoral systems, transition literature and other current work (e.g. Bugge, Hansen & Klitkou, 2016; Rotolo, Hicks & Martin, 2015). We also refer to ongoing work in the SusValueWaste project (see note 1) using project and CV data to explore empirical ways of getting a handle on the question of knowledge and competencies. Improving the measurement of an emerging sector or meta-sector like the bioeconomy boils down to evolving metrics along the following dimensions:

- Coverage of supply and demand side measures for resources, activities and actors;

- Compatibility across countries and across time;
- Granularity in terms of the actors involved and their activities;
- Timeliness; and
- Replicability and legitimacy.

The following section will present a first estimate of formal R&D activities, the population and the survey based on this population. Two of the lenses featured here are based on collecting data from the actors themselves: the Agrifood R&D carried out by NIFU (Rørstad & Sundnes, 2017), and the survey of Norwegian firms engaged in organic waste activities carried out by TIK in 2017 (Normann, 2018). These two collection rounds are census-based activities rather than sample surveys. Each is based on an established (notionally) complete list of entities (for the defined categories); established criteria have been applied and respondents and non-respondents have been validated. This provides a point of departure that is distinct from other efforts (such as Biosmart (2018)) and that can reveal something new about the extent and direction of R&D allocations and innovative activities respectively.

We address the following questions:

- Who are active 'bioeconomy' actors in Norway?
- What activities do the different subcategories report?
- How do these activities square with related activities within the sector?

11.3.1 Baseline: R&D activity in the circular bioeconomy

NIFU, which produces the official R&D statistics for the higher education sector (HES) and the research institute sector in Norway, conducts extended census work to look more deeply into thematic areas of specific interest such as polar research, climate change or 'agriculture and food' research. We utilise the results and the population frame from the study of 'agriculture and food' R&D to set the baseline for sizing up the circular bioeconomy in Norway. The term 'agriculture and food' corresponds to the following categories in the Web of Science database: agriculture economics and policy, agricultural engineering, agriculture, dairy and animal science, agriculture multidisciplinary, agronomy, food science and technology, forestry, and veterinary science.

Conducted in 2015/2016 (reference year 2015), this national survey mapped the allocations of Norwegian actors to R&D in the area of agricultural and food-related R&D. Agriculture and food research is admittedly not a perfect proxy for the research area we wish to capture. However, agriculture and food research does provide an instrumental foundation from which to start to size up actors and activities: it spans a number of important industries that focus on organic matter and waste streams; it involves a range of commercial activities; and these activities are relatively research intensive within our scope of interest.

The population frame of the survey includes all departments in the higher education sector (84) and the research institute sector (47) who report R&D expenditures in this area and/or publish in this area. A further 462 private sector actors were included. This population represents a full count of firms that receive research, development and innovation (RD&I) grants from the Research Council of Norway in relevant fields. Over 80% of the entities canvassed replied. Information from 230 research active entities is used in the analysis.

The basis of our analysis is thus all entities who receive public money to finance research, development and innovation activities in the area of agriculture and food. This represents a quasi-totality of research activity, and the selected lens provides a census for this important part of the bioeconomy. The university (HEI) and the research institute (RI) sectors are known to be key in this research area. R&D resource allocation, including expenditures, is reported by the institutions themselves, based on a breakdown of in-house activities. The departments in the higher education sector (HES) and research institutes reported a percentage of their R&D activity which was defined as agricultural and food-related R&D. Firms in the industrial sector reported on actual amount spent on R&D in that particular research field. The questionnaire then asked all respondents to break down the agricultural R&D into thematic sub-fields which included circular bioeconomy. The numbers on R&D in circular bioeconomy are therefore estimates made by the R&D performers themselves.

Agriculture and food R&D totalled NOK 2.4 billion and has grown at an annual rate of about 2.4% in real terms between 2007 and 2015. A total of 2,900 researchers were reported to be involved in R&D activity in this area in 2015 (Rørstad & Sundnes, 2017). It should be noted that this approach does not include the important activities of ocean fisheries and aquaculture.

Table 11.1 provides a breakdown of R&D expenditures that were allocated to the area of 'circular bioeconomy' in 2015. The study defines

Table 11.1 R&D expenditures on circular bioeconomy (million NOK) per sector of performance and number of institutions/firms in 2015

Sector of performance	*Circular bioeconomy (million NOK)*	*Circular bioeconomy, share of total R&D (%)*	*Circular bioeconomy, share of agricultural R&D[1] (%)*	*Number of institutions/firms with circular bioeconomy R&D*
Higher education sector	29	0.2	9	16[2]
Research institutes	164	1.2	16	17
Industry	291	1.0	27	84
Total	485	0.8	20	117

Source: NIFU Report, 2017, p. 2.

Notes

1 Total agricultural R&D expenditures in 2015 were 2.4 billion NOK.

2 The 16 HEI departments were located at seven higher education institutions.

knowledge about the 'circular bioeconomy' (Rørstad & Sundnes, 2017, p. 12) as 'knowledge that contributes to the efficient utilisation of bio-based resources, products and residual-inputs so that they remain in the economy through multiple stages (of production and utilisation)' (translation by the authors).

The R&D study illustrates that roughly 120 actors carried out R&D for the circular bioeconomy for 485 million NOK in 2015. The private sector accounted for 60% of this activity, the research institute sector for 1/3 and the rest was carried out by the higher education sector. Activity is not evenly distributed through the RD&I system. Instead, there are a handful of dominant actors that account for the lion's share in each sector. Although seven institutions in the higher educational sector and 17 research institutes performed R&D in this field, the clear majority was carried out by organisations located at or close to the Norwegian University of Life Sciences.

Compared to the total R&D volume in Norway, the circular bioeconomy is a minor field and accounts for less than 1% on average and varies from 0.2% in the HES to around 1% for both the institute sector and industrial sector. However, the bioeconomy R&D volume is not negligible compared to the agricultural R&D. In total, bioeconomy accounts for 20% of total agricultural R&D, but the shares vary across the sectors. The highest share of bioeconomy is in the industrial sector, with 27%, followed by the research institutes with a share of 16%, while only 9% of HES agricultural R&D occurs within the bioeconomy. These findings imply that R&D within bioeconomy is a type of applied research that is likely to be conducted at firms and research institutes rather than at universities. Moreover, the research performers in each sector are not evenly distributed. Around 80 firms conducted R&D in this field, while the numbers of research performers in the other sectors were 17 research institutes and 16 university departments.

11.3.2 Population frame: establishing the NIoBE inventory of active bioeconomy actors

This R&D expenditure data provides a valuable starting point from which to take stock of the actors that are active in the circular bioeconomy in Norway. The effort to create a stable and robust population that can be used in different empirical exercises is dubbed the Norwegian Inventory of Bioeconomy Entities (NIoBE). It was initiated at NIFU in 2015 (Iversen, 2018). An earlier iteration of NIoBE is presented in Iversen and Rørstad (2017). A current version is now being finalised as a reference tool. NIoBE is designed to address the overriding question, 'Who is involved in bioeconomy innovation in Norway?', from which it can focus on more specific areas of the circular economy.

We go on to outline the identification procedure behind NIoBE before presenting some key dimensions of the resulting population. This stage is then used as a population frame for the questionnaire-based exercise carried out by the University of Oslo in 2017 that focused on mapping organic waste-related activities in the following subsectors: forestry, aquaculture and seafood processing, beer brewing, meat processing, dairy, and organic waste processing. This study will be presented in section 11.3.3.

11.3.2.1 Identification strategy of the Norwegian Inventory of Bioeconomy Entities (NIoBE)

The approach used in this research to identify a target population of active organisations in Norway included two main stages. The *first stage* involved collecting and collating a first estimation of the population. We used three types of data to open the population. The inclusion rules moved from the more stringent (the entity is involved in RD&I activities in the area, as in the example above), to an intermediate level of accuracy (the entity has been identified by another systematic project), to a more generic association (the entity is a member of a population that is nominally associated with the bioeconomy).

This first stage, which is akin to using three nets with different meshes, was designed to include as many of the true population as possible (i.e. maximise 'recall'). It is clear, however, that as we progress from the narrow to the more broad-meshed nets, we risk including considerable bycatch in the form of entities that are not a part of the true population. It proved difficult to weed out these 'false positives' from our population as the 'true' population is not known. Therefore, a *second stage* was undertaken. In this stage we set out to increase precision by using other standardised information to exclude entities that were clearly not part of the population. In particular, we used industrial classifications (NACE, which is the starting point of other studies) and other firm-level information, such as 'trade descriptions' found in financial data sources. In the following, we briefly present the three components of our approach before fleshing out the resulting population.

11.3.2.1.1 CONFIRMATION BY ACTIVITY

The first inductive stage of the approach establishes a stable foundation. It identifies Norwegian actors – universities, research institutes and firms – using data on RD&I activities that are recognised to advance the 'bioeconomy'. In this stage, recognised definitions are employed by impartial authorities in three contexts:

1 The R&D survey that demonstrates that the entity is actively involved in innovative bioeconomy activities as described above in section 11.3.1.
2 Research and innovation projects funded by the Bionær programme at the Research Council of Norway as described in Chapter 10.
3 Patenting activity in the bioeconomy area based on the Cooperative Patent Classification (CPC) (particularly the taggings under Y02W,

targeting climate change mitigation technologies related to wastewater treatment or waste management), but augmented by the work of WIPO, the OECD and other work that links patenting to the bioeconomy. The approach is elaborated on in the EU report (Frietsch, Neuhausler, Rothengatter & Jonkers, 2016) and by Kreuchauff and Korzinov (2017).

The external authorities that delimit the activities include patent examiners, funding organisations, university administrations and other researchers. They use recognised criteria to determine what constitutes the 'bioeconomy' in relation to innovative activities. Entities that conduct R&D, that engage in research and innovation activities and/or that patent novel products or processes according to clearly relevant criteria are strong candidates as innovative contributors to the Norwegian bioeconomy. The narrow definition of this first phase yields 900 firms and other actors.

11.3.2.1.2 CONFIRMATION BY EXISTING STUDIES

A second phase uses a broader identification procedure to help eliminate false negatives, and reduce the likelihood that we were excluding members of the 'true' population. In this phase, the identification strategy was loosened to include other sources where the tie to the bioeconomy had been confirmed either by other studies and/or by some form of explicit self-identification with the bioeconomy.

The sources include two earlier studies that have tried to establish populations of bioeconomy firms, one primarily focusing on the primary industries and the other primarily focusing on waste and recycling. The first study we used to firm up the bioeconomy population was the Biosmart survey (Biosmart, 2018). Given the lack of a pre-established population of bioeconomy firms, this survey was sent out by another Bionær project to many actors in the primary industries (Bjørkhaug, Hansen & Zahl-Thanem, 2018). BioSmart's wide net approach yielded a small set of respondents (650 firms) who confirmed involvement in the bioeconomy according to the definition that was provided by the survey. This form of self-identification arguably provides a strong, although more subjective, signal.

The second study takes a complementary approach and is focused on a complementary section of the bioeconomy. The study was conducted by Menon (Espelien & Sørvig, 2014), and was sponsored by Oslo Renewable Energy and Environment Cluster (OREEC, 2018), and set out to map Renewable Energy and Environmental Technologies in Norway in 2014. This study was primarily deductive: it used industrial categories (NACE) to select entities from national register data that fell into environmental technology categories: renewable energy, environmental technologies and services, and relevant parts of the electricity distribution industry (largely related to hydroelectric power in Norway). A supervised review of these entities led to a final list based on input from the branch expertise of the OREEC team.

This population was further pared down (removing 200 entities) in order to exclude the non-organic sections of the population and to focus on the area of bioenergy and circular economy related to organic waste.

11.3.2.1.3 CONFIRMATION BY ASSOCIATION

To ensure that we had not excluded any target entities, a final analysis of the population was conducted. Here we use two registers of entities that have a strong but more nominal association with the bioeconomy. The so-called 'Biodirectory', which originated in 2016, showcased 80 entities that had been involved in research programmes into sustainable technologies including those funded under the Centres for Research-Based Innovation (SFI) and the BIOTEK2021 Programme (BIOTEK2021, 2018). Half of these entities overlap with either the project data or the patent-data already presented.

The other Biodirectory entities overlap with our final source, namely the relevant branch organisations from the Norwegian Confederation of Companies (NHO) and the Federation of Norwegian Industries (Norsk Industri, 2018). Branch organisations were included in dialogue with the organisations themselves and include those dedicated to wood processing, recycling, seafood and aquaculture, as well as the broad category of food and beverages. More than half of the members fit other identifiers in our approach. The remaining entities are less certain and can be excluded depending on what NIoBE is being used for.

11.3.2.2 The Norwegian Inventory of Bioeconomy Entities (NIoBE)

In the first stage, we once again focused on improving the 'recall' of the identification procedure by casting our nets wide enough that we did not prematurely exclude potential candidates of the 'true population'. This stage yielded a gross population of 2,792 entities, which can be considered an upper bound for the population. The overall population entities may be narrowed according to the focus of the analysis. For example, firms that are identified through more than one lens arguably yield the most robust identifier and could be the focus. In other cases, a broader population may be useful.

In the subsequent step, we collected a variety of information on this gross population. The industrial affiliations of the entities by NACE or by other markers of activities, such as the trade descriptions found in financial data in the AMADEUS dataset that Bureau van Dijk harvests from company annual reports (Nelson & Rosenberg, 1993), provide information about the activities of the firms: this is instrumental information that can be used to make informed decisions about which types of firms fall outside the boundaries of the circular economy. On this basis the firms were first graded by their apparent relevance to the circular bioeconomy (core, secondary, periphery) and then arranged according to the six categories studied in this project: breweries, aquaculture, dairy, meat processing, waste processing, as well as a seventh category, residual population consisting of other research activities.

Table 11.2 breaks down the resulting population of 2,369 firms by type of sector (based on NACE) and by the mode by which the entity was identified. We have excluded actors from our last identification phase if they feature in the NIoBE population solely due to membership in an interest organisation within the broad area of food and drinks or in generic industries that do not directly involve biomass. We found that 419 firms allocated to the bioeconomy population were based on more than one stage of the identification procedure. More than half of the entities from the R&D data overlap with other bioeconomy markers, while one third of the patenting firms also do so. This overlapping category (first column) is arguably the most robust population for framing the bioeconomy population, although it is biased towards larger firms which have a higher probability of appearing in multiple firm populations.

In general, entities in the BioSmart Survey are sole proprietor companies (i.e. very small firms) while other categories such as membership in relevant Federation of Norwegian Industries (NHO) and patenting firms tend to characterise larger firms. Table 11.3 illustrates the breakdown of the NIoBE population based on size classes in terms of the locations of the entities, which are spread throughout the country. We note a larger concentration of sole proprietorships (e.g. farms or forestry companies) in more rural areas of the country. The large population centres of Oslo and Akershus account for many of the larger firms in the population, as do the other population centres of Trondheim (Sør Trøndelag), Bergen (Hordaland) and Stavanger (Rogaland).

Some areas of Norway are more rural and rely more on primary industries; others are more urban and service-oriented; while still other localities are mostly dependent on manufacturing. These differences in the economic landscape influence the question of where circular economy activities take place. Table 11.4 classifies the location of the entities using Statistics Norway's 'classification of municipality groups' (SSB, 2018); it illustrates how the different economic activities are distributed across different parts of the country.

Primary industries – forestry, aquaculture and seafood processing, brewing, meat processing, dairy, and waste processing – are seen here to be spread throughout the country, as are the entities that produce food and beverages. The utilities classification includes recycling firms as well as bioenergy entities. These are more concentrated in population centres, as are the universities (education, etc.) and R&D service companies. Non-private services include interest-organisations and government organisations – whose involvement is qualitatively different from that of other entities in the inventory.

The Norwegian Inventory of Bioeconomy Entities (NIoBE) can thus be used to provide a systematic look at the 'circular bioeconomy' in terms of the actors who actually work with organic resources. The inventory of firms provides a great deal of information about who is involved in the circular economy in Norway. However, it does not in itself provide information

Table 11.2 Population of bioeconomy actors in Norway: by sector and identification source ($n = 2,369$)

Sector	*Overlap*	*R&D agricult. & food 2015*	*Patents*	*Projects*	*BioSmart survey*	*OREEC register*	*Confederation members*	*Grand total*
Primary	**37**	**8**	**12**	**64**	**483**		**29**	**633**
Farm	22	7	1	48	363			441
Fish	10	1	5	9	17		29	71
Forest	3			6	103			112
Other resources	2		6	1				9
Manufacturing	**110**	**25**	**72**	**79**	**48**	**15**	**28**	**377**
Food & drinks	70	15	4	38	20	1	22	170
Other production	40	10	62	41	28	14	6	201
Other resources			6					6
Utilities	**164**		**13**	**11**	**20**	**430**	**5**	**643**
Construction	2		7	6	1	8		24
Electricity	26		5	4	6	72		113
Recycling	136		1	1	13	350	5	506
Services	**105**	**21**	**145**	**207**	**43**	**85**	**24**	**630**
Other Activities	7		12	54	1			74
Other Services	12	5	20	29	2			68
R&D services & education	38	7	95	71	22	25	4	262
Wholesale & retail	48	9	18	53	18	60	20	226
Public & other activities	**3**		**6**	**76**	**1**			**86**
Other activities	3		4	54				61
R&D services & education			2	22	1			25
Grand total	**419**	**54**	**248**	**437**	**595**	**530**	**86**	**2,369**

Source: NIoBE, 2018.

Table 11.3 Reduced NIoBE population of bioeconomy entities by employment class and county (n=2,113)

County	*Sole proprietorships*	*Micro-firm*	*Small firm*	*Medium-sized firm*	*Large firm*
AKERSHUS	38	68	43	26	16
AUST-AGDER	17	16	11	2	0
BUSKERUD	28	53	32	14	1
FINNMARK	6	13	13	3	3
HEDMARK	35	47	29	18	5
HORDALAND	32	50	41	23	15
MØRE OG ROMSDAL	19	55	41	14	4
NORD-TRØNDELAG	25	29	14	15	2
NORDLAND	10	36	29	15	3
OPPLAND	35	45	17	11	1
OSLO	16	79	68	46	49
ROGALAND	20	54	39	19	12
SOGN OG FJORDANE	13	25	21	7	2
SØR-TRØNDELAG	26	49	31	16	17
TELEMARK	11	31	19	8	3
TROMS	9	33	34	8	6
VEST-AGDER	7	29	16	3	0
VESTFOLD	15	35	19	10	3
ØSTFOLD	15	35	25	10	2
Total	377	782	542	268	144

Source: NIoBE, 2018.

Notes

1 Reduced population as defined above; 112 entities that lacked information about location and/or employment were not included here.

2 Employment classes are based on maximum annual numbers of employees between 2009 and 2016. Micro-firms have fewer than 10 employees, small firms between 10 and 49, medium between 50 and 249, and large over 250 employees. The national VoB database is used (Brønnøysundregistrene), supplemented by the stock-value of the firm and firm type (e.g. sole proprietorships).

about how these different types of actors that are located in different parts of the country actually contribute to innovation in the circular-economy. In the final section, NIoBE is used to target a questionnaire that was directed at firms whose activities appeared to be linked to organic waste streams.

11.3.3 Mapping of innovation in the Norwegian circular economy

In this final empirical section we present the results of a mapping exercise that was carried out in Norway in the spring of 2017 to better understand what firms do to derive value from different organic waste streams. This mapping exercise targeted firm-level activities involving organic waste in the six focal subsectors of the SusValueWaste project: (i) forestry, (ii) aquaculture and seafood processing, (iii) beer brewing, (iv) meat processing, (v) dairy and (vi) organic waste processing. In addition, a seventh category, consisting of

Table 11.4 NIoBE population of bioeconomy entities by sector (NACE) and by type of municipality ($n = 2{,}239$)

	Primary Industry	*Mixed agriculture and manufacturing*	*Manufacturing*	*Less central service industry*	*Less central, mixed service industry and manufacturing*	*Central, mixed service and manufacturing*	*Central service industry*	*Total*
Agriculture, forestry, fishing	77	90	76	78	179	34	87	621
Education, health and other public services	1	1	3	5	12	7	50	79
Manufacture food products and beverages	20	13	16	23	29	19	49	169
Manufacture other	8	15	27	15	56	8	68	197
Non-private services	1	0	2	2	4	2	37	48
Professional, scientific and technical activities	8	8	19	8	50	11	178	282
Utilities, transport and diverse services	39	36	69	88	247	68	296	843

Note
Sectors are classified according to NACE divisions. Classification of municipality groups according to Statistics Norway at the lower level of aggregation of municipalities ('kommuner').

R&D-oriented service firms, was also included, echoing our original focus on R&D expenditure described above (an overlap of 14 entities).

The purpose of the exercise was not to perform a representative survey of activity in the circular bioeconomy (cf. Biosmart) but to get a better idea of how biomass and organic waste is used by different entities in different markets. The instrument was therefore addressed to entities in the NIoBE population to increase the likelihood that respondents in fact hosted (or planned) activities involving such organic resources. We go on now to briefly introduce the design of the non-probabilistic sample procedure and of the questionnaire, before finishing by reviewing some of the results. A more in-depth account of the data collection process can be found in a separate report by the TIK Centre at the University of Oslo (Normann, 2018).

11.3.3.1 Approach and population

A questionnaire was used to collect data about the extent and orientation of organic waste activities, the sources of feedstock used, the distribution of innovation in different contexts and other questions such as barriers to innovation activities or the importance of collaboration. In this sense, the instrument modelled some of its questions on items in the Community Innovation Survey (CIS), as well as adding in other questions (about feedstocks, etc.).

The questionnaire consisted of nine sets of questions, including a control question about current activities by type of organic waste. The sections collected information about the types of feedstocks, technology and knowledge sources, drivers and barriers, the importance of public measures, costing and financing relevant activities, collaboration, innovation activities, as well as generic information about the firm. A pilot round was used to calibrate the questionnaire before a new version was sent to a population of 304 entities. The survey was sent by email. The relevant contact points at the individual entities were identified in advance either by accessing publicly available information (website) or by phoning the entity.

A census-type survey approach was applied to collect data in this mapping exercise. Given the noted problems when identifying target firms, data collection utilised the NIoBE dataset (above) as a population frame for the Norwegian circular bioeconomy. The sample for this exercise included about 12% of the total NIOBE data current at the time of sampling. The design for this subpopulation was based on a number of clear criteria. Selection criteria included the following:

1 the entity was a private sector firm;
2 the firm was linked to at least one of the six targeted activities, namely (i) forestry, (ii) aquaculture and seafood processing, (iii) beer brewing, (iv) meat processing, (v) dairy and (vi) organic waste processing; or a seventh category consisting of R&D-oriented service firms;

3 the firm was drawn from the overlapping category of NIoBE (i.e. the subpopulation of bioeconomy actors present in more than one data source); and
4 the firm was not a sole proprietorship and it was registered as active in the underlying databases at the time of the survey.

The sample constituted a full count of entities that fulfilled these criteria in NIoBE. Following an initial drawing of the sample, a round of validation was conducted to exclude defunct or misreported entities (especially in waste processing).

11.3.3.2 Results

The questionnaire, which was sent to the 304 actors, resulted in 133 responses, of which 85 were complete responses confirming ongoing activities in the area (see Normann, 2018). The completed questionnaires provide the main focus of the review presented here. A further 48 reported no current activity and were only asked generic questions, e.g. about unexploited opportunities related to organic waste activities.

Which types of organic waste activities do the companies carry out? Figure 11.1 breaks the population of 85 entities down by size, subsector and the type

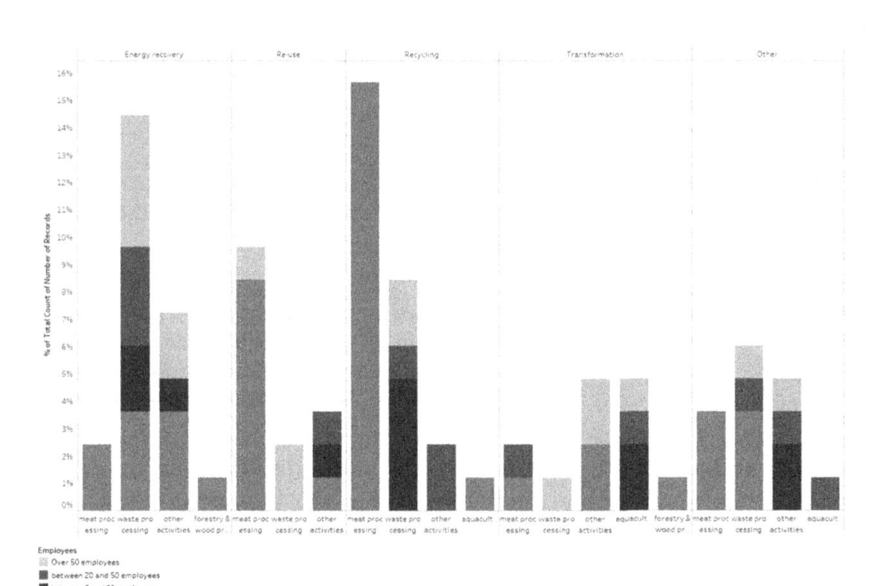

Figure 11.1 TIK questionnaire: 85 respondents by size class, subsector and type of activity.

Source: TIK, 2017, collated by NIFU.

of activity the firm is involved in: the recovery of energy from different organic feedstocks, their re-use or their transformation, as well as recycling or other uses. With more than one type of activity allowed for each respondent, we observed that recycling appeared to be the main activity type, while re-use and transformation, which aim to upgrade to a different type of product, were less frequent. Energy recovery is pursued by 31 firms, i.e. by more than one third of the firms involved in organic waste activities. Simultaneously, we observed the predominance of small firms, not least in the meat-processing area. The waste-processing firms, on the other hand, tend to be larger.

Of the 85 firms that have declared activity connected to organic waste, 64 consider it to be the core activity of the firm, while 21 define it as a supplementary activity. When asked specifically about activities involving transforming bio-based feedstock into new intermediate products, less than half of the firms involved in the activity defined the activity as 'core'. A similar proportion is observed for involvement in the development of new OW technologies to be sold to other companies. In contrast, activities devoted to the development of new products for end users are almost as frequently defined as 'core' as 'supplementary'. An intermediate case is represented by activities of selling and/or delivering to other companies without transformation.

The orientation and intensity of organic waste-oriented activities differ by type. Figure 11.2 demonstrates that the branch of firms involved in recovering energy from organic waste reports that roughly a third of their business activities (and turnover) are related to this activity. The proportion is higher for transformative activities. Noting that the activities might overlap, we see that firms involved in recycling-oriented activities report on average a fourth of their activity in this category.

What is the source of bio-based feedstock? Noting that there may be more than one source, half of the companies responded that their feedstock was produced as a by-product of the company's own production activities. Significantly fewer companies obtained bio-based feedstock from other companies, for free (10) and/or by purchase (14).

Thirty-one firms have invested in R&D linked to its organic waste activities in the last three years. A minority of seven of the 31 firms reported spending more than 80% of the R&D budget on activities related to organic waste. Table 11.5 breaks down the RD&I activity by the share that report R&D expenditure and the share that also report product and/or process innovation. A final category indicates whether the firm has acquired new machinery expressly to process organic waste. The same firm can report multiple types of waste-related activities (e.g. energy recovery and re-use).

Those firms that report ongoing activities in transforming organic waste are the most active innovators in this area. Eighteen firms have introduced and commercialised a new product related to organic waste in the last three years. For most firms, the new products relate to recycling and transformation. Four of the firms have new products related to energy recovery, and

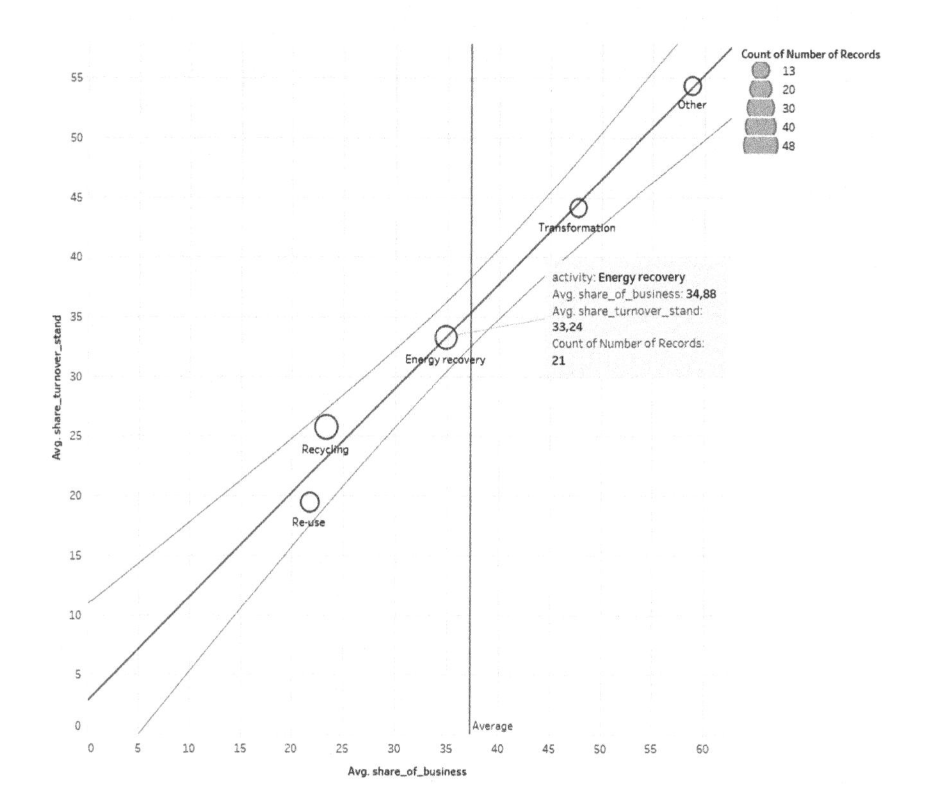

Figure 11.2 TIK questionnaire: 85 respondents by importance of organic waste in terms of share of business activities (x-axis) and turnover (y-axis).

Source: TIK, 2017, collated by NIFU.

Table 11.5 TIK survey: firms reporting R&D investments, reporting R&D or innovation activities, and purchase of machinery to process organic waste: by type of organic waste-related activity

Organic waste-related activity	*Number of firms*	*R&D activity (%)*	*RD&I activity (%)*	*New machines (%)*
Recycling	23	22	39	39
Energy recovery	35	29	33	43
Re-use	22	33	33	33
Transformation	48	77	85	69
Other	59	38	62	31
Average shares	**35**	**36**	**47**	**42**

Source: TIK, 2017.

only two firms have new products related to re-use. On the other hand, 25 firms have introduced a new process related to organic waste in the last three years. Pre-treatment processes and fermentation/biochemical processes are the most frequent, while extraction and separation processes have been introduced by seven firms.

11.4 Preliminary conclusions

There are two preliminary conclusions that can be drawn from our study.

The first one relates to the difficulties of approaching the 'bioeconomy' meta-sector empirically, since even the theoretical definitions of the meta-sector are still evolving within the current scientific literature. When we wanted to target a specific survey about organic waste at the population of Norwegian bioeconomy firms, we faced discrepancies between the theoretical directions which we ideally wanted to explore, and the empirical possibilities we had according to the information available. Indeed, the data sources we looked at, concerning the reconstruction of the population of 'bioeconomy' firms in Norway, were based on firms' innovation inputs and outputs (funded RD&I projects; R&D surveys; patents), on the economic context the firms belong to (industry classification; affiliation to industry networks) and on more subjective judgements made internally by the firm or externally by experts (self-identification vs. supervised reviews). In order to cover the different conceptual approaches to the bioeconomy meta-sector, researchers need to navigate through the available data sources and make a series of decisions about how to intersect or merge different data layers, each of which connects to one or more theoretical approaches to the bioeconomy.

A second conclusion relates to the specific role of organic waste activities within bioeconomy firms. On the one hand, organic waste activities are often core activities, thus constituting a distinctive characteristic of a firm. On the other hand, the development of new products related to organic waste is not a central concern for such firms. Moreover, unless there is waste transformation involved, the activities related to organic waste seem to attract a low share of the firms' total R&D budget. Therefore, in order to reach out to firms who actively seek to realise value from organic waste streams, an identification based solely on RD&I indicators may not be sufficient. R&D&I data sources can provide an important first indication of organic waste activities, e.g. by highlighting R&D allocated to the areas of 'circular economy' and 'ecology', or by recording patents on biological treatment of waste (CPC class 'Y02W'). However, complementary sources, such as trade descriptions in financial data or self-identification in response to specific survey questions, can become necessary to detect other relevant firms, whose active role in organic waste activities may not be supplemented by corresponding activities in research and innovation.

Note

1 This section is based on earlier background work reported in Iversen (2016) and Iversen and Rørstad (2017).

References

BIC Consortium. (2018). Retrieved from https://biconsortium.eu.

Biosmart. (2018). Hvordan fremme overgangen til en smart bioøkonomi? Retrieved from https://biosmart.no/?lang=en.

BIOTEK2021. (2018). Retrieved from www.forskningsradet.no/prognett-biotek 2021/Home_page/1253970728140.

Bjørkhaug, H., Hansen, L. & Zahl-Thanem, A. (2018). *Sektorvise scenarier for bioøkonomien.* Trondheim: Ruralis – Institutt for rural- og regionalforskning.

Bugge, M. M., Hansen, T. & Klitkou, A. (2016). What is the bioeconomy? A review of the literature. *Sustainability, 8*(691), 1–22. doi:10.3390/su8070691.

CEPI. (2018). Confederation of European paper industries. Retrieved from www.cepi.org.

Espelien, A., & Sørvig, Ø. S. (2014). *Fornybar energi og miljøteknologi i Norge. Status og utvikling 2004–2013* (Vol. 34). Oslo: MENON.

EurObserv'ER. (2018). Retrieved from www.eurobserv-er.org.

Frietsch, R., Neuhausler, P., Rothengatter, O. & Jonkers, K. (2016). Societal grand challenges from a technological perspective: Methods and identification of classes of the International Patent Classification IPC. *Fraunhofer ISI Discussion Papers Innovation Systems and Policy Analysis, 53.*

Iversen, E. J. (2016). *The market-square in the circular bio-economy: Empirical strategies. Conference Poster.* Paper presented at the Nordic Research and Innovation Pathways towards a Circular Bio-economy NoRest Conference Copenhagen.

Iversen, E. J. (2018). *The Norwegian Inventory of Bioeconomy Entities (NIoBE)* [database]. Oslo: NIFU, Nordic Institute for Studies in Innovation, Research and Education [producer]. Unpublished (publication expected in 2019).

Iversen, E. J., & Rørstad, K. (2017). *Empirical strategies to analyze innovative firms in the bioeconomy.* Paper presented at the European Association for Evolutionary Political Economy (EAEPE), Bremen, Germany.

Kreuchauff, F., & Korzinov, V. (2017). A patent search strategy based on machine learning for the emerging field of service robotics. *Scientometrics, 111*(2), 743–772.

LUKE Natural Resources Institute Finland. (2018). Retrieved from www.luke.fi/en.

Nelson, R. R., & Rosenberg, N. (1993). Technical innovation and national systems. In R. R. Nelson (Ed.), *National innovation systems: A comparative analysis* (pp. 3–22). New York, Oxford: Oxford University Press.

Normann, H. (2018). *Mapping activities in the Norwegian organic waste sector: A survey of Norwegian firms engaged in organic waste activities.* Oslo: SusValueWaste project, University of Oslo/TIK Centre for Technology, Innovation and Culture.

Norsk Industri. (2018). Norsk Industri. Retrieved from www.norskindustri.no/bransjer.

OECD. (2018). *Meeting policy challenges for a sustainable bioeconomy.* Paris: OECD.

OREEC. (2018). Oslo renewable energy and environment cluster. Retrieved from http://oreec.no/about.

Rotolo, D., Hicks, D. & Martin, B. R. (2015). What is an emerging technology? *Research Policy*, *44*(10), 1827–1843. doi:10.1016/j.respol.2015.06.006.

Rørstad, K., & Sundnes, S. L. (2017). *Kartlegging av landbruks- og matrelatert FoU i 2015: Ressurser og vitenskapelig publisering* (Vol. 2017: 2). Oslo: NIFU.

SSB. (2018). Classification of municipalities. Retrieved from www.ssb.no/en/klass/klassifikasjoner/131.

Part IV
Policy implications

12 Directionality and diversity

Contending policy rationales in the transition towards the bioeconomy

Lisa Scordato, Markus M. Bugge, and Arne Martin Fevolden

12.1 Introduction

One of the pressing societal challenges today relates to climate change and the need to replace fossil-based inputs with renewable resources in the production of fuel, energy, and chemical compounds. This has resulted in the development of biofuels, such as bioethanol, biodiesel, and biogas; bio-products, such as bio-plastics, bio-chemicals, and bio-pharmaceuticals; and bioenergy, such as electricity and district heating generated at biogas or combustion plants. The magnitude and diversity of these initiatives have led scholars, commentators, and policy-makers to talk about a "bioeconomy" and, subsequently, to call for a more comprehensive policy framework to support and direct this emerging field of the economy. The bioeconomy concept has been embraced by many governments around the world with a view to responding to diverse societal challenges, including not only solving issues related to climate change, but also dealing with areas such as food security, resource efficiency, and health problems (German Bioeconomy Council, 2015; Staffas, Gustavsson, & McCormick, 2013). Nevertheless, it remains unclear what the bioeconomy is, and how it can contribute to achieving these broad and potentially contending policy objectives.

In recent years, a growing body of academic literature has emerged that aims to understand the roles of the bioeconomy in mitigating the challenges of climate change, and which also tries to disentangle the notion of the bioeconomy and its implications for governance. A recent contribution from Bugge et al. shows that the notion of the bioeconomy is multifaceted and covers several sectors and meanings, including different "rationales or visions of the underlying values, directions and drivers of the bioeconomy" (Bugge, Hansen, & Klitkou, 2016). Opposing rationales may also reflect the diversity of the sectors and policy areas involved, which stresses the need for horizontal policy mixes across sectors (Bugge et al., 2016).

Theorising on the governance of socio-technical transitions has emphasised the need for an active state, formulating societal needs and establishing the direction in socio-technical transitions.

The urgent need for green innovation requires policies that are not merely designed to improve coordination and fix market failures, but that are based on clear strategies aimed at reducing the risks and uncertainties in the field (Mazzucato, 2013; Weber & Rohracher, 2012). At the same time, it is increasingly acknowledged that, when compared with traditional policy fields, challenge-oriented policy measures require more demand-side policies, such as public procurement stretching across policy domains. Finally, it is assumed that the coordination and reflexivity needed in such societal transitions can be perceived as a form of meta-governance. The involvement of a broad range of actors in agenda setting is perceived as crucial. A pertinent question, however, is what the implications of such participative governance suggest, in terms of the possibilities for policy-makers to direct transitions in an effective and efficient way.

Against this background, the aim of this chapter is to improve our understanding of whether and in what way the bioeconomy consists of contending rationales for governance and policy-making. In order to do this, we apply a typology of three visions of the bioeconomy onto the policy discourse on the bioeconomy. This typology distinguishes between (i) bio-technology visions, emphasising the importance of biotechnology and its commercial applications; (ii) a bio-resources vision, focusing on processing and upgrading biological raw materials, as well as establishing new value chains; and (iii) a bio-ecology vision, which highlights sustainability and ecological processes, including biodiversity. This typology was created by Bugge et al. (2016) through an extensive review of the scientific literature that dealt explicitly with conceptual aspects of the bioeconomy, focusing on areas such as innovation and value creation, driving forces, governance, and the spatial implications of the bioeconomy.

The chapter applies these visions to a number of submissions in a public inquiry process on the development of a national strategy for the bioeconomy in Norway. Through the analysis, the chapter seeks to depict (a) the types of actors involved in shaping the direction of the new bio-based economy and (b) their positions on this emerging field. Based on this analysis, the chapter discusses the implications and possibilities for governance in setting the direction for the current socio-technical transition.

The chapter is structured as follows: in section 12.2, we describe the conceptual framework for the analysis. Section 12.3 will then outline the approach and methods used to analyse the empirical material. In section 12.4, some background on the Norwegian economy is outlined. The results from the analysis are presented in section 12.5. Finally, section 12.6 concludes the chapter by summing up the findings and reflecting upon their implications.

12.2 Conceptual framework

Over the last couple of decades, there has been increasing interest among innovation scholars and policy-makers in grand challenges and socio-technical transitions (European Commission, 2011, 2012; Geels, 2002; Kemp, Schot, & Hoogma, 1998; Kuhlmann & Rip, 2014; Schot & Steinmueller, 2016). Such societal challenges and system transformations are seen as open-ended and constantly redefined and renegotiated across several sectors and stakeholders (Kuhlmann & Rip, 2014). In the search for possible solutions to these highly integrated and complex societal challenges, it has been pointed out that there is a need for an "opening up" of decision-making processes, in order to include participation from a broader array of societal stakeholders. One concrete example of such an open approach is the use of public hearings to develop policy strategies, alongside consensus conferences and foresight exercises, as well as other approaches, in order to make decision-making processes more open and inclusive (Martin, 2015; Stirling, 2008).

Such integrative approaches to policy making can be demanding. They require the coordination and processing of complex and often conflicting inputs from a broad array of actors. Weber and Rohracher (2012) have developed a framework for legitimising policies addressing grand challenges, and operating with four required roles for governance in societal transformations: directionality, demand, coordination, and reflexivity. Directionality failure refers to a deficit in the pointing of innovation efforts and collective priorities in a certain direction to meet societal challenges. Demand articulation failure refers to a deficit in anticipating and learning about user needs, resulting in inappropriate and misleading specifications guiding development through, e.g. procurement or policy programmes. Policy coordination failure refers to a deficit in managing and synchronising the inputs from different policy areas to meet societal challenges. Such coordination might include coherence between policies at international, national, regional, and municipal levels (vertical coordination failure), or across different sectors (horizontal coordination failure). Reflexivity failure refers to a deficit in the learning feedback loops and in the ability to continuously monitor the progress of ongoing innovation processes and to adjust the course of action. Alongside the existing categories of market and system failures, such forms of transformational system failures constitute a more comprehensive framework and legitimacy for policy intervention and formulation. In general, the role for governance in addressing socio-technical transitions is seen as more proactive and entrepreneurial than as has traditionally been regarded the norm for state intervention, in terms of fixing market failures or system failures (Klein Woolthuis, Lankhuizen, & Gilsing, 2005; Schot & Steinmueller, 2016).

Still, although the state is expected to take a leading role in these processes of societal transformation, there is reason to question whether and how these four roles (i.e. setting the direction, formulating demand, coordinating various stakeholders, and ensuring continuous learning and reflexivity) only constitute an extension of the former technocratic policy framework associated with systems of innovation in terms of fixing system failures. Rather, one may argue that societal shifts like the transition into a sustainable bioeconomy represent conflicting rationalities and perspectives that transcend the coordination of various inputs sharing the same societal ontology and objectives. In this sense, there are potential challenges related to navigating in a landscape consisting of diverse stakeholders. Governance across heterogeneous interests poses challenges for policy-makers and complicates strategy-making processes. Various stakeholders might express diverging and conflicting interests which, again, might lead to power struggles and negotiations.

A prominent approach to understanding socio-technical transitions involving diverse stakeholders is the multi-level perspective (MLP), which sees systemic transitions as co-evolutionary processes that unfold through an interplay between three interrelated analytical levels: regimes, niches, and landscapes (Geels, 2002, 2004, 2005; Geels & Schot, 2007; Schot & Geels, 2008). A socio-technical regime refers to the existing configurations of technologies, infrastructures, production processes, practices, and consumption patterns. Niches are seen as the locus for the development of disruptive innovation to supplement or replace existing socio-technical regimes. Finally, landscapes refer to the contextual and long-term societal trends that create pressures on existing socio-technical regimes, thus opening windows of opportunity for innovative niches.

Although the MLP perspective has advanced our understanding of socio-technical transitions, it has also been criticised for putting too much emphasis on the emergence of niches as the principal locus for regime change (Geels & Schot, 2007). Much of the MLP literature has also tended to focus upon the emergence of new regimes, and less is said about the decline of existing and old regimes (Geels, 2014; Turnheim & Geels, 2013). Moreover, it has been pointed out how theorising on socio-technical transitions has traditionally downplayed the role of power relations and politics in many ways (Geels, 2014). First, there has been a lack of focus on how power relations affect the development of policies in socio-technical transitions. Second, there has been a tendency to focus on the development of innovative niches rather than on the destabilisation of existing regimes. Third, the stability of existing regimes is often understood and explained in terms of socio-technical configurations and user practices related to notions, such as lock-in and path dependence, rather than political priorities and deliberate decisions (Geels, 2014). In sum, there is a need to improve our understanding of how (political) power relations and negotiations affect directionality in the processes of socio-technical transitions.

In an effort to address these shortcomings, throughout the last decade there has been increased interest in better understanding the power struggles and institutional underpinnings involved in socio-technical transitions (Geels, 2014; Markard, Raven, & Truffer, 2012; Meadowcroft, 2011; Shove & Walker, 2007; Smith & Raven, 2012; Smith, Stirling, & Berkhout, 2005). Some of these contributions have started to pay attention to the stability of existing socio-technical regimes, and to how the incumbent actors and stakeholders of existing regimes show resistance to niche innovations and developments that threaten the status quo (e.g. Geels, 2014). In this sense, power and politics have increasingly started to be seen in connection with the multi-level perspective, with the actors of existing socio-technical regimes treated as actively resisting change and protecting their current positions in the existing regime.

12.2.1 Contending visions on the bioeconomy

The notion that grand challenges transcend sector boundaries is especially relevant in the case of the bioeconomy. The development of a bioeconomy represents a move from fossil-based to bio-based products, fuel, and energy, and can, therefore, be seen as a way to address the grand challenge of climate change. However, the notion of the bioeconomy can also be seen to address other grand challenges related to food security, health, industrial restructuring, and energy security (Ollikainen, 2014; Pülzl, Kleinschmit, & Arts, 2014; Richardson, 2012). The bioeconomy can thus be seen as a generic phenomenon spanning a broad range of technologies and sectors of the economy, such as the agriculture, marine, forestry, bioenergy, chemicals, materials, and health sectors. Nevertheless, it is important to stress that a transition to a sustainable bioeconomy does not constitute a predetermined path, but still remains an open, future possibility. It is not something that will necessarily happen, and it will only occur through considerable and coordinated efforts, which will involve a wide range of actors. The bioeconomy has been conceptualised by previous scholars as a particular policy ambition and framework (Birch, Levidow, & Papaioannou, 2010; Levidow, Birch, & Papaioannou, 2012), representing "a techno-economic imaginary of the future that is co-produced with certain policies, institutions, and infrastructures that are framed as desirable and possible, while others are framed as undesirable and problematic" (Birch, 2016).

Richardson (2012), among others, found considerable difference of opinion between "farmers and agribusiness, between those convinced and those sceptical of environmental technofixes, and between pro-corporate and anti-corporate NGOs" with regards to the use of biotechnology (Richardson, 2012). The tensions arising from biotechnology innovations have also been emphasised by De Witt, Osseweijer, and Pierce (2015). Conflicting lines exist around the genetic modification of food, bio-based products, and pharmaceuticals (De Witt et al., 2015). Levidow et al. (2012) argue that the concept of the bioeconomy is still rather new and is not yet explicitly integrated into policy-making. In this regard, they suggest that policy strategies which declare the intentions and visions for the development of a bioeconomy may have an important role in achieving a transition towards the bioeconomy, and in determining the direction of the transition, in terms of the funding, instruments, and involved organisations (Birch et al., 2010; de Besi & McCormick, 2015; Levidow et al., 2012). On the other hand, Bosman and Rotmans found that the governments that have adopted bioeconomy policies already have approached this policy area quite differently – with the Dutch government acting as a "facilitator" and the Finnish government as a "director of transition" for the bioeconomy, for instance (Bosman & Rotmans, 2016). The novelty of the bioeconomy as a concept has also been stressed by Hilgartner (2007). In a critical analysis of the definition of the bioeconomy promoted by the Organisation for Economic Co-operation and Development (OECD), he argues that this has created "a new policy-oriented machinery suited to the work of ongoing technoeconomic anticipation" (Hilgartner, 2007, p. 385). At the same time, he raises critical concerns for how the bioeconomy definition of the OECD takes "economic operations as its distinctive focus", as this places other policy fields, such as the environment, health, and agro-food, in secondary positions. He therefore calls for a "reflexive examination" of a highly political concept, as the bioeconomy as a policy field has emerged to be (Hilgartner, 2007). In sum, this suggests that the notion of the bioeconomy has developed as a complex and highly contested policy field.

Although subject to increasing interest in the last decade, there is, thus, still little clarity in terms of what the notion of the bioeconomy implies and means. In an attempt to improve our understanding of the different perspectives on the bioeconomy, Bugge et al. (2016) have distinguished between three visions of what the bioeconomy constitutes. These are (1) the biotechnology vision; (2) the bio-resource vision; and (3) the bio-ecology vision.

The bio-technology vision emphasises the importance of the application and commercialisation of bio-technology in different sectors. The objectives of the bio-technology vision relate to economic growth and job creation (Pollack, 2012; Staffas et al., 2013). Value creation is based on the application of biotechnologies in various sectors, as well as on the commercialisation of research and technology within the framework of a globalised economy. As such, the bio-technology vision is in many ways similar to the so-called linear model of innovation, where a science push is seen as the primary driver of innovation and economic growth. Within this vision, close interaction between universities and industry is needed in order to commercialise relevant research (Zilberman, Kim, Kirschner, Kaplan, & Reeves, 2013).

The bio-resource vision focuses on the processing of bio-based resources as the primary driver and objective for innovation and economic growth. Whereas economic growth in the bio-technology vision is based on capitalising on biotechnologies, growth in the bio-resource vision is expected to come from capitalising on bio-resources. Value creation in the bio-resource vision emphasises the processing and conversion of bio-resources into new products. In addition to an optimisation of land use and existing value chains, waste management and the development of new value chains are also important in this vision. Moreover, the role of research, development, and demonstration (RD and D) is central. Whereas the bio-technology vision takes a point of departure in the potential applicability of science, the bio-resource vision emphasises the potential of upgrading and converting the biological raw materials.

The bio-ecology vision highlights the importance of ecological processes that optimise the use of energy and nutrients, promote biodiversity and avoid monocultures and soil degradation. While the first two visions are technology-focused and assign a central role to RD and D in globalised systems, this vision emphasises the potential for locally or regionally integrated circular processes and systems. In contrast with the importance of external linkages in the first two visions, the bio-ecology vision calls for the development of locally embedded economies in the form of "place-based agri-ecological systems" (Marsden, 2012), as a central characteristic for ensuring a sustainable bioeconomy.

The three visions are seen as analytical categories, and should, thus, not be considered as completely distinct from each other, but rather as ideal-type visions of the bioeconomy. Similar analytical categories have also been used in previous studies of the national discourses and narratives of the emerging bioeconomy. Birch (2016), for instance, analyses policy visions and frameworks in the Canadian bioeconomy by applying four distinct definitions of the bioeconomy: (1) product based; (2) substitution; (3) renewable-versus-sustainable; and (4) societal transition. Like the visions framework used in this chapter, Birch finds that one of the definitions (societal transitions) represents a competing alternative to the others, and this represents an example of the tensions and conflicts that exist in developing the bioeconomy (Birch, 2016). In our case, this alternative is represented by the bio-ecology vision.

In order to test the relevance of this conceptual framework, we wish to apply it to a number of submissions to a public hearing process that was part of the development of a national strategy for the bioeconomy in Norway. By doing so, we wish to see whether the different visions on the bioeconomy that were earlier identified in a review of the research literature can also be found among other types of civic and business stakeholders in the bioeconomy. This exercise will, thus, test the analytical framework applied, as well as improving our understanding of the power struggles and politics in socio-technical transitions.

12.3 Materials and methods

The method for the analysis is based on a discourse analysis of a recent national public inquiry process for a bioeconomy strategy in Norway. The public inquiry was initiated by the Ministry of Trade, Industry and Fisheries and the Ministry of Food and Agriculture in 2015. The inquiry was launched with the aim "to identify overall priorities for a national strategy within the field and formulate goals and instruments in a long-term perspective". Parties were invited to submit their opinions by sending in written submissions. The public inquiry material comprises 41 written submissions made by as many different actors representing private companies, industry associations, universities and university colleges, research institutes, interest organisations, municipalities, and NGOs. Most of the written submissions were about two pages long, whereas a few were detailed reports of up to 13 pages, with annexes.

Although a considerable number of written statements were submitted, it remains an open question whether or not these statements reflect the opinion of all the stakeholders in the bioeconomy. Some stakeholders might not have learned about the public inquiry process and others might not have felt sufficiently competent to express their opinion publicly. Nevertheless, we believe that both the variety of submissions and the extensive participation of interest organisations and NGOs suggest that the public inquiry process gathered a fairly broad and balanced selection of different stakeholders' opinions.

The text analysis was carried out in two main steps. First, the submissions were categorised by actor groups and sectors (Figure 12.1). Second, the text elements were coded according to the applied predefined bioeconomy visions (Bugge et al., 2016) and a corresponding set of sub-topics (Table 12.1). In this way, the text corpus was systematically analysed, allowing us to identify emerging discursive patterns.

In sum, the discourse analysis was carried out through a bottom-up and iterative process of the identification, interpretation, and categorisation of

Table 12.1 Key characteristics of the bioeconomy visions (Bugge, Hansen, & Klitkou, 2016)

	The bio-technology vision	*The bio-resource vision*	*The bio-ecology vision*
Aims and objectives	Economic growth and job creation	Economic growth and sustainability	Sustainability, biodiversity, conservation of ecosystems, avoiding soil degradation
Value creation	Application of biotechnology, commercialisation of research and technology	Conversion and upgrading of bio-resources (process oriented)	Development of integrated production systems and high-quality products with territorial identity
Drivers and mediators of innovation	R&D, patents, TTOs, research council funders (Science Push, linear model)	Interdisciplinary, optimisation of land use, include degraded land in the production of biofuels, use and availability of bio-resources, waste management, engineering science and market (interactive and networked production mode)	Identification of favourable organic agro-ecological practices, ethics, risk, transdisciplinary, ecological interactions, re-use and recycling of waste, land use, (circular and self-sustained mode)
Spatial implications	Global clusters/ Central regions	Rural/Peripheral regions	Rural/Peripheral regions

(a) actors; (b) their statements and advice; and (c) their visions on the bio-economy. (At the time of writing, the Norwegian bioeconomy strategy was not yet published.)

We use three ideal visions of the bioeconomy to identify the prevailing and potentially contending understandings of what the bioeconomy constitutes, and the diverse bases and perspectives that actors may have on this field. The ideal types are used as a method of interpretative analysis for understanding the way actors and organisations view a defined context, and to facilitate comparisons. In this sense, ideal types do not conform completely to reality, but are simplified models of interpretation (Thornton & Ocasio, 2008). Moreover, these visions should not be considered to be completely distinct from each other, but interrelated (Bugge et al., 2016). Table 12.1 illustrates the key characteristics of the three bio-economy visions, with their respective implications in terms of overall aim, value creation, drivers and mediators of innovation, and spatial implications. Hence, we expect that these visions also co-exist across heterogeneous bioeconomy stakeholders.

Although Bugge et al.'s (2016) bioeconomy visions constitute a useful tool for classifying the written submissions to the public inquiry process, there are also some methodological challenges related to applying these visions to new contexts. First, the visions were primarily created based on analysing academic articles, and these might not express ideas and ideals that are prevalent among NGOs, industry associations, and private companies. Second, the visions reflect the scientific discourse through the last decade and, thus, refer to ideas that were sometimes expressed many years ago, which might not be equally relevant for processes that are ongoing today. Nevertheless, we believe that these potential methodological challenges are minor, and that we were able to deal with them effectively by carefully reading and cataloguing the written inputs to the public hearing.

12.4 Background

Over the past century, Norway has developed a strong resource-based economy. It has established strong industries within forestry, aquaculture, and petroleum, and these resource-based industries have typically accounted for about 80%–90% of the country's exports. Nevertheless, the relative importance of these natural resource sectors has varied over time (Ville & Wicken, 2012). In the early 1900s, the prominent export products were fish and wood, in addition to relatively smaller quantities of paper and minerals (Ryggvig, 1996). Today, the most prominent export products are oil and gas and related petroleum products, in addition to relatively smaller quantities of fish and metals (see Statistics Norway). If we look specifically at bio-resources, fish was already an important export product in the early 1900s and is, today, by far the largest bio-based export product. Wood products were important in the early 1900s, but have lost much of their market share in the past decade. Agricultural products, on the other hand, have never been exported in large quantities. Seen together, this development path has created an imbalance in the Norwegian bioeconomy, in which the seafood sector can be described as an export leader that is seeking to expand its markets abroad; the agricultural sector can be described as an export lightweight that is seeking to protect its domestic markets from foreign import; and the forestry sector can be seen as a struggling has-been that is seeking to find new ways of regaining some of its former glory.

Although many resource-based economies have become victims of the "resource curse", Norway has – along with a few other countries, such as Australia – been able to reach modern levels of development while relying extensively on the extraction and refinement of natural resources. Ville and Wicken argue that what distinguishes the successful from the less successful resource-based economies is their ability to diversify into new "resource products and industries", through a dynamic interplay between the natural resource industries and the knowledge-producing and disseminating sectors within their societies (Ville & Wicken, 2012). Ville and Wicken describe these knowledge-producing and disseminating sectors as "enabling sectors" that are typically composed of capital goods suppliers and R and D institutions, and they maintain that a healthy interplay between these enabling sectors and the natural resource industries leads to both improved productivity in old resource-based sectors and the development of new resource-based industries. For instance, they found that the development of a strong mechanical engineering industry played a crucial role in establishing a vibrant Norwegian wood processing industry, and that Norwegian marine biologists helped the fisheries develop and make use of new fishing methods, thereby improving their productivity (Ville & Wicken, 2012).

Today, this dynamic interplay between the enabling sectors and the natural resource industries defines the working of Norway's bioeconomy. In terms of R and D institutions, most of the Norwegian bio-industries rely on a range of well-developed scientific institutions. The seafood industry benefits from research carried out in as many as 20 semi-public and private research institutes, of which some of the most important institutions include the Norwegian Veterinary Institute and the Norwegian College of Fishery Science (Doloreux, Isaksen, Aslesen, & Melançon, 2009). The agriculture and forestry industries also rely on strong academic institutions – such as the Norwegian University of Life Sciences (NMBU) and the Norwegian University of Science and Technology (NTNU) – both of which carry out a considerable amount of relevant R and D (Klitkou, 2010). In terms of capital goods suppliers, the seafood industry has a much more developed industrial base to draw upon nationally than the agriculture and forestry industries. The seafood industry – and in particular the aquaculture sector – can rely on a large group of mostly small- and medium-sized suppliers that carry out a substantial amount of R and D, and it is generally considered to be at the technological forefront of this area globally (Doloreux et al., 2009). The agricultural and forestry sectors, on the other hand, rely on national suppliers that import much of the equipment from foreign companies and carry out a limited amount of development work at home (Klitkou, 2010). The natural resource industries and enabling sectors can be described as the prime carriers and promoters of the three bioeconomy visions described above (Bugge et al., 2016). Together, they comprise the core of the Norwegian bioeconomy and the policy positions they advocate reflect, to a large extent, their position within the wider Norwegian economy.

12.5 Findings

Based on the discourse analysis of the different texts submitted to the public inquiry on Norway's bioeconomy strategy, we have been able to identify discursive patterns which relate to the three bioeconomy visions (Bugge et al., 2016). The three bioeconomy visions are crucial elements in this analysis, as they reveal the tendency for both conflicts and alignments between the participating actors. As we discussed above, the various bioeconomy visions differ in terms of aims and objectives, as well as over which might or might not be compatible as guideposts for a future bioeconomy. It is, therefore, interesting to see to what extent the different actors commit themselves to these bioeconomy visions and how the visions become manifested in terms of specific policy suggestions. It is also interesting to see whether certain actor groups commit themselves to specific visions and if these groups are large and powerful enough to influence the direction or pace of the transition process towards the bioeconomy.

Regarding the participating actors, we find that more than half of the submissions were from industry associations or private firms, followed by an even distribution of other actor groups representing public authorities, academia, and environmental and social NGOs. When further dissecting the private sector group, we find that they represent a multitude of industrial sectors ranging from forestry to bioenergy, agriculture, waste management and recycling, meat and poultry, marine, health, food, chemicals, and aquaculture (Figure 12.1). Among these sectors, the forestry sector clearly dominates, followed by bioenergy and agriculture.

Figure 12.2 shows that topics relating to the bio-resource vision, such as RD and D in agriculture, forestry, bioenergy, and new bio-based materials, the establishment of new value chains, resource management, and conversion technologies, all receive considerable attention (62% of the submissions in the public inquiry). While these issues are discussed across different actor groups, the industry actors and public authorities lead these discussions. The considerable involvement of private actors within the forest sector may reflect the interests these stakeholders have in influencing the direction of the bioeconomy discourse. This sector advocates increased harvesting and the exploitation of biomass resources from the forests, and the expanded use of biomass resources to create, among others, bio-materials, bioenergy, and biofuels. This position can be interpreted as an outcome of the uncertainties associated with

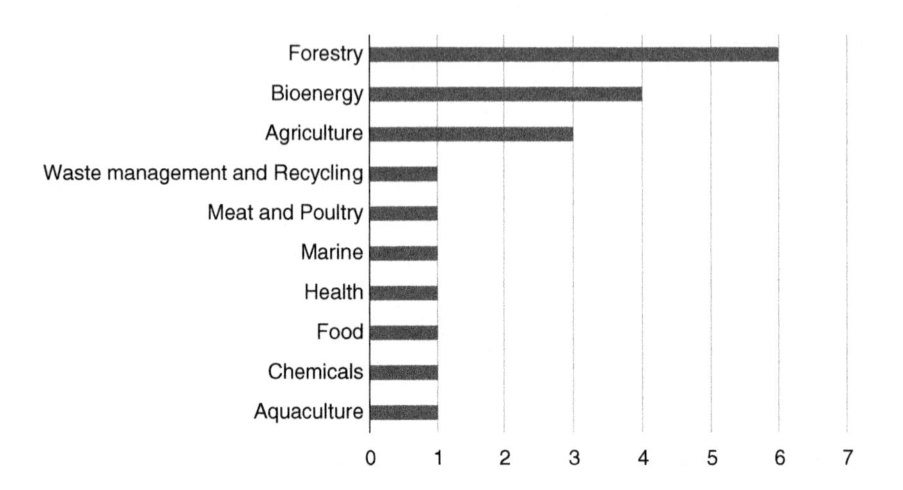

Figure 12.1 Industrial sectors represented by private firms and industry associations in the public inquiry.

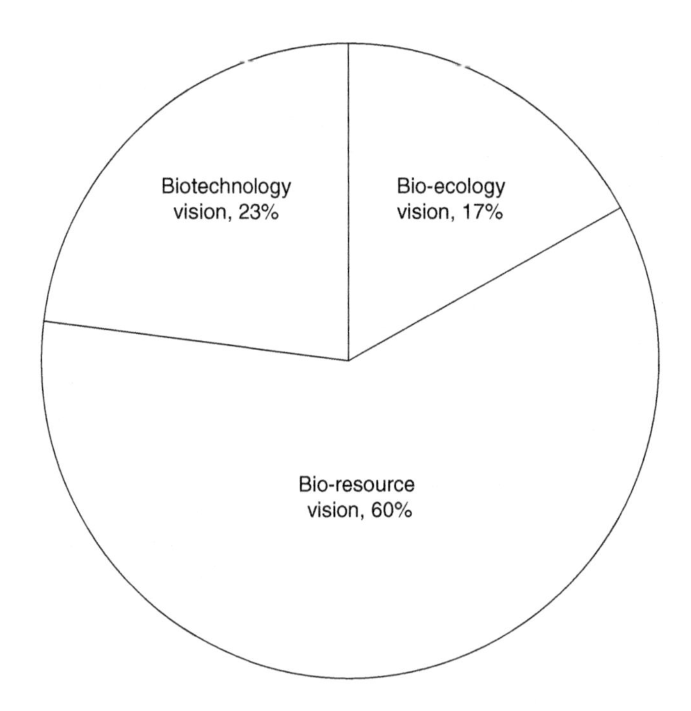

Figure 12.2 Relative share of bioeconomy visions in the public inquiry.

the crisis in the Norwegian pulp and paper sector, a sector which has practically collapsed in recent years. The industry's main interest is to find alternative uses for timber and forest residuals, and to encourage policy-makers to support and foster investments in the sector. The submissions from the energy sector advocate the increased use of bioenergy, and public investments in biofuel development. Some exemplar quotes from the submission of an energy company illustrate the argument:

> Bioenergy will be part of the bioeconomy. A bioeconomy strategy hence needs to include a further development of those parts in the value chain that are already active and commercially available today. It is important that stationary bioenergy and 2G biofuels are among the building blocks in a Norwegian bioeconomy strategy.... Forest resources are the best raw material for biofuels.... Biofuels will increase value creation from the forest by inverting the export of timber and will strengthen the wood manufacturing industry in general.
>
> (Energy company)

Biogas technology suppliers focus on competition and market structure and argue that the public authorities need to take a more active role in creating better market conditions for green and bio-based products. They, therefore, push for a more active and strategic use of demand-side policies, such as public procurement.

> There is a need for markets and cost levels which can compete with fossil-based solutions. In order to create new value chains based on biomass stable framework conditions, coordination and support from the public authorities are needed.... There is enormous potential to produce food and products from the sea.... We need to find alternative products that can exploit forest resources.... The solutions build on combining known technologies, but the challenges are related to both technology and profitability.
>
> (Industry association)

In addition, public authorities, such as local and regional governments, promote issues associated with the bio-resource vision. Their main focus is on research and innovation, and on the capitalisation of a wide range of bio-resources, which will presumably lead to economic growth and employment opportunities. In particular, they view their role as the promoters of cross-sector collaboration amongst regional actors, typically between industry and universities and research institutes. While positive effects related to sustainability and environment are portrayed implicitly, they are not emphasised as the main outcomes of the development of the bioeconomy. Thus, sustainability aspects receive limited attention from the public policy actors. See the following exemplar quote:

> The development of new value chains requires investments in research and innovation. A national strategy for research and innovation will create the basis for an innovative industrial sector based on national natural resources.... The role of the local municipality will be to connect existing competences within biomass utilisation (R&D, industrial production, market) and other knowledge sectors/networks (e.g., oil, shipping, finance, defense, and ICT) with the objective to identify new opportunities for value creation and job creation.
>
> (Local government)

The biotechnology vision is represented by 23% of the submissions, emphasising issues related to biotechnology research and the commercialisation of R and D within the life sciences and health. Unsurprisingly, it is essentially the academics (universities and research institutes) who are leading the biotechnology/science push discourse. Notably, the issue of life sciences and health-related biotechnology is discussed as having the potential to become a growing field. Their arguments clearly stress the need to develop biotechnology to prevent and treat different diseases, and at the same time, they highlight the many new business opportunities, which may come as a result of the commercialisation of biotechnology. It is assumed that the market for biotechnology products is promising, and that it would be a missed opportunity to neglect this field in a national bioeconomy strategy. A quote from the submission of an industry association illustrates this point:

> Medical and health-related biotechnology needs to be a part of the national bioeconomy strategy. New technology based on gene- and biotechnology has the potential to bring us large opportunities for treatment and prevention of diseases, and at the same time create new business opportunities.... The market for biotechnology products is large and will become even larger in the years to come.
>
> (Industry association)

Overall, these submissions express disappointment with the way the government at the outset has defined the scope of the bioeconomy, leaving health-related biotechnology and life sciences out of the definition. An example quote in this regard can be traced back to the submission from a university, which states that:

> Microbal biotechnology is a decisive research field if Norway is to develop economically sustainable and competitive bioprocesses based on Norwegian biomass in the future. Such investment will contribute to increased industrial activity and create new jobs within the bioeconomy.
>
> (University)

While the two first visions (i.e. the biotechnology vision and bio-resource vision) share many similar aspects (the focus on technology development,

R and D and economic growth, employment, market development, etc.), the bio-ecology vision has sustainability as the primary objective of the bio-economy. The actors within this vision express strong concerns related to sustainability and environmental issues and, at the same time, criticise the capitalisation of public goods, warning about risks related to the over-extraction of biological resources. A quote from an environmental NGO illustrates this argument:

> It is important to have a realistic assessment of how much biomass can be harvested from forests.... If the forest is to contribute with biomass resources to more than a small part of the potential application areas, much more needs to be harvested than just the forest waste (GROT). And this fact must make us listen to the alarm signals: what implications will this have for biodiversity and recreation? What are the implications for the climate and carbon storage function of the forest? ... We need to set strict requirements for the application of harvested forest biomass, so that it is used as effectively as possible and with the highest possible conversion rates. From our viewpoint it is hence not interesting to produce liquid biofuels from forest biomass.
>
> (NGO)

The analysis of the material shows that a minority of submissions (14%) emphasise questions such as environmental preservation, biodiversity, ecosystem services, and a circular approach to the bioeconomy. These issues are important predominantly for NGOs and a few industry associations. This perception radically contrasts with arguments from actors that have vested interests in, for instance, forest resources. In this sense, the bio-ecology actors favour the need to protect forest resources and pursue a careful assessment of the actual available biomass resources. They argue against the utilisation of forest biomass resources for the purpose of energy use, and contend that estimates of biomass extraction from forests need to be carefully assessed and managed. Moreover, they argue for the need to preserve forest resources, which are seen as public goods, serving important functions in terms of preventing further losses of biodiversity, preserving the essential ecosystem, and delivering recreational services:

> There are several opportunities for R&D and business development within an increased investment in the bioeconomy, but it requires a holistic perspective on the limitations that exist with regards to the exploitation of raw materials, the quantity of accessible raw material, the need to stop the loss of biodiversity, preservation of landscape qualities, recreational life and other interests in the same areas, and the real consequences of climate change.
>
> (NGO)

In addition, we see that perspectives on a circular production mode are represented by new actors from the waste management and recycling industry. This is linked to how the idea of waste has been transformed from a disposable pollutant to an important raw material in manufacturing and energy production. Concepts such as recycling and re-use are increasingly being redefined in terms of waste prevention, future material use, and opportunities for the circular economy. The point is illustrated by an industry association:

> It is important from a resource and climate perspective that the food production and waste from society is reintegrated in the life cycle through the reutilisation of bio-manure from biogas production in agriculture and the production of new food. The EU's vision on a circular economy and our own bioeconomy strategy will be important drivers.
>
> (Industry Association)

This last statement suggests that environmental sustainability may increasingly become integrated into new business models in the sector. However, the empirical material at hand is too limited to make any general conclusions on this aspect.

12.6 Conclusions and reflections

The aim of this chapter has been to improve our understanding of the politics of socio-technical transitions. The study has been based on an analysis of the different visions and contending rationales of different actors shaping the policy discourse on the bioeconomy. The analysis has been accomplished by applying three visions on the bioeconomy to analyse the content of a recent public inquiry process that sought to inform the direction of a national policy strategy for the bioeconomy in Norway. Although the findings have revealed a substantial diversity in visions and interpretations of the bioeconomy, they also show that the policy positions advocated by the stakeholders, to a large extent, reflect their roles and positions within the dominant regimes of the wider national economy.

Among the three bioeconomy visions, the bio-resource vision dominates the discourse. This finding reflects the traditionally important role of natural resource industries in the Norwegian economy and it is, hence, not very surprising that these sectors' positions on the bioeconomy are central in the material analysed. Still, this may serve as an illustration of how power structures are manifested in the existing socio-technical regime of the resource-driven Norwegian economy, and how these actors actively try to position themselves within the emerging bioeconomy.

It is primarily industry actors and public authorities that promote this vision. Overall, there seems to be a consensus among this group of actors regarding what needs to be prioritised and included in a national bioeconomy strategy. These actors advocate policies enabling the bioeconomy with attention to creating new markets for bio-based products and hence rebalancing the playing field between bio-based products and products based on fossil resources. They promote an increased use of policy instruments, such as the public procurement of bio-based products, in order to stimulate market demand. Their position often highlights the large unexploited potential for utilising biomass resources extracted and harvested nationally. In this context, biomass from forests is seen as having huge potential for the development of a national bioeconomy. According to this view, the state should take a more proactive role in developing new value chains based on biomass from forestry. As we have seen, the Norwegian forestry sector has been struggling in recent years and is seeking new ways of exploiting wood resources. In addition to the industrial players, regional governments seem to have a similar vision, focusing on the role of research and innovation and on the capitalisation of a wide range of bio-resources. This position emphasises the capitalisation of natural bio-resources in order to make Norwegian industries more competitive, and to create jobs nationally.

Overall, this position dedicates limited attention towards issues related to sustainability. Sustainability is seen as an effect of the bioeconomy rather than as a main starting point for it or outcome from it. To some extent, these positions hence reflect a "business as usual" approach to the bioeconomy, rather than presenting alternative ways to develop it, or counter-framings to contemporary industrial production practices.

Interpreted through the lens of our conceptual framework, these findings reflect how the incumbent actors of existing socio-technical regimes often try to resist change (e.g. Geels, 2014). However, we do not only find that the incumbent actors try to resist change brought about by niche-level actors; we also find that some of the incumbents pro-actively take part in the shaping of the future socio-technical regime of the bioeconomy.

The biotechnology vision, emphasising the application and commercialisation of science and technology, is most frequently advocated by the academic community. Similarly to the arguments put forward by the natural resource industries, they view the bioeconomy as an opportunity to create new businesses based on biotechnology products, and at the same time as an opportunity to make important advances within the treatment and prevention of diseases.

However, these visions based on the development and application of technology and the industrial exploitation of biomass resources are contrasted by sustainability concerns from NGOs arguing for a more careful use of biomass resources. These perspectives, reflecting the bio-ecology vision, highlight how the national bioeconomy strategy should take into account the relationship between biomass utilisation and sustainability, and make sure that the activities within the bioeconomy minimise negative environmental impacts. This group of actors represents a minority of the submissions to the public inquiry. The bio-ecology vision, hence, represents a contending alternative to the other two visions in how they portray and bring sustainability into the bioeconomy discourse.

Overall, the analysis has shown that the conceptual framework consisting of the three visions of the bioeconomy, applied to the submissions in the national hearing process, has proved to be a relevant and appropriate tool. The study has categorised the different submissions into the three respective visions, and the conceptual framework has thereby helped clarify how the different submissions express and represent diverging perspectives on the bioeconomy. This said, there seems to be extensive agreement across the various submissions in terms of seeing the bioeconomy as an opportunity to address societal challenges such as climate change. The divergence of perspectives rather relates to the means by which these societal challenges should be addressed.

The contending visions observed among the public submissions illustrate how socio-technical transitions often comprise competing points of view and values. The chapter has illustrated how the different responses to the public inquiry may contribute to a destabilisation of the existing (fossil) regime; whereas some of the actors in the existing regime oppose the new possibilities of the circular and sustainable bioeconomy, others embrace these and wish to contribute to the shaping of an alternative regime. This may cause a shift in the power balance between the various stakeholders involved. In particular, the policy strategy on the bioeconomy needs to deal with emerging tensions, such as the balance and relationship between economic growth and sustainability.

In this sense, the chapter has illustrated the relevance of the transformational policy framework of Weber and Rohracher (Geels, 2004), in terms of how giving direction to a socio-technical transition can be complicated by coordinating and balancing the different interests and stakeholders involved. Still, it remains crucial to ensure broad and democratic involvement and reflexivity across different stakeholders and interests in the process of shaping the bioeconomy of tomorrow.

References

Birch, K. (2016). Emergent imaginaries and fragmented policy frameworks in the Canadian bio-economy. *Sustainability*, *8*(1007), 1–16.

Birch, K., Levidow, L., & Papaioannou, T. (2010). Sustainable capital? The neoliberalization of nature and knowledge in the European "knowledge-based bio-economy". *Sustainability*, *2*(9), 2898–2918. doi:10.3390/su2092898.

Bosman, R., & Rotmans, J. (2016). Transition governance towards a bioeconomy: A comparison of Finland and the Netherlands. *Sustainability*, *8*(10), 1–20.

Bugge, M. M., Hansen, T., & Klitkou, A. (2016). What is the bioeconomy? A review of the literature. *Sustainability*, *8*(7), 1–22.

de Besi, M., & McCormick, K. (2015). Towards a bioeconomy in Europe: National, regional and industrial strategies. *Sustainability*, 7(8), 10461–10478.

De Witt, A., Osseweijer, P., & Pierce, R. (2015). Understanding public perceptions of biotechnoogy through the "Integrative Worldview Framework". *Public Understanding of Science*, *26*(1), 70–88.

Doloreux, D., Isaksen, A., Aslesen, H. W., & Melançon, Y. (2009). A comparative study of the aquaculture innovation systems in Quebec's coastal region and Norway. *European Planning Studies*, *17*(7), 963–981.

European Commission. (2011). *Horizon 2020: The Framework Programme for Research and Innovation Communication from the European Commission*. Brussels: European Commission.

European Commission. (2012). *Responsible research and innovation: Europe's ability to respond to societal challenges*. Brussels: European Commission.

Geels, F. W. (2002). Technological transitions as evolutionary reconfiguration processes: A multi-level perspective and a case-study. *Research Policy*, *31*(8–9), 1257–1274.

Geels, F. W. (2004). From sectoral systems of innovation to socio-technical systems: Insights about dynamics and change from sociology and institutional theory. *Research Policy*, *33*(6–7), 897–920.

Geels, F. W. (2005). Processes and patterns in transitions and system innovations: Refining the co-evolutionary multi-level perspective. *Technological Forecasting & Social Change*, *72*(6), 681–696.

Geels, F. W. (2014). Regime resistance against low-carbon transitions: Introducing politics and power into the multi-level perspective. *Theory, Culture & Society*, *31*(5), 21–40.

Geels, F. W., & Schot, J. (2007). Typology of sociotechnical transition pathways. *Research Policy*, *36*(3), 399–417.

German Bioeconomy Council. (2015). *Synopsis of national strategies around the world*. Berlin: German Bioeconomy Council.

Hilgartner, S. (2007). Making the bioeconomy measurable: Politics of an emerging anticipatory machinery. *BioSocieties*, *2*(3), 382–386.

Kemp, R., Schot, J., & Hoogma, R. (1998). Regime shifts to sustainability through processes of niche formation: The approach of strategic niche management. *Technology Analysis & Strategic Management*, *10*(2), 175–198. doi:10.1080/09537329808524310.

Klein Woolthuis, R., Lankhuizen, M., & Gilsing, V. (2005). A system failure framework for innovation policy design. *Technovation*, *25*(6), 609–619. doi:10.1016/j.technovation.2003.11.002.

Klitkou, A. (2010). *Innovasjon i matvare- og skogsektoren i Norge*. Retrieved from Oslo.

Kuhlmann, S., & Rip, A. (2014). *The challenge of addressing Grand Challenges: A think piece on how innovation can be driven towards the "Grand Challenges" as defined under the prospective European Union Framework Programme Horizon 2020*. Retrieved from Twente.

Levidow, L., Birch, K., & Papaioannou, T. (2012). EU-agri-innovation policy: Two contending visions of the bio-economy. *Critical Policy Studies*, *6*(1), 40–65.

Markard, J., Raven, R., & Truffer, B. (2012). Sustainability transitions: An emerging field of research and its prospects. *Research Policy*, *41*(6), 955–967.

Marsden, T. (2012). Towards a real sustainable agri-food security and food policy: Beyond the ecological fallacies? *The Political Quarterly*, *83*(1), 139–145. doi:10.1111/j.1467-923X.2012.02242.x.

Martin, B. (2015). *Twenty challenges for innovation studies*. Retrieved from Brighton, UK.

Mazzucato, M. (2013). *The entrepreneurial state: Debunking Public vs private sector myths*. London and New York: Anthem Press.

Meadowcroft, J. (2011). Engaging with the politics of sustainability transitions. *Environmental Innovation and Societal Transitions*, *1*(1), 70–75. doi:10.1016/j.eist.2011.02.003.

Ollikainen, M. (2014). Forestry in bioeconomy: Smart green growth for the humankind. *Scandinavian Journal of Forest Research, 29*(4), 360–366. doi:10.1080/02827581.2014.926392.

Pollack, A. (2012, 26 April 2012). White House promotes a bioeconomy. *New York Times*. Retrieved from www.nytimes.com/2012/04/26/business/energy-environment/white-house-promotes-a-bioeconomy.html?_r=0 (accessed on 9 March 2016).

Pülzl, H., Kleinschmit, D., & Arts, B. (2014). Bioeconomy: An emerging meta-discourse affecting forest discourses? *Scandinavian Journal of Forest Research, 29*(4), 386–393. doi:10.1080/02827581.2014.920044.

Richardson, B. (2012). From a fossil-fuel to a biobased economy: The politics of industrial biotechnology. *Environment and Planning C: Government and Policy, 30*(2), 282–296.

Ryggvig, H. (1996). Statoil, Stoltenberg og den nye norske imperialismen. *Internasjonal Sosialisme* (1).

Schot, J., & Geels, F. W. (2008). Strategic niche management and sustainable innovation journeys: Theory, findings, research agenda, and policy. *Technology Analysis & Strategic Management, 20*(5), 537–554.

Schot, J., & Steinmueller, E. (2016). *Framing innovation policy for transformative change: Innovation policy 3.0*. Retrieved from Brighton, UK.

Shove, E., & Walker, G. (2007). CAUTION! Transitions ahead: Politics, practice, and sustainable transition management. *Environment and Planning A, 39*(4), 763–770.

Smith, A., & Raven, R. (2012). What is protective space? Reconsidering niches in transitions to sustainability. *Research Policy, 41*(6), 1025–1036.

Smith, A., Stirling, A., & Berkhout, F. (2005). The governance of sustainable socio-technical transitions. *Research Policy, 34*(10), 1491–1510.

Staffas, L., Gustavsson, M., & McCormick, K. (2013). Strategies and policies for the bioeconomy and bio-based economy: An analysis of official national approaches. *Sustainability, 5*(6), 2751–2769. doi:10.3390/su5062751.

Stirling, A. (2008). "Opening up" and "closing down" power, participation, and pluralism in the social appraisal of technology. *Science, Technology & Human Values, 33*(2), 262–294.

Thornton, P. H., & Ocasio, W. (2008). Institutional logics. In R. Greenwood, C. Oliver, K. Sahlin, & R. Suddaby (Eds.), *Organizational institutionalism* (pp. 99–129). London: Sage Publications.

Turnheim, B., & Geels, F. W. (2013). The destabilisation of existing regimes: Confronting a multi-dimensional framework with a case study of the British coal industry (1913–1967). *Research Policy, 42*(10), 1749–1767.

Ville, S., & Wicken, O. (2012). The dynamics of resource-based economics development: Evidence from Australia and Norway. *Industrial and Corporate Change, 22*(5), 1341–1371.

Weber, K. M., & Rohracher, H. (2012). Legitimizing research, technology and innovation policies for transformative change: Combining insights from innovation systems and multi-level perspective in a comprehensive "failures" framework. *Research Policy, 41*(6), 1037–1047. doi:10.1016/j.respol.2011.10.015.

Zilberman, D., Kim, E., Kirschner, S., Kaplan, S., & Reeves, J. (2013). Technology and the future bioeconomy. *Agricultural Economics, 44*(S1), 95–102. doi:10.1111/agec.12054.

13 Multi-level governance of food waste

Comparing Norway, Denmark and Sweden

Julia Szulecka, Nhat Strøm-Andersen, Lisa Scordato and Eili Skrivervik

13.1 Introduction

While numerous political initiatives and civil society efforts have, for years, focused on undernourishment and hunger in vulnerable regions, an astonishing one third of all edible food produced globally is wasted (FAO, 2011). Studies commissioned by the United Nations Food and Agriculture Organization (FAO) estimate annual global food loss and waste by quantity for root crops, fruits and vegetables (40–50%), fish (35%), cereals (30%), oilseeds, and meat and dairy products (20%) (FAO, 2015, p. 2). Without a doubt, the alarming scale of food waste indicates a substantial market failure. Global food waste translated directly to economic terms has a value of US$1 trillion annually (FAO, 2015, p. 3). Another FAO study of total global costs of food waste estimated that apart from the direct economic loss, indirect environmental costs can be translated to some US$700 billion, and social costs of around US$900 billion per year (FAO, 2014). What is more, food waste is also a huge environmental challenge. Uneaten food is not only a loss within its own production chain, but squanders other valuable resources (land, water, energy and labour). It has a direct impact on the global climate – estimates suggest that global food loss and waste generate 4.4 Gt CO_2eq annually, which is some 8% of total anthropogenic GHG emissions (FAO, 2015, p. 3). The European Commission's analyses conclude that the food sector, together with housing and transport, has the largest environmental impacts in Europe, and that food waste has negative impacts due to the production burdens of uneaten food and waste treatment (Stenmarck, Hanssen, Silvennoinen, Katajajuuri & Werge, 2011). Lastly, but no less importantly, the scale of food waste is a significant moral issue. All the starving and malnourished people around the world could be fed using only a portion of the food we see being wasted.

Given the alarming scale of food waste, with edible food landing in composts, landfills and incineration plants, it is somewhat surprising that the problem stayed below the radar of most politicians, scientists and civil society activists for a long time. Scholars note that 'until recently, food waste has been largely ignored' (Halloran, Clement, Kornum, Bucatariu & Magid, 2014, p. 295), and that the issue started to appear on political agendas and in public debates only in the 2010s. Although reducing food waste seems a win-win situation for consumers, our planet and industries, it is a very complex issue, requiring diverse and well-tailored governance measures. It is known that the scale and occurrence of food waste in the value chain depend on the economic situation, climate, local culture and consumer habits. There are also significant tensions between the food security and resource efficiency perspectives (Hartikainen, Mogensen, Svanes & Franke, 2018, p. 509). In medium- and high-income countries, food waste is mainly caused by consumer behaviour and a lack of coordination between different actors in the supply chain. While the first relates to poor purchase planning, food labelling ('best-before dates') and customers' expectations regarding a wide range of fresh products and discounts for buying more, the latter is a systemic failure. This can be illustrated by farmer–buyer agreements that specify certain product characteristics related to appearance (FAO, 2011), retailer rights to return unsold products or a lack of knowledge on how to handle certain products (Stenmarck et al., 2011).

In the last decade, many international, national and local initiatives were created in order to address the food waste problem. The UN's Sustainable Development Goal 12, aiming to 'ensure sustainable consumption and production patterns', lists several targets related to food waste. It foresees that by 2030 we ought to be able to 'halve per capita global food waste at the retail and consumer levels and reduce food losses along production and supply chains, including post-harvest losses' and 'substantially reduce waste generation through prevention, reduction, recycling and reuse', but also 'encourage companies, especially large and transnational companies, to adopt sustainable practices and to integrate sustainability information into their reporting cycle' (SDG, 2015).

Many initiatives exist on the supra-national level, e.g. the multi-stakeholder Save Food Initiative (of FAO, UNEP, Messe Düsseldorf and Interpack packaging company), with their Think Eat Save campaign (Save-Food, 2018). Different solutions for tackling waste are being explored upstream and downstream of the production chain, in high- and low-income countries.

However, no single blueprint for tackling the problem exists, and this is especially visible on the national level, where a number of domestic political, economic and cultural factors influence the way food waste is tackled. This chapter provides a comparative analysis of the governance and the resulting policies and tools introduced by three high-income Nordic countries: Denmark, Sweden and Norway. In all three cases, the food waste issue has been relatively high on the political agenda for almost a decade, and the different governance frameworks that have emerged to reduce food waste provide interesting insights into food waste reduction strategies. Even though the three share many similarities in terms of political, cultural and economic features, such as a consensual parliamentary political culture, traditionally strong labour unions and social democratic parties, as well as a distinct regional identity, they also vary in some respects, and as our research finds, governance of food waste is one of them. The chapter describes these national governance arrangements and explains this observed variation. We first briefly discuss the theoretical and methodological foundation for our research, and then present food waste governance from a historical perspective, with a particular focus on the actors and milestones involved. We then move to a comparative analysis of the three countries, from which we draw conclusions and offer possible policy recommendations. We further categorise the observed governance systems and highlight important actors, who, in each of these cases, have proven particularly dedicated to food waste reduction.

13.2 Theoretical approach and method

We apply a multi-level governance framework to organise our comparative analysis of the three country case studies. Multi-level governance is a concept initially proposed to capture the changing nature of policymaking and policy implementation in the European Union (Hooghe & Marks, 1996). The approach builds on the growing complexity of tailoring accurate policy measures in modern states, especially in light of 'subsidiarity' and the emergence of sub-national authority levels as part of the European Union. Marks was one of the first to introduce the phenomenon in the 1990s (Stephenson, 2013, p. 818), and the concept has picked up a great deal of academic currency since, mostly because it has been shown to best reflect the nature of modern European politics.

The diffusion of authorities and competences, the creation of transnational regimes and the proliferation of public–private partnerships are all trends observed in the last three decades, challenging traditional forms of hierarchical authority and undermining traditional centralised forms of government (Hooghe & Marks, 2010, p. 17). 'Multi-level' refers to different levels of 'governance', i.e. 'above' national (European), national and sub-national, but it also implies the involvement of both public and private actors at these levels (Van Kersbergen & Van Waarden, 2004, pp. 149–150). Multi-level

Table 13.1 Two types of multi-level governance arrangements

Type I	*Type II*
General purpose	Task specific
Well ordered	Fluid, intersecting memberships
Clear lines of accountability	Accountabilities less clear
'Russian doll set'	Puzzle of many units, providing services, solving problems

Sources: Bache, 2012; Marks & Hooghe, 2004; Smith, 2007; Stephenson, 2013.

governance studies investigate the move from centralised forms of authority to more fluid, problem-focused networks (Smith, 2007) and are well suited to comparative analyses (Stephenson, 2013, p. 830). Marks and Hooghe operationalise this term by distinguishing between two ideal types of multi-level governance: type I and type II (Marks & Hooghe, 2004).

Scholars applying multi-level governance need to be careful and reflective in the way they employ the approach. As already noted, it is both a descriptive metaphor and an analytical framework for studying change in public policy making. At the same time, multi-level governance is simultaneously a framework for analysis, a *solution* for collaboration between stakeholders and a strategy for better policy implementation, improved information and consultation, and should improve the quality of decision-making and implementation (Gollata & Newig, 2017). Empirical evidence indicates that polycentric governance systems produce greater environmental outputs than monocentric ones (Newig & Fritsch, 2009). In the context of food waste, multi-stakeholder collaboration and public–private partnerships are necessary to find sustainable solutions to reducing food waste (Halloran et al., 2014).

In this analysis, we try to provide an overview of existing food waste policies and governance arrangements, and so we employ multi-level governance as an analytical framework, operationalising the two ideal-typical arrangements – type I and type II – in a comparative analysis. In the conclusion, drawing lessons and policy recommendations from our research, we reflect on food waste multi-level governance as a strategy or benchmark for good governance and effective policy for tackling the problem.

Multi-level governance gives us an effective tool with which to conduct a structured comparison of the three cases. We identify the relevant actors, i.e. all stakeholders and institutional actors who influence the food issue area, and we look at their sectoral category (public, industry and third sector), as well as the levels of governance on which they operate. Further, we analyse the different policy outcomes, and categorise them according to the ideal-typical distinction proposed by Hooghe and Marks, identifying characteristic elements of type I and type II arrangements (Table 13.1).

The empirical part of the chapter builds on a wide-ranging review of secondary scientific literature, government documents, public and industry statistics and applied grey literature. This is combined with the data gathered at an expert workshop on food waste, hosted by the Nordic Institute for Studies in Innovation, Research and Education (NIFU) in Oslo on 23 November 2017, and supported by ten expert interviews conducted by two of the authors in the spring of 2018 with food waste researchers, NGOs and industry and policy representatives.

13.3 Background

13.3.1 Definitions

The first challenge in any food waste-related research and action is defining the problem, since no internationally agreed definition of food waste exists (Hartikainen et al., 2018). Food waste can be *edible* and *non-edible*, *avoidable*, *possibly avoidable* and *unavoidable*. It can appear at different stages of the value chain (primary production, handling and storage, in processing and packaging, distribution or at the consumption stage). It can be distinguished from food loss in primary production, e.g. during harvesting. It is also debatable whether goods produced for human consumption but used for animal feed or biogas production should also be considered food waste.

Another important factor is the cultural definition of *edibility* as only parts defined as edible can become food waste or food loss (Klitkou & Iversen, 2016). This culturally constructed category (e.g. some parts of animals are considered inedible and discarded in certain societies while they are eaten elsewhere) may also change over time, influencing the measured scale of food lost or wasted. Edibility is also an important category at the end of the chain, as certain volumes of food will be deemed inedible for health and safety reasons (e.g. due dates). Departing from this food input–food output perspective, a systemic approach to food chains allows us to introduce more nuanced categories relating to food quality loss or possible degradation of products (Klitkou & Iversen, 2016, pp. 9–10).

Food waste data may be collected in tons, calories, economic values, etc. In Denmark and Sweden industrial food waste data is more confidential than in Norway[1] and some European Union (EU) data is not available for Norway. For instance, Eurobarometer surveys for food waste are not conducted in European Economic Area (EEA) states. All these questions translate to direct difficulties for research, including methodological challenges when comparing the three Scandinavian countries. In response to this problem, the Nordic Council of Ministers initiated a process aiming to develop common definitions of food waste for all Nordic countries (Swedish EPA, 2013).

To illustrate the different definition types and their implications, we can refer to the Norwegian industry agreement on reduction of food waste, signed on 23 June 2017, which stated that

> food waste includes all edible parts of food produced for humans, but which is either disposed of or removed from the food chain for purposes other than human consumption, from the time when animals and plants are slaughtered or harvested.
>
> (Regjeringen, 2017)

This is considered a 'strict' definition of food waste because it also categorises food produced for humans but used for animal feed or biogas as food waste.

In comparison, FUSIONS (Food Use for Social Innovation by Optimizing Waste Prevention Strategies), a European project in which Norway is an active participant, used a different and notably less 'strict' definition, understanding food waste as 'any food, and inedible parts of food, removed from the food supply chain to be recovered or disposed (including composted, crops ploughed in/not harvested, anaerobic digestion, bio-energy production, co-generation, incineration, disposal to sewer, landfill or discarded in the ocean)' (FUSIONS, 2014).[2]

13.3.2 Food waste hierarchy

The waste pyramid is a framework created to define, prevent and manage waste (see also Chapter 3) which has since been adopted for surplus food and food waste concerns. Taking a combined sustainability perspective, it lists strategies for dealing with food waste from 'best' to 'worst' (Bugge, Dybdahl & Szulecka, 2018). According to the hierarchy, the best possible option is always to prevent food waste through minimising food surpluses and avoidable waste. The second best option is to re-use the food that would otherwise be wasted through redistribution, food banks and all possible forms of charity and redistributing organisations. This is followed by food waste recycling, as animal feed and through composting. The fourth option is energy recovery, while the least favourable option is waste disposal (Papargyropoulou, Lozano, Steinberger, Wright & Ujang, 2014). The pyramid is also useful when considering the definition problem. The Norwegian definition puts food waste as anything below re-use, while the FUSIONS definition sets the boundary closer to the recycling stage (see Figure 13.1).

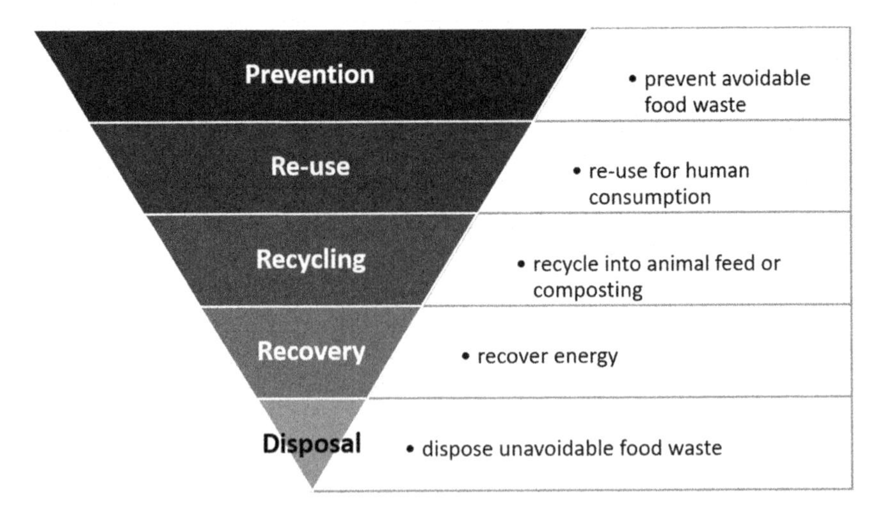

Figure 13.1 The food waste hierarchy.

13.3.3 Comparative food waste assessments in Scandinavia

The scale of food waste in the three analysed countries can be divided according to position on the value chain. Food waste quantification in *primary production* shows that Denmark is in the lead (288,000 tons), followed by Sweden (277,000 tons) and Norway (85,000 tons) (Hartikainen et al., 2018, p. 508). This might be explained by the fact that utilised agricultural areas cover only 3.1% of Norway's land territory, equalling one million hectares (EUROSTAT, 2012b), while they cover 7% of land and three million hectares in Sweden (EUROSTAT, 2012c), and as much as 61% and 2.5 million hectares in Denmark (EUROSTAT, 2012a). Taking into account these agricultural areas, relative food waste from primary production is quite comparable, yet Denmark is still in the lead.

Data from the *retail* sector shows that Sweden had double the amount of absolute food waste at this stage (83,500 tons per year in the retail sector) but considering the fact that Sweden has twice the population of Norway (43,000 tons) and Denmark (40,000–46,000 tons), the data seems very comparable (Stenmarck et al., 2011).

Some *combined value-chain* data on food waste from households, retailers and wholesalers, presented in kilograms of food waste per capita, shows that Denmark has the highest volume of food waste per citizen (above the EU average) and a very large share of food waste in groceries, while Norway and Sweden are more comparable, and are slightly under the EU food waste average (Figure 13.2) (Stensgård & Hanssen, 2018). However, the data is from different years, and the Norwegian data is most up-to-date, especially compared to EU data, making a direct comparison difficult. This will be further discussed in the comparative analysis. In section 13.4, we present the

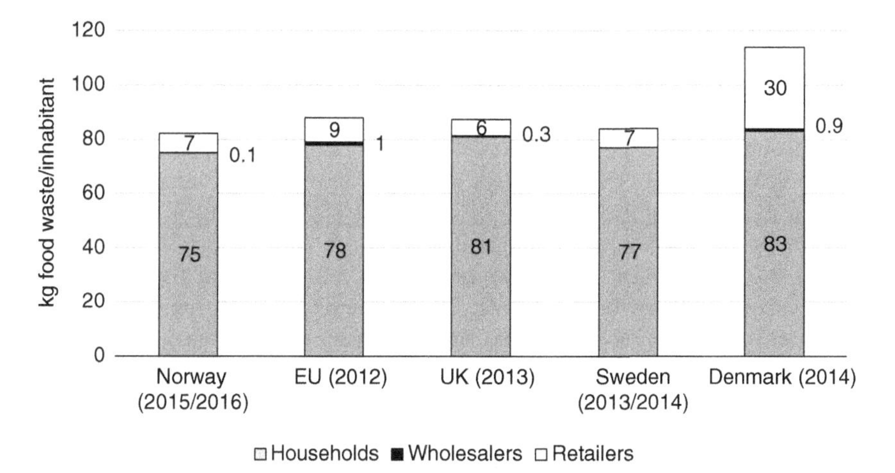

Figure 13.2 Combined value chain data on food waste in kg per inhabitant, including inedible parts, excluding primary production and waste that is used for animal feed (adapted from Stensgård & Hanssen, 2018).

most recent numbers for each country. Because of the differences in years and changing food waste definitions and units of measurement, the numbers can differ slightly from the comparative synthesis in 13.3.3.

13.4 Analysis

13.4.1 Food waste governance in Norway

Food waste has recently been identified as a very significant and pressing problem in Norway. The ForMat project (2010–2015), a collaborative endeavour involving the food industry, research organisations and state agencies to prevent food waste, estimated that food was wasted at all four stages of the value chain (industry, wholesale, retail and households) and amounted to 355,000 tons in 2015, worth about NOK 20 billion (Stensgård & Hanssen, 2016). Food wasted annually in Norway could potentially feed 900,000 people, and creates emissions equal to those of 375,000 cars (Lindahl, 2016), totalling 978,000 tons of CO_2 (Boffey, 2017). The distribution of food waste in the value chain is very uneven, with 61% generated by households, 21% by producers, 1% by wholesalers and 17% by retailers (Stensgård & Hanssen, 2016).

The ForMat project reshaped the food waste debate in Norway with extensive surveys of food waste in the food chain. Matvett, a non-profit multi-stakeholder hub and the administrator of ForMat, worked with the business sector, five ministries and Østfold Research (Østfoldforskning) to put food waste on the agenda in Norway. This was a very important first step as no statistics on food waste existed in Norway prior to ForMat. Although the target of 25% reduction in food waste by the end of the project was not reached (food waste levels in the project period fell by 12% (Stensgård & Hanssen, 2016)), the ForMat project drew international interest, due to its openness, methodologies and strong public–private collaboration. The data collection was unique in the entire European context, with documented measuring methods, and details on food waste quantities in the different stages of the value chain.

The work of the ForMat network led the industry to prepare an *Agreement of Intent to reduce food waste*, signed on 7 May 2015 (Regjeringen, 2015). The Agreement was further developed and finalised on 23 June 2017 in the form of an *Industry Agreement on the reduction of food waste* between five ministries and a dozen industry organisations. The agreement had a main reduction target of 50% by 2030, which was further subdivided into two sub-targets: 15% by 2020 and 30% by 2025 (Regjeringen, 2017). In parallel, the narrower Cut Food Waste 2020 (*KuttMatsvinn2020*) is another initiative and result of the ForMat collaboration, targeting the hospitality sector. It aims to cut food waste by 20% by 2020 (NorgesGruppen, 2016).

Alongside the Industry Agreement, a parallel law-making and governance design process with the aim of reducing food waste took place in the Norwegian Parliament. In 2016, a parliamentary committee asked the government to evaluate the introduction of a potential food waste law (similar laws had recently been introduced in France (Mourad, 2015) and Italy), meant to strengthen the bottom-up voluntary process with binding state regulation. The final evaluation issued in September 2017 by the Ministry of Agriculture and Food concluded that the Industry Agreement was a sufficient first step in emerging Norwegian food waste governance, putting the binding public regulation on hold. Because of the strong food waste reduction commitment of the opposition, a food waste law is still on the political agenda. The Parliament backed by the Labour Party (Ap), the agrarian Centre Party (Sp), the Socialist Left (SV), the Christian Democrats (KrF) and the Greens (MDG) asked the government to draft a proposal for a food waste law with emphasis on the food industry (NTB, 2018). The draft was expected at the end of 2018.

One Norwegian NGO was particularly active in food waste reduction campaigns and support for the state regulations: Future in our Hands (*Framtiden i våre hender*) wrote many press articles regarding food waste and delivered a petition to the Norwegian Parliament. It also recently started to work with municipalities and consumers through its 'MatVinn' project (Riise Jenssen, 2017).

A number of bottom-up initiatives appeared, e.g. a mobile app TooGoodToGo brought to Norway from Denmark. Online shopping websites and traditional shops started selling food that would be discarded in traditional supermarkets due to nearing best-before dates (e.g. the Best Før supermarket, as well as Holdbart stores and online shopping, etc.). Another interesting initiative is a student restaurant concept offering gourmet lunches made from food that is approaching its expiration date. Since its start on the Blindern campus at the University of Oslo in 2015, it has further expanded to the Oslo Metropolitan University (in 2017).

Summing up, it can be said that in Norway the main driver of food waste governance is the industry, collaborating with civil society actors and public administration.

Box 13.1 Milestones in food waste governance in Norway

- **2010:** Launching of ForMat, a business-driven project involving the collaboration of supermarkets, food industry organisations and public authorities (industry driven)
- **Spring 2012:** The NGO 'Future in our Hands' launched a series of food waste reduction articles, events and campaigns (civil society driven)
- **August 2013:** Norway launched a national waste management strategy 'From Waste to Resources' (public agency driven)
- **Autumn 2013:** First food bank in Oslo emerges, orchestrated by the Church Mission, the Blue Cross and the Salvation Army. The authorities, industry (NorgesGruppen) and the ForMat project all supported the idea (civil society driven)

- **Spring 2015:** First food waste restaurant (KUTT Gourmet) at the University of Oslo (civil society driven)
- **7 May 2015:** The 'Agreement of Intent to reduce food waste', signed by the food industry and authorities (industry driven)
- **4 January 2017:** 'KuttMatsvinn2020' cut food waste by 20% by 2020 (industry driven)
- **23 June 2017:** Finalisation of 'Industry Agreement on reduction of food waste' (industry driven)
- **22 September 2017:** The evaluation of the food waste law was negative because of the *Industry Agreement* (public agency driven)

13.4.2 Food waste governance in Sweden

Food waste in Sweden is defined as 'avoidable food waste – food that is discarded but could have been eaten had it been handled properly' (Naturvårdsverket, 2013, p. 29). Examples of avoidable food waste are bread, food leftovers, fruit and vegetables. Avoidable food waste was sometimes referred to as *food wastage*. In 2014, 1,278,000 tons (equivalent to 134 kg/person[3]) of food waste was generated in Sweden (see Table 13.1 in Elander, Sternmarck & Ostergren (2016)). The total volume of food waste was calculated along the food value chain from primary production, food producers, distributors, retailers and consumers.

In the last few years, Sweden has paid increasing attention to the food waste issue. The EU Directive (2008/98/EC) on waste (the Waste Framework Directive) required European Union member states to establish waste prevention programmes. As of June 2018, Sweden has not yet implemented a specific strategy or national plan for food waste management and reduction. Nevertheless, it embedded proposed milestone targets for food waste in several national plans such as the Swedish Waste Management Plan 2012–2017 and the Swedish Waste Prevention Programme 2014–2017 (Elander et al., 2016). The first had important provisions on energy and nutrient recovery from food waste as it assumed that by 2018, 50% of food waste from consumption would be collected separately and treated biologically (Naturvårdsverket, 2012). The other assumed a waste reduction by 2020, at least 20% compared to 2010, throughout the entire food value chain (except for primary production) (Naturvårdsverket, 2014). A separate action plan has been developed for food wastage reduction in primary production. The Swedish Environmental Protection Agency (EPA) was commissioned by Sweden's government to create 'interim objectives for decreasing avoidable food waste, and to suggest measures for reaching the target' (Naturvårdsverket, 2013, p. 31).

Food waste reduction in Sweden was also targeted together with CO_2 emission reductions at the municipal level, for example through the Klimatsmart campaign, which aims to reduce food waste in school canteens (Klimatsmart, 2018).

An important milestone in Swedish food waste reduction governance was the creation of SaMMa (Swedish Collaboration Group for Reduced Food Waste) in 2012. The Swedish EPA initiated SaMMa as a liaison group for food waste prevention, information exchange and to assist in reducing food waste. It includes the National Food Agency (*Livsmedelsverket*) and the Stockholm Consumer Cooperative Society (*Konsumentföreningen Stockholm*). It is coordinated by the Swedish EPA, Swedish Board of Agriculture and Sweden's National Food Agency, all taking turns to host meetings, which are held several times a year. The network was designed to help reduce food waste in Sweden by promoting collaboration throughout the food supply chain. By collaborating with political representatives, researchers, authorities, organisations and businesses can discuss issues and share experiences and knowledge (Naturvårdsverket, 2013).

In 2013 a public campaign called 'Stop food waste' (*Stoppa matsvinnet*) was launched by Sweden's National Food Agency, the Swedish EPA and the Swedish Board of Agriculture to inform and inspire people about how to reduce food waste (StoppaMatsvinnet, 2018).

Governmental bodies such as Sweden's National Food Agency, the Swedish EPA and the Swedish Board of Agriculture played a key role in food waste governance in Sweden. These public agencies set up targets, launched programmes and initiated different campaigns and initiatives to fight food waste.

Box 13.2 Milestones in food waste governance in Sweden

- **2009:** The Campaign 'Climate Smart' was first launched by Halmstad municipality, aiming to decrease food waste in school canteens to reduce carbon footprint (public agency driven)
- **2012:** A target to collect 50% of food waste from consumption separately and treat it biologically by 2018 was included in the Swedish Waste Management Plan 2012–2017 of the Swedish Environmental Protection Agency (public agency driven)
- **2012:** SaMMa Swedish Collaboration Group for Reduced Food Waste was founded as a broad network involving research bodies, businesses and civil society (public agency driven)
- **2013:** The public campaign 'Stop food waste' (*Stoppa matsvinnet*) was launched (public agency driven)
- **2013:** A target was set for reducing food waste along the entire value chain by 20% in 2020 compared to 2010 in the Swedish Waste Prevention Programme 2014–2017 developed by the Swedish EPA, Swedish Board of Agriculture and the National Food Agency of Sweden (public agency driven)
- **2013–2015:** The campaign Stop Food Waste Now – an initiative to inform and inspire people to reduce food waste – was launched by the National Food Agency, the Swedish EPA and the Board of Agriculture (public agency driven)

13.4.3 Food waste governance in Denmark

According to the Danish Ministry of Environment, Danes throw away 700,000 tons of edible food annually (Gadd, 2017). Danish retailers create 45,676 tons of food waste per year (Halloran et al., 2014), while households are responsible for about half of the total amount (Kjær & Werge, 2010).

Actions aiming to reduce food waste in Denmark started largely as a consumer movement, led by the influential activist Selina Juul and the movement 'Stop Wasting Food' (*Stop Spild af Mad*), established in 2008 (Stop Spild Af Mad, 2018). It is considered the largest European non-profit consumer movement against food waste, and is strongly supported by Danish citizens, as well as Members of the European Parliament, Members of the Danish Parliament, top Danish chefs and food personalities (Halloran et al., 2014).

Danish food waste figures were higher than the EU average (Figure 13.2) but have fallen sharply in the last decade. The trade magazine *Dansk Handelsblad* reported that Denmark is the leading country in the EU regarding the number of initiatives to limit food waste (FUSIONS, 2015). NGOs, retailers, the hospitality sector, authorities, industry and consumers are all participating in various actions to cut food waste (FUSIONS, 2016). In response to this development, Denmark's Prime Minister put the food waste problem on the political agenda. In 2011 the Ministry of the Environment established a public–private partnership called 'Initiative Group Against Food Waste' to exchange information and work together on food waste reduction. In the same year, a 'Charter on Less Food Waste' was signed by different ministries and industry representatives. This can be seen as an example of industry self-regulation through a voluntary agreement on food waste. In 2013 the Danish Consumer Council initiated a consumer awareness campaign on food waste (Halloran et al., 2014).

The civil movement's success was based on its wide societal resonance, partnerships and cooperation with both the authorities and the private sector. It also expanded its links with academia, with Selina Juul initiating the first think tank in the world dedicated to food waste prevention. Many public and industry initiatives were developed together with NGOs (ThinkEatSave, 2015). In recent years, innovative applications have been developed in Denmark to fight food waste (including YourLocal, ReFood, Green Menu Planner, etc.). A Danish humanitarian NGO opened Wefood, the first food waste supermarket.

Between 2010 and 2015, Denmark's food waste was reduced by 25% (translating to an economic saving of DKK 4.4 billion). This was considered a European record, followed only by the United Kingdom's reduction of 21% between the years 2008 and 2013 (FUSIONS, 2016).

Box 13.3 Milestones in food waste governance in Denmark

- **2008:** Activist Selina Juul established the leading organisation 'Stop Wasting Food', the largest such movement in Europe, with the aim of raising awareness about food waste (civil society driven)
- **2008:** Major food store chain REMA 1000 stopped 'buy 2 get 3' sales and only sold by the piece to reduce consumer food waste (industry driven)
- **2011:** The Ministry of the Environment established a voluntary 'Initiative Group Against Food Waste' with stakeholders representing the public and private sectors (public agency driven)
- **2011:** The 'Charter on Less Food Waste' created and signed by nineteen major stakeholders such as various supermarket chains, restaurants, ministries and hotel chains (cross sectoral)
- **2011:** The Ministry of the Environment established the 'Initiative Group Against Food Waste' (public agency driven)
- **2013:** Ministry of Environment and Food introduced a strategy on growth and resource-efficiency on food and a fund to support food waste reduction actions (public agency driven)
- **2013:** The Zero Waste initiative was launched by municipalities with the aim of raising public awareness and making citizens the pivot of anti-waste initiatives (public and civil society driven)
- **2013:** The Danish Consumer Council initiated a campaign to increase awareness of food waste (civil society driven)
- **2014:** Launch of the Danish Government Resource Plan for Waste Management 2013–2018 (public agency driven)
- **2015:** YourLocal was the first mobile app to help small businesses and supermarkets fight food waste (civil society driven)
- **2016:** The ReFood initiative, in partnership with Stop Wasting Food, united 800 restaurants, cafes, food producers and institutions. Partners could use the ReFood label to show their commitment to reducing food waste (civil society driven)
- **2016:** The world's first food waste think tank – ThinkEatSave – was established to gather knowledge and develop action plans to combat food waste (civil society driven)

13.4.4 Comparative analysis of governance pathways

The country-level analysis shows very different governance pathways that can be identified in the three countries with regard to food waste reduction. Although a variety of stakeholders take action to reduce food waste in each country, certain leaders can be highlighted. We identify them as drivers for food waste action, establishing powerful collaborations between different actors, levels and sectors in all three countries. We notice a 'spill-over' effect of sorts: where food waste reduction initiatives and the leading actors should be given credit for shaping the agenda, pushing for action and proposing innovative solutions, but not for taking unilateral action.

Governance differences may also be linked to citizens' expectations. The Eurobarometer report on food waste and date marking (unfortunately only available for Sweden and Denmark, not conducted in Norway) points to different actors that should take responsibility for preventing food waste. In Denmark and Sweden, consumers are indicated as being responsible (85% and 88% respectively), followed by shops and retailers (68% and 75%), the hospitality sector (59% and 66%), food manufacturers (51% and 58%), public authorities (44% and 58%) and farmers (27% and 28%) (EUROBAROMETER, 2015). This shows that the biggest difference is regarding the role of public authorities, where significantly more Swedes expect a state intervention compared to Danes. Interestingly, among the twenty-eight analysed countries in this report, the highest responsibility of shops and retailers is perceived in France (90%) and the highest responsibility of public authorities in Spain and France (77% and 74% respectively). Both these factors can explain France's lead in European food waste legal regulation (EUROBAROMETER, 2015).

13.4.5 Comparative analysis from a multi-level governance perspective

It is worth noting that food waste governance in Sweden is closer to the traditional centralised forms of steering, with the Swedish EPA taking a clear lead in tackling the food waste problem. It is therefore in line with the type I multi-level governance characteristics presented in Table 13.1. Open collaboration with industry and civic society partners also signals that the state is trying to coordinate all efforts to reduce food waste.

While in Sweden, initiatives were worked out 'mostly in the public sector',[4] in the Norwegian case, initiatives to prevent food waste 'always came from the industry'. In many countries, food waste regulations started with e.g. food banks; in Norway 'it started with the industry'.[5] Voluntary self-regulation occurred in the shadow of hierarchy (Héritier & Eckert, 2008; Newman & Bach, 2004) and the role of the traditional state authorities was limited in the early agenda setting and food waste policy formulation phase. It was, however, understood that if the industry tackled the problem 'before the politicians', it would 'bring cheaper and more effective solutions'.[6] It is nevertheless clear that public agencies are paying close attention to the tangible effects of self-regulation and a state-led food waste law proposal might centralise food waste governance in Norway in the future. Therefore, it can be stated that Norway initiated food waste governance with a type II multi-level governance framework with the industry taking the lead. The industry agreements from 2015 and 2017 were very task specific, but with quite open memberships and not very clearly defined accountability of the actors in the common goals. We also observe that the Norwegian ministries, having an observer status in the agreement of intent, and becoming full members in the final agreement, brought public type I legitimacy to business-driven type II activities. It can also be seen that the industry's lead in food waste governance created a counter proposal of a type I state-led hierarchical response in the form of a food waste law proposal debated by the Parliament. The Norwegian case illustrates that there might be significant learning processes between type I and type II activities; the food waste law suggestions were built on solid experience from the industry collaboration and could even incorporate this learning process.

The Danish case can also be seen as an illustration of type II multi-level governance of fluid, problem-focused networks but with the civic sector taking the lead. Here even more type I legitimacy was given to those type II activities as state actors began participating in various initiatives as observers, funders and project partners. This might be caused by a very strong collaboration between the leading NGO and various public entities (municipalities, ministries and legislative bodies). The high effectiveness of the food waste reduction effort in Denmark (although Denmark could initially be seen as a laggard in the comparative analysis; see Figure 13.2) also increased the output of legitimacy to this particular governance arrangement. Another key aspect might be the strong ownership displayed towards this arrangement by Danish civil society, which seems to be a subject rather than an object of food waste reduction action. As consumers are responsible for most food waste, civic engagement might also be crucial for reaching more ambitious goals after the 'lowest-hanging fruits' have been collected.

The multi-level governance analysis shows that the three countries can be put on an axis with Sweden on one end, with a state regulations and type I arrangement, Norway stretched in between with some competition between type I and type II, and Denmark being the most liberal, with strong civic activism and fluid, public partnerships governance and a strong type II legitimacy.

It is important to note that the state-centric approach is not enough to see the full food waste governance framework. All analysed countries are strongly influenced by the UN Sustainable Development Goals on food waste and their incorporation into EU regulations. The goals are the same, but Denmark, Sweden and Norway illustrate that different governance pathways are possible. There is a great deal of information exchange between all the Scandinavian and other European countries (through the Nordic Council of Ministers in Scandinavia and through the FUSIONS project in Europe). Norway looks to Sweden and Finland for policy inspiration and lessons on food waste initiatives in the hospitality sector,[7] while Danish innovative mobile phone applications and food waste supermarkets are replicated in many European countries. Similarly, the Norwegian industry's self-regulation of food waste should also receive more scholarly and policy attention.

13.5 Conclusions and policy implications

In this chapter, we have presented the food waste governance efforts in Norway, Denmark and Sweden. We illustrate that three experimentalist governance approaches emerged in the analysed countries: a civil society-driven framework in Denmark, a public policy-driven one in Sweden and finally an industry-driven arrangement in Norway. This shows the diverging origins of food waste regulations in the three countries, as all three sectors are in different ways involved in food waste reduction.

Learning and collaboration between the three sectors (public policy, industry and civil society initiatives), at various levels and between countries, is an important and necessary component for tackling the food waste challenge. This means a departure from the analytical mode of multi-level governance (with type I and type II arrangements) to the normative mode. Multi-level governance should also be seen as a solution for better policy implementation, public participation and involvement in decision-making, as a strategy or benchmark for good governance and effective policy for tackling food waste. Multi-stakeholder collaboration is particularly important in addressing complex social, economic and environmental problems that span across the value chains and particular ministerial competences. Our three national cases show how such collaboration might look in practice.

Denmark, with its well-coordinated mass civil society engagement to reduce food waste, has been very successful in recent years: its reductions were the largest and exceeded the pre-set targets, but it also initially had higher food waste than the EU average. Now in all countries the 'low-hanging fruits' have been collected, meaning the next targets might be more difficult to meet. There is no one-size-fits-all solution to tackle the global food waste problem, but possibilities for learning between actors and countries, to overcome weaknesses and enhance their strengths, are very important for the next stages of food waste reduction. Further studies should concentrate on best practices and lessons that could be drawn from industry self-regulation to reduce food waste in Norway, from the effectiveness of public food waste policies in Sweden, and from the successful engagement of civic society in the case of Denmark.

Notes

1 In Norway industry actors report food waste data but only aggregated numbers are made publicly available. In Denmark in 2016 Netto was the first supermarket chain to disclose the amount of food waste it produced. See also: Stenmarck et al., 2011.
2 FUSIONS was an EU-funded project with twenty-one partners in thirteen countries, running from 2012 to 2016, in which all the analysed countries actively participated. The project's aim was to reduce food waste, and to harmonise food waste monitoring in Europe. It provided the first detailed national statistics covering the entire value chain, accompanied by suggested methodology for data collection and analysis.
3 Lower number in Figure 13.2 in the comparative analysis does not include food waste in primary production.
4 Expert interview with a food waste research expert, 12 March 2018.
5 Expert interview with a food industry representative, 13 February 2018.
6 Expert interview with a food industry representative, 15 March 2018.
7 Expert interview with a food waste research expert, 12 March 2018.

References

Boffey, D. (2017). How Norway is selling out-of-date food to help tackle waste. Retrieved from www.theguardian.com/environment/2017/aug/17/how-norway-is-selling-out-of-date-food-to-help-tackle-waste (accessed 4 February 2018).

Bugge, M. M., Dybdahl, L. M. & Szulecka, J. (2018). *Recycling or preventon of waste? Addressing the food waste challenge in the right end. Policy Brief.* Oslo: NIFU.

Elander, M., Sternmarck, A. & Ostergren, K. (2016). *Sweden: Country report on national food waste policy.* Retrieved from www.eu-fusions.org/phocadownload/country-report/SWEDEN%2023.02.16.pdf.

EUROBAROMETER. (2015). Food waste and date marking. Retrieved from http://data.europa.eu/euodp/en/data/dataset/S2095_425_ENG.

EUROSTAT. (2012a). Agricultural census in Denmark. Retrieved from http://ec.europa.eu/eurostat/statistics-explained/index.php?title=Agricultural_census_in_Denmark.

EUROSTAT. (2012b). Agricultural census in Norway. Retrieved from http://ec.europa.eu/eurostat/statistics-explained/index.php/Agricultural_census_in_Norway.

EUROSTAT. (2012c). Agricultural census in Sweden. Retrieved from http://ec.europa.eu/eurostat/statistics-explained/index.php?title=Agricultural_census_in_Sweden.

FAO. (2011). Global food losses and food waste: Extent, causes and prevention. Retrieved from www.fao.org/docrep/014/mb060e/mb060e00.pdf.

FAO. (2014). *Food wastage footprint: Full-cost accounting.* Final report. Rome: FAO.

FAO. (2015). Global initiative on food loss and waste reduction. Retrieved from www.fao.org/3/a-i4068e.pdf.

FUSIONS. (2014). FUSIONS definitional framework for food waste. Full report. Retrieved from www.eu-fusions.org/phocadownload/Publications/FUSIONS%20Definitional%20Framework%20for%20Food%20Waste%202014.pdf.

FUSIONS. (2015). Food waste in Denmark reduced by 25% and 4,4 billion DKK [Press release]. Retrieved from www.eu-fusions.org/index.php/14-news/238-food-waste-in-denmark-reduced-by-25-and-4-4-billion-dkk.

FUSIONS. (2016). Food waste in Denmark reduced by 25% and 4,4 billion DKK. Retrieved from www.eu-fusions.org/index.php/about-fusions/news-archives/238-food-waste-in-denmark-reduced-by-25-and-4-4-billion-dkk.

Gadd, S. (2017). Netto weighs in to cut food waste. *CPH Post Online.*

Gollata, J. A. M., & Newig, J. (2017). Policy implementation through multi-level governance: Analysing practical implementation of EU air quality directives in Germany. *Journal of European Public Policy*, *24*(9), 1308–1327. doi:10.1080/13501763.2017.1314539.

Halloran, A., Clement, J., Kornum, N., Bucatariu, C. & Magid, J. (2014). Addressing food waste reduction in Denmark. *Food Policy*, *49*, 294–301. doi:10.1016/j.foodpol.2014.09.005.

Hartikainen, H., Mogensen, L., Svanes, E. & Franke, U. (2018). Food waste quantification in primary production: The Nordic countries as a case study. *Waste Management*, *71*, 502–511. doi:10.1016/j.wasman.2017.10.026.

Héritier, A., & Eckert, S. (2008). New modes of governance in the shadow of hierarchy: Self-regulation by industry in Europe. *Journal of Public Policy*, *28*(1), 113–138. doi:10.1017/S0143814X08000809.

Hooghe, L., & Marks, G. (1996). 'Europe with the regions': Channels of regional representation in the European Union. *Publius*, *26*(1), 73–90.

Hooghe, L., & Marks, G. (2010). Types of multilevel governance. In H. Enderlein, S. Wälti & M. Zürn (Eds.), *Handbook of multilevel governance* (pp. 17–31). Cheltenham: Edward Elgar.

Kjær, B., & Werge, M. (2010). Forundersøgelse af madspild i Danmark. Retrieved from www2.mst.dk/udgiv/publikationer/2010/978-87-92617-88-0/pdf/978-87-92617-89-7.pdf.

Klimatsmart. (2018). Retrieved from www.klimatsmart.se.

Klitkou, A., & Iversen, E. (2016). *Inventory of organic waste resources*. Oslo: SusValueWaste.

Lindahl, H. (2016). Derfor trenger Norge en matkastelov [Press release]. Retrieved from www.framtiden.no/201608317029/aktuelt/mat/derfor-trenger-norge-en-matkastelov.html (accessed 4 February 2018).

Marks, G., & Hooghe, L. (2004). Contrasting visions of multi-level governance. In I. Bache & M. Flinders (Eds.), *Multi-level governance* (pp. 15–30). Oxford: Oxford University Press.

Mourad, M. (2015). France moves toward a national policy against food waste. Retrieved from www.nrdc.org/sites/default/files/france-food-waste-policy-report.pdf.

Naturvårdsverket. (2012). Från avfallshantering till resurshushållning. Sveriges avfallsplan 2012–2017. Retrieved from www.naturvardsverket.se/Documents/publikationer 6400/978-91-620-6502-7.pdf.

Naturvårdsverket. (2013). Together we will gain from a nontoxic, resource efficient society: The Swedish Waste Prevention Programme for 2014 to 2017. Retrieved from http://naturvardsverket.se/upload/miljoarbete-i-samhallet/miljoarbete-i-sverige/avfall/avfallsforebyggande-programmet/Together-gain-rom-non-toxic-resource-efficient-society-2017-05-22.pdf.

Naturvårdsverket. (2014). Förslag till etappmål för minskad mängd matavfall. Redovisning av ett regeringsuppdrag den 16 december 2013. Retrieved from www.naturvardsverket.se/upload/miljoarbete-i-samhallet/miljoarbete-i-sverige/regeringsuppdrag/2013/etappmal2013forslag/matavfallsrapport-reviderad.pdf.

Newig, J., & Fritsch, O. (2009). Environmental governance: Participatory, multi-level – and effective? *Environmental Policy and Governance*, *19*(3), 197–214. doi: 10.1002/eet.509.

Newman, A., & Bach, D. (2004). Self-regulatory trajectories in the shadow of public power: Resolving digital dilemmas in Europe and the United States. *Governance*, *17*(3), 387–413.

NorgesGruppen. (2016). Bærekraftsrapport. Retrieved from www.norgesgruppen.no/globalassets/finansiell-informasjon/rapportering/ng_barekraftsrapport_2016_ok.pdf.

NTB. (2018). Flertallet på Stortinget for matkastelov. *Aftenposten*. Retrieved from www.aftenposten.no/norge/i/J1l4V4/Flertallet-pa-Stortinget-for-matkastelov.

Papargyropoulou, E., Lozano, R., K. Steinberger, J., Wright, N. & Ujang, Z. b. (2014). The food waste hierarchy as a framework for the management of food surplus and food waste. *Journal of Cleaner Production*, *76*, 106–115. doi:10.1016/j.jclepro.2014.04.020.

Regjeringen. (2015). Intensjonsavtale om reduksjon i matsvinn. Retrieved from www.regjeringen.no/contentassets/e54f030bda3f488d8a295cd0078c4fcb/matsvinn.pdf.

Regjeringen. (2017). Industry agreement on reduction of food waste. Retrieved from www.regjeringen.no/contentassets/1c911e254aa0470692bc311789a8f1cd/industry-agreement-on-reduction-of-food-waste_norway.pdf.

Riise Jenssen, E. (2017). Går sammen med kommunene for å kutte matkasting [Press release]. Retrieved from www.framtiden.no/201702107103/aktuelt/mat/gar-sammen-med-kommunene-for-a-kutte-matkasting.html.

SaveFood. (2018). Initiative SaveFood: Solutions for a world aware of its resources. Retrieved from www.save-food.org.

SDG. (2015). 17 Sustainable Development Goals (SDGs) of the 2030 Agenda for Sustainable Development. Retrieved from https://sustainabledevelopment.un.org/?menu=1300.

Smith, A. (2007). Emerging in between: The multi-level governance of renewable energy in the English regions. *Energy Policy*, *35*(12), 6266–6280. doi:10.1016/j.enpol.2007.07.023.

Stenmarck, Å., Hanssen, O. J., Silvennoinen, K., Katajajuuri, J. & Werge, M. (2011). *Initiatives on prevention of food waste in the retail and wholesale trades*. Retrieved from Copenhagen.

Stensgård, A. E., & Hanssen, O. J. (2016). Food waste in Norway 2010–2015: Final report from the ForMat Project. Retrieved from https://ec.europa.eu/food/sites/food/files/safety/docs/fw_lib_format-rapport-2016-eng.pdf.

Stensgård, A. E., & Hanssen, O. J. (2018). Matsvinn i Norge. Rapportering av nøkkeltall 2016. Retrieved from www.ostfoldforskning.no/media/1788/or0618-matsvinn-i-norge-rapportering-av-noekkeltall-2016.pdf.

Stephenson, P. (2013). Twenty years of multi-level governance: 'Where does it come from? What is it? Where is it going?'. *Journal of European Public Policy*, *20*(6), 817–837.

Stop Spild Af Mad. (2018). Stop wasting food. Retrieved from www.stopspildafmad.dk.

StoppaMatsvinnet. (2018). Retrieved from http://stoppamatsvinnet.nu.

Swedish EPA. (2013). Food waste volumes in Sweden. Retrieved from www.naturvardsverket.se/Documents/publikationer6400/978-91-620-8695-4.pdf.

ThinkEatSave. (2015). The world's first think tank on food waste is Danish [Press release]. Retrieved from www.thinkeatsave.org/index.php/multimedia/in-the-news/79-in-the-news/359-the-world-s-first-think-tank-on-food-waste-is-danish.

Van Kersbergen, K., & Van Waarden, F. (2004). 'Governance' as a bridge between disciplines: Cross-disciplinary inspiration regarding shifts in governance and problems of governability, accountability and legitimacy. *European Journal of Political Research*, *43*(2), 143–171. doi:10.1111/j.1475-6765.2004.00149.x.

14 Life cycle assessment

A governance tool for transition towards a circular bioeconomy?

Andreas Brekke, Kari-Anne Lyng, Johanna Olofsson and Julia Szulecka

14.1 Introduction

There is broad agreement that human activities should be sustainable. Although most activities have an impact on the environment, no human activity should restrict the possibility for other people to meet their needs. Sustainability is easy to agree upon on a general and abstract level, but it is harder to judge what it means in practice. When is an activity sustainable and when is it unsustainable? Which products are sustainable and which are not? While those are not easily answerable questions, policymakers must have a measure allowing them to grade sustainability, helping them develop long-term strategies, current regulation or appropriate policy incentives. Similarly, companies need to know that the manufacturing of their products does not harm people or the environment unnecessarily and that they can sustainably create value for the company, its shareholders and the society at large.

Life cycle assessment (LCA) is often referred to as a method or tool to answer questions about what is more or less sustainable, at least in relation to environmental sustainability (Finnveden et al., 2009). The International Organization for Standardization (ISO) defined LCA as "compilation and evaluation of the inputs, outputs and the potential environmental impacts of a product system throughout its life cycle" (ISO, 2006). Recent years have seen a development in both the breadth and depth of LCA, where dimensions of sustainability other than the direct impact on natural environment have been included and where impacts are modelled more specifically.

Already by the beginning of the 1970s, the Environmental Protection Agency in the USA was considering the use of LCA for all products as part of public policy (Reed, 2012). Following the United Nations Conference on the Human Environment held in Stockholm in 1972 and the 1973 oil crisis – both drawing attention to finite resources, LCA – or Resource and Environmental Profile Analysis as it was called – was seen as a way to understand different products' impact on the environment. For a number of reasons, for instance the "cancellation" of "peak oil", the interest in LCA declined during the second half of the 1970s and the method was deemed impractical as a regulatory tool (ibid.). At the end of the 1980s, a joint initiative by the

Society of Environmental Toxicology and Chemistry and the United Nations Environmental Programme led to the revival of LCA. The following decades have witnessed a large and growing inclusion of LCA in the governance and the testing of Product Environmental Footprint (PEF) in the European Union (EU). A possible transformation to a circular economy might call for even greater use of LCA.

This chapter explores how LCA is currently employed in public policy and how its use can guide the governance of a transition towards a circular bioeconomy. In order to distinguish between different uses of LCA, the chapter presents LCA along three dimensions: (1) as results; (2) as a method; and (3) as a mindset. Existing legislation is used to exemplify each of the dimensions. The focus is to demonstrate the advantages and pitfalls of using LCA in governance and the consequences of choices of results, method and mindset. Throughout the chapter, most information is related to the use of LCA for assessment of environmental performance (also referred to as environmental LCA), and it will be emphasised whenever LCA as a term is applied in connection to other aspects of sustainability.

14.2 Life cycle assessment as results

Governance processes need access to information about the topics to be governed. LCA promises to bring results to include in governance processes, all the way from large processes of societal change to small processes such as the procurement of single items. This section describes the various results from LCAs and how they are used in governance.

14.2.1 Being good from just performing life cycle assessment

Historically, there are examples where performing an LCA has been a result in itself. Environmental Product Declarations (EPDs) are based on an LCA of a material or product and should give purchasers information about the environmental performance of products. There has been a large increase in the number of EPDs available, especially relating to the built environment. The Building Research Establishment Environmental Assessment Method (BREEAM) certification scheme awards points to buildings which have collected EPDs for important materials or components (Norwegian Green Building Council, 2012). Similarly, public and private purchasers have used EPDs as a binary environmental criterion in projects. If an EPD is made, the manufacturer of a product can tick the appropriate box and be qualified for a tender process.

A result from an LCA is, however, not proof of something being environmentally friendly. This is of course because an assessment of a product's environmental profile alone says nothing about how it compares to the environmental profiles of similar products or the usability of the product. This might seem obvious, but there is still a reason to stress this point. Since many eco-labels, such as the Green Dot or the Nordic Swan, are awarded to environmentally preferable products or services, EPDs are easily confused as proof of a product being greener. The main merits of LCA and EPD results are not in the simple, binary answer that something is good or bad, but rather the ability to compare different products, services or product systems fulfilling the same function. Whereas an environmental label can guide decision-makers in the choice of environmentally friendlier products, it does not quantify the product's environmental impact. If one wants to compare functionally equivalent products or services by, for instance, calculating the total emissions from construction projects, LCAs and EPDs are needed.

14.2.2 Product benchmarking

An EPD contains results for a defined set of environmental impacts. Presently, the BREEAM scheme has included extra points if the EPDs that are actually used to choose among alternatives and tenders also start to include numerical benchmarks based on LCAs. The construction sector has been an early mover in applying environmental information in development and procurement processes (Frischknecht, Wyss, Knöpfel & Stolz, 2015), but the use of LCA for governance purposes is not confined to this sector. For instance, the Renewable Energy Directive states that biofuels' greenhouse gas emissions are to be compared to fossil fuels' emissions of 83.8 grams of CO_{2e} per MJ throughout their life cycle – from extraction of raw materials to combustion when used as a fuel (EU, 2009), or, if revised sustainability criteria are agreed upon, 94 grams of CO_{2e} in 2021 (European Commission, 2017). Any biofuel producer or importer who wants to certify their fuel for a market in Europe must relate to these numbers, at least if they want their biofuel to be part of the targeted use of biofuel.

The aforementioned Product Environmental Footprints (PEFs) (European Commission, 2013) take the comparison of product life cycles one step further. The European Union has initiated several pilot projects on PEFs for different product categories with a methodology based on LCA. Every product should have a label with information on the environmental impacts related to its manufacturing, much like a label on nutritional facts, to give consumers the possibility to demand environmentally friendlier products.

Environmental Product Declarations, the sustainability criteria in the Renewable Energy Directive, and Product Environmental Footprints, are all examples of tools for informing on environmental performance for different product categories. There are also many examples of companies and governments commissioning LCAs for specific areas to make environmental decisions. The first LCA was made for comparing different beverage containers for Coca Cola, and LCA has helped decisions on packaging alternatives for both companies and governments ever since. For instance, this includes the Danish Environmental Protection Agency commissioning an LCA on packaging systems for beer and soft drinks in 1995 to support regulations on beverage containers.

Probably more than any other issue, LCA has been guiding governance on waste handling, also relating to organic material. Waste handling companies have used LCA results to aid decisions on which treatment methods to choose, where to place facilities or which vehicles to purchase for the transport fleet. In Norway, national authorities have used results from LCAs to support policies on source separation and recycling of organic waste (Raadal, Stensgård, Lyng & Hanssen, 2016; Syversen et al., 2018), and regulations on food waste or the proposed revision of Norwegian Fertiliser Ordinance which refers to two LCA studies (Hanserud, Lyng, Vries, Øgaard & Brattebø, 2017; Modahl et al., 2016). Local authorities have used results from LCA to support overall strategies for waste handling as well as decisions on the construction of biogas facilities (Councilor of Vestfold, 2012).

There are several easy-to-use models for waste management based on LCA, such as Orware and EaseTech (Gentil et al., 2010). These models are based on the decision-maker providing numbers for a few important parameters, such as transport distances, waste composition and amounts of waste, and getting results for various environmental impacts in return.

14.2.3 Detailed insights from life cycle assessment as results

The numerical results of the environmental impacts from a product, product system or service are the obvious outcomes of an LCA. In many instances, however, numbers themselves are not the most important results. The knowledge of the life cycle of a product or service is often sparse before the LCA is conducted. LCA commissioning parties suddenly learn how their suppliers are connected to global production systems and are often surprised to find out which energy or material inputs turn out to be the most challenging with respect to sustainability, or to find out that the real challenges in manufacturing are not part of their own product's life cycle, but rather how other products' life cycles are affected. Similarly, companies learn more about the use and the waste handling of their products. They may find out that by increasing the environmental load from their own production – for instance, through selecting a different material – other product life cycles may be positively affected to an even greater extent, resulting in an overall lower environmental load.

LCAs as results give benchmarks on which to base policies and strategic targets. A single number gives a target or threshold that actors can agree upon and use to steer activities. There is, however, a danger that LCA results may conserve existing systems. For instance, results from LCAs comparing different waste treatment options can lead to waste prevention disappearing from sight. An even greater risk when presenting a number as an absolute value is that all the information about how the number has been produced gets lost. How the world is modelled determines the results, as we shall explore more thoroughly in the following section on LCA as a method.

14.3 Life cycle assessment as method

LCA does not only bring ready-made results for decision-makers to compare alternatives. When speaking of LCA, one is normally referring to the method, or procedure, to understand the environmental impacts of products, product systems or services. Notwithstanding, speaking of "the method" may blur the issue, since LCA is rather an umbrella term for several methods needed to translate both the uses of natural resources and the emissions and wastes to the environment into information about potential environmental harm. This involves, on the one hand, multiple methods and models connecting elements in nature; such as wind, chemical composition, uptake of compounds in organisms and the effect of different compounds on various organisms and physical systems. On the other hand, it involves guidance on how to model societal systems such as industrial production, consumers' use of products, waste handling after end-of-useful life and on what data are needed. In the following section, different method issues in LCA are presented, with an emphasis on those that are relevant to guide governance or that may influence governance if the actors are unaware of their consequences.

14.3.1 The functional unit

One of the first methodological choices of the LCA practitioner is to define a functional unit. Every LCA is based on an implicit or explicit comparison of the environmental performance of how a production system is able to fulfil a certain function. The functional unit should be quantified and enable comparison of different ways to fulfil the same function. This might be difficult when governing the use of natural resources. The Renewable Energy Directive, for instance, applies a unit of energy – MJ – to compare different fuels. On the one hand, this unit does not tell how the fuels are able to fulfil a function, as different engine technologies might need a different amount of energy in order to produce the same amount of work. On the other hand, the sole focus on using biomass for fuels does not provide any input on the ability of the same resources to be put to better use as a food or feed or input to the manufacture of a material.

14.3.2 Comparability and standardisation

An important aspect of a decision-informing tool is the comparability of different alternatives. From the great bargainings over selecting global pathways for the future, to the small everyday choices of purchasing an item, rightful comparison of alternatives is key to correct decisions. This explains why LCA development work has been intrinsically linked to development of standards, guidelines and handbooks such the ISO standards, the European International Reference Life Cycle Data System (ILCD) Handbook (European Commission – Joint Research Centre – Institute for Environment and

Sustainability, 2010) and the Nordic guidelines on life cycle assessment (Finnveden & Lindfors, 1996). Comparability between LCAs has been an important issue as studies following different method rules and using different assumptions could suggest different, and even contradictory, conclusions for action. It also links to the development of two different types of LCA, namely attributional LCA and consequential LCA.

14.3.3 Different life cycle assessments for stable systems and systems in change

A standardised and harmonised method is not sufficient if the method is considered to produce unreliable results. Developers of regulations or business strategies, or purchasers of specific items, are rightfully confused about the merits of LCA when results for the environmental performance of what seems to be the same product vary by several orders of magnitude. This has been shown within the waste community (Lazarevic, 2015), and, as another example, a significant stir was caused in the field of bioenergy in 2008 when studies showed that the assumptions regarding land use change for corn ethanol completely overturned the idea of bioethanol as a climate-friendly fuel (Fargione, Hill, Tilman, Polasky & Hawthorne, 2008; Searchinger et al., 2008). Still, there might be good reasons for different results for the same production system. In the first stage of performing an LCA, the LCA practitioner should clearly state the goal of the study. The goal affects the method choices and data to be used since an assessment of the environmental performance of last year's production of a well-known commodity requires a different set of assumptions and method tools than the assessment of the environmental performance of the transportation system in 2030.

The LCA community has recognised the difference in assessing stable systems with few consequences for global economic markets and assessing systems that lead to changes in demand of products and, consequentially, changes in production systems. These two different approaches are coined attributional LCA and consequential LCA. There are heated debates in the LCA community as to which of these approaches are better, not least for policy applications (Bento & Klotz, 2014; McManus & Taylor, 2015; Plevin, Delucchi & Creutzig, 2014). While no consensus has been reached, the majority of practitioners adhere to the view that they are both useful, just for different purposes.

One of the main characteristics speaking in favour of consequential LCA's potential to inform policy is the focus on assessing impacts as a result of change, including indirect impacts (Bento & Klotz, 2014; McManus et al., 2015). McManus and Taylor (2015) go so far as to say that consequential LCA "is essentially a policy tool, rather than a technology assessment tool", while Suh and Yang (2014) criticise an overly optimistic view on consequential LCA and point out that the methodology is still developing and to a large extent untested within the policy field. There is, however, little doubt that consequential LCA is a useful tool for exploring alternative states of the world, while attributional LCA is a useful tool to assess the environmental (or sustainability) performance of different stable product systems. Thus, attributional LCA is used as a basis for the calculation method in the Renewable Energy Directive, although there might be controversies related to how stable the different production systems are.

14.3.4 Dividing impacts on several inputs or outputs

The chapter started by asking the question of how sustainable a product or service is and it has presented LCA as a useful tool to provide answers. However, whenever a process produces more than one output or a waste handling process transforms more than one input, there are difficulties in relating the environmental performance of the system to single inputs or outputs.

As part of a strategy for a circular bioeconomy, the concept of biorefinery is often mentioned as a technical approach to making the most of organic resources and diversifying product portfolios. Whereas the idea of a biorefinery is to diversify output and minimise waste, a much-debated attribute in LCA is how to deal with systems which involve more than one output. As examples, burning wood can result in both heat and power, and anaerobic digestion of food in both biogas and biofertiliser (digestate). In order to produce results for one output only, e.g. the environmental impacts of biogas, it is necessary to divide the impacts from anaerobic digestion between the biogas and the digestate. If possible, such divisions should be avoided, but in other cases the ISO standard 14044 prescribes two different method options: primarily to use system expansion, and secondarily to allocate impacts between the two outputs (biogas and biofertiliser).

In the biogas system, system expansion could entail expanding the function studied to include both outputs, which makes it impossible to separate the environmental impacts of the biogas from those of the biofertiliser. An option which is often regarded as a case of system expansion (e.g. in the ILCD handbook) is to consider the biofertiliser as substituting another type of fertiliser, and therefore the impact of that fertiliser is subtracted from the anaerobic digestion system. The idea is that the biofertiliser is dealt with, and the resulting environmental impact can be attributed to the biogas. Such a subtraction requires two conditions to be met: (1) an alternative product can clearly be defined, and (2) life cycle inventory data for the alternative product exists. The choice of which product is substituted and how can be of high importance to the final results, biogas included (Hanserud, Cherubini, Øgaard, Müller & Brattebø, 2018; Lyng et al., 2015), and the approach of substitution is not undisputed (Heijungs & Guinee, 2007). With an increasing number of outputs, such as in a biorefinery, the complexity of the model and its results may increase.

As an alternative, the total environmental impacts of the biogas system can be divided between, and allocated to, each co-product. The division can be based on mechanisms in the production process, or on different characteristics of the outputs, such as the mass of the different flows, their energetic or nutritional content or their economic market value. Choosing the most suitable approach to deal with multiple outputs is not necessarily easy, and since a division can be considered an artefact without equivalent in reality, the choice can be considered arbitrary or subjective (Heijungs & Guinee, 2007). To further complicate the picture, the choice often affects the LCA results, particularly in a system such as a biorefinery (Cherubini, Stromman & Ulgiati, 2011; Sandin, Røyne, Berlin, Peters & Svanström, 2015). The ISO standards state a general prioritisation scheme, and, although widely used, it has also been criticised as incoherent (Schrijvers, Loubet & Sonnemann, 2016) and not providing enough guidance (Weidema, 2014). Since a universally correct and unproblematic approach is not likely to be agreed upon, the tailoring of methods to the study purpose is essential. Nevertheless, in order to function as a more direct policy instrument, the EU Renewable Energy Directive has set this type of parameter in order to minimise incomparability among results for biofuels.

14.3.5 *What data and data for what?*

Intrinsically connected to the method is the data used as input to perform an LCA. Standards specify how a life cycle inventory should be collected and provide example data sheets for practitioners to use. The methods for impact assessment define the lists of components that must be studied. In a business perspective, the specificity of the data used is an important aspect. The more data that can be collected from the specific production, use and end of life of a product will affect how realistic the results are. The specificity of the results is one of the possible strengths of LCA.

When LCA is used for policy development, however, generalisation or use of average data is necessary. When using generalised or average data, the specificity may be lost, resulting in non-representativeness for some producers. It can also lead to frontrunners not being identified (Modahl, Askham, Lyng, Skjerve-Nielssen & Nereng, 2013). Challenges related to data also exist in dimensions other than the distinction between general and specific data. The quality of data, including age, geographical location, representativeness and other aspects, will determine the results from assessments.

14.3.6 *From environment-only to "holistic" sustainability*

Within the framework of LCA, all sorts of impacts connected to a product life cycle can potentially be studied. In the environmental sphere, impact categories have grown from mainly including energy to waste and water quality in the 1990s, to greenhouse gases, land use and ecosystem services (Sala, Reale, Cristóbal & Pant, 2016).[1] Adding to the scope is the development of social LCA and attempts at coupling LCA and life cycle costing which could potentially result in a life cycle sustainability assessment, where all aspects of sustainability are ideally covered (Sala et al., 2016, p. 9).

In reality, however, social LCA (S-LCA) is in an early stage of development and there is still a large need to define the relevant impact categories and indicators to quantify the social dimension. There is also a question of whether all social issues should be quantified, or if qualitative information is sufficient for governance. Still, the LCA framework can be useful to identify which people in which locations are affected by a decision.

Life cycle costing has followed a different path of development than LCA and although many LCA practitioners are trying to merge the two, there are also other scientists and practitioners developing life cycle costing in other directions. This means that system boundaries and data might be incompatible and the two cannot be put in the same equation proposed by Kloepffer (2008).

Here, the emphasis is placed on possible pitfalls related to social LCA and life cycle costing. One should be aware, however, that the models used to assess various environmental impacts in LCA are also in different stages of development. While the impact assessment models for climate change and acidification, for example, are mature and give results with small uncertainties, other impact assessment models, for instance for toxicity from heavy metals or ocean acidification, are in an early stage of development.

Scientific communities will most likely continue to expand models to include more possible impacts, and public and private policymakers who, for instance, work with the UN Sustainable Development Goals (United Nations, 2015) back this trend.

LCA as a method can provide a calculation procedure to give a result as an output. Just as important in terms of governance are, however, the discussions relating to specific methodological issues. These discussions reveal important aspects of society where science is unable to produce a clear-cut answer. There is, however, also the opportunity for life cycle principles to guide governance on a more general level than as results and methods – namely as a mindset or worldview.

There is, for instance, no single correct way to define the functional unit, but one might learn something about what functions we ask for and how they can be fulfilled when trying to put the functional unit into words. Similarly, there is no correct way to perform allocation, but the attempt at allocating displays features of the industrial system of relevance for governance. How much credit should the production of biofuels get for the simultaneous production of by-products? Furthermore, the choice between attributional and consequential LCA is not a choice between the better type of method to apply but rather the outcome of a conscious process towards defining the information one wants.

14.4 Life cycle assessment as a mindset

There is a third dimension – in addition to results and method – connected to the use of LCA in governance. This is the underlying mindset or worldview and how LCA depicts the function and connectedness of different elements in the world.

14.4.1 Life cycle thinking and industrial ecology

Constructs like extended producer responsibility, circular economy and product environmental footprint have all made their way into legislation and the everyday reality of companies. They are constructs with large influence from life cycle thinking which can be defined as "going beyond the traditional focus on production site and manufacturing processes to include environmental, social and economic impacts of a product over its entire life cycle" (UNEP-SETAC, 2018). Life cycle thinking is an umbrella concept for various strategies rooted in the life cycle mindset. It contains both qualitative and quantitative approaches, such as life cycle assessment, life cycle costing, social LCA or life cycle sustainability assessment but also integrated analysis with combined tools (Petit-Boix et al., 2017), function-oriented assessment paradigms or simple considerations of up- and downstream impacts (Bidstrup, 2015). The first policy promoting the life cycle thinking and Ecodesign arose from the waste management perspective and from the recognised relevance of consumers' awareness (Sala et al., 2016). Life cycle thinking is visible in certain policy actions plans, as eco-labels and eco-innovation strategies (Petit-Boix et al., 2017), and in the Communication on Sustainable Consumption and Production, the Communication on Circular Economy (Sala et al., 2016) and the EU waste hierarchy (Lazarevic, Buclet & Brandt, 2012).

Life cycle thinking is connected to the scientific field of Industrial Ecology which depicts the industrial system as "a certain kind of ecosystem" (Erkman, 1997) and tries to provide real solutions for sustainable development. It constitutes a broad framework intended to guide the industrial system in its transformation towards sustainability, shifting from linear industrial processes to closed loop systems (Saavedra, Iritani, Pavan & Ometto, 2018), and strives for an optimal circulation of materials and energy that limits damage in both industrial and ecological systems (Cohen-Rosenthal, 2004). One of the key concepts in Industrial Ecology is Industrial Metabolism. The term, partly borrowed from biology, focuses on the environmental impact of natural resources' use and how industrial systems can be organised to utilise all of companies' (energy and material) waste streams by mimicking nature (Saavedra et al., 2018).

Industrial ecology frameworks with LCA and ecological footprints are common methods for designing transitions from unsustainable to sustainable business models (Korhonen, 2003). Extended producers' responsibility, product stewardship and work with upstream and downstream chains become increasingly demanded by governments and accepted by industries (EEA, 1997).

Life cycle thinking and LCA as a mindset are more than just small corrections to business as usual. If the concepts are taken seriously, they have widespread consequences for both private and public governance, especially related to how the world is viewed. This is presented in two sections; one that discusses the view of time, space and connectedness in LCA, and one on cyclic models in opposition to linear models.

14.4.2 Time, space and connectedness

Most national policies are related to processes taking place within the state's boundaries and often they are also confined to a single sector. Whenever a product, product system or service is studied in an LCA, other sectors within other states are making their presence known immediately.

The political implications of life cycle thinking are related to its spatial and temporal outlook. If the product is not only what exists in the here and now, but is also seen as an outcome, entangled in a long network of relationships and processes – from resource extraction, through manufacturing, the labour involved, the transport route, to its marketing, consumption and further recycling – it also implies that the circle of stakeholders and actors involved is much wider. There are two kinds of normative issues that LCA can help bring to the surface: moral and political ones. Life cycle thinking can make it possible to identify and incorporate many ethical issues such as equity (possible unequal and unsustainable features of international trade), futurity (taking the future generations and their access to the resources into account) and locality (inter-regional material and energy flows) (Korhonen, 2003). This becomes particularly important when the question of responsibility for environmental externalities is posed. Who is responsible for the water pollution involved in manufacturing Product X? Is it only the company owning the plant, or perhaps also the company marketing the product on the other side of the world, or the consumer choosing it over some alternative? The political implications of a broadened scope are currently very visible in global climate negotiations, where the focus on emissions at point-sources is being gradually replaced by a life cycle approach which tracks emissions along the value chains.

Despite the problems with setting system boundaries and data requirements, space and time dimensions in LCA are potentially infinite. If resources circulate, a product can have a life cycle which starts in a distant past and theoretically never ends. The life cycle perspective revolutionised and extended the traditional approach in environmental management (Korhonen, 2003). LCA applications in business, academia and by policy actors bring a broader philosophy and way of thinking that go beyond the particular calculations. LCA can be seen as a constant learning process between the sectors, and for both practitioners and their audience. Therefore, LCA as a mindset provides a possibility to avoid the myopia of solely focusing on single sectors or nations.

14.4.3 Circularity

What are the implications of the expansion of life cycle thinking in different contexts? The results of the life cycle mindset's mainstreaming may be new concepts and ideas formed in business, academia and policy. Industrial Ecology's concepts and tools further evolving in life cycle thinking brought in applied concepts such as the circular economy (Saavedra et al., 2018, p. 1519) and the cascading use of biomass, functioning as autonomous and powerful social and political discourses. While LCA and circular economy share a view of the flow of the resources in the world as cyclic or circular, most LCAs are made for an industrial reality that is seen as linear, for instance as a supply or value chain. The recycling of materials from a product system is therefore often treated as a deviation from the normal operating system. In combination with the need to define the life cycle and its system boundaries, similar issues of division and burden-partitioning between co-products emerge for recycled materials (Ekvall & Tillman, 1997).

For instance, the environmental impacts from the recycling processes may be attributed to only one life cycle, or divided into multiple ones. Similarly, the benefits which arise from recycling and reuse of a material instead of relying on virgin production must be credited to either one or both life cycles – unless subsequent life cycles (and their functions) can be studied as an aggregate system. In theory, LCA always allows for expanding the studied system in order to avoid division issues, but the questions posed to LCA often require such division as they concern a single function, such as environmentally preferable transportation, food, heat, etc. In systems where outputs fulfil fundamentally different functions, such as the use of slaughterhouse waste to produce transportation fuel, an aggregated study may not be considered helpful to the decisions at hand.

Further challenges arise from the ideas of material loops and cascading material use, implying fuzzy borders between products, by-products and waste outputs. In 2008 some controversial papers relating to bioethanol pointed to residues as one safer option for biomass supply. But are organic residues per definition better resources from an environmental point of view? Waste materials have often been considered "free" as resources in terms of environmental impact (Oldfield, White & Holden, 2018; Pradel, Aissani, Villot, Baudez & Laforest, 2016), and such an assumption is currently implemented in the EU Renewable Energy Directive calculation rules, where agricultural and processing residues are to be considered burden-free at the point of collection (Annex V, part C, point 18). There is, however, no rule or definitive logic in LCA which states that residual materials have zero environmental impact, especially not when wastes are increasingly considered as valuable resources (Olofsson & Börjesson, 2018). The role of residual materials in a circular economy calls for a full life cycle perspective for residues (Oldfield et al., 2018), but the reasoning for how to include the environmental impacts of residual materials can vary depending on the study scope and intended use (Olofsson & Börjesson, 2018; Pelletier, Ardente, Brandão, De Camillis & Pennington, 2015; Svanes, Vold & Hanssen, 2011). In order to study the most sustainable uses of residues, several pathways and alternative uses must also be considered (Tonini, Hamelin & Astrup, 2016). In this way, both new production systems and the mental shift of wastes as resources have implications as to method considerations in LCA.

LCA as a mindset does not give direct input to governance as the specific content of policies or regulations such as LCA as results or LCA as a method might do. Instead, LCA as a mindset provides a frame for governance at all levels. It challenges every decision-maker to broaden the perspective both in time and space and keep in mind that no material or energy flows will vanish but possibly change into something else.

14.5 The different dimensions of life cycle assessment used in governance

Viewing LCA as results, method and mindset has indicated that there are several possible dimensions for applying LCA in relation to governance – each with different possibilities and pitfalls, as displayed in Table 14.1.

When performing LCA of a product or a service, the result is one or several numbers representing the quantified environmental impacts of the system(s) under study. This represents the first dimension of using LCA in governance, presented in Table 14.1. Results for different alternatives can be compared in order to evaluate which option is preferable, but they should not be used without keeping in mind the purpose of the study and the underlying assumptions. LCA results can also provide qualitative information about each life cycle step and which ones contribute the most to the total impact, as well as the function of the product and the kind of environmental problems of which one should be aware. LCA results may also reveal important trade-offs between types of environmental impact, or between optimisation options in different life cycle stages. To develop effective policy, policymakers should be informed about possible trade-offs based on life cycle information.

When using LCA as a method in governance (the second dimension presented in Table 14.1), it might be a ready-made procedure to produce results for a certain application such as the computational rules stated in the Renewable Energy Directive. It can also be the information stemming from discussions about methodological issues, such as what the real function of the product is and whether there are other ways to fulfil the same function, or how to view by-products from a production system when allocation is discussed. These methodological discussions thus give input to how specific production systems should be governed.

Table 14.1 Three dimensions of life cycle assessments (LCAs) (own elaboration)

	LCA as results	*LCA as method*	*LCA as a mindset*
Description	Using the quantified environmental impacts from life cycle assessment to determine which solution is the most environmentally beneficial	Policy makers can utilise life cycle assessment methodology when evaluating the consequence of different choices	Value chain perspective and quantification of environmental impacts in relation to the function of the product or the service
Example	Using Environmental Product Declarations in public procurement Sustainability criteria in renewable energy directive	Assessment of impacts from implementing a certain policy instrument	The philosophy of including the entire life cycle: circular economy
Opportunities	Standardised results enable a quantitative comparison between products/services	Enables a holistic perspective on the environmental effects	Complex thinking, including larger time horizons, more geographical connections, circularity
Pitfalls	Complexity can be lost if only the final results are presented	Type of life cycle assessment and assumption can largely affect the results and may be a matter of value choice	Mistrust towards data and results manipulation

The third dimension of use of LCA in governance shown in Table 14.1 is LCA as a mindset. The philosophy behind LCA and life cycle thinking is that it is not sufficient to look at only one life cycle phase or to address only one environmental problem. Rather, to make informed decisions when it comes to reducing environmental impacts, a holistic approach is required. LCA represents the idea that the environmental impact of a product or a service can be quantified, and that this should be quantified in relation to the function that the product or service provides to the user.

14.5.1 Cross-sectoral policy development

While the national emissions reporting to the United Nations Framework Convention on Climate Change is sector based, applying LCA when developing policies enables policymakers to see across sectors to make more informed decisions. A measure implemented in one sector can lead to impact reductions or increases in another sector. A high level of sector integration in the value chain (waste, agricultural and transport sectors) has been beneficial for biogas production in Norway in terms of reduction of greenhouse gas emissions (Lyng, Stensgård, Hanssen & Modahl, 2018). At the same time, the demand for biogas in the transport sector is limited (Sund, Utgård & Christensen, 2017) and biogas has lost in public tenders in competition with traditional diesel engines. Without targeted political instruments, the actors involved may lack incentives to choose the options that give the largest overall reduction of environmental impacts and instead focus on a single sector (see the case of liquefied biogas in Chapter 4).

14.5.2 Pitfalls and possibilities

LCA is particularly useful for regulation (Lowi, 1972), framing conditions and constraints for individual and collective behaviours (environmental or consumer protection), and it can be applied at various stages of the policy cycle: agenda setting and problem definition, policy formulation and impact assessment, as well as implementation and evaluation (Sala et al., 2016, p. 11). With these opportunities, however, come certain pitfalls. LCA provides information on environmental impacts, but their prioritisation lies in the policy domain, including qualitative or quantitative weighting and prioritisation of impacts of higher societal importance (as climate change versus eutrophication). This division is not always clear-cut, with differing expectations for LCA to, on the one hand, deliver robust results (Wardenaar et al., 2012) as input to policymaking in different forms, and, on the other, provide broader knowledge on trade-offs, uncertainties, risks, etc. (Herrmann, Hauschild, Sohn & McKone, 2014).

As previously discussed, using LCA for policy without careful distinction and understanding of attributional and consequential LCA can cause various problems such as inappropriate method selections, unfair results, misinterpretation of results, etc. (Brander, Tipper, Hutchison & Davis, 2009). In addition, LCA studies can rely on different types of data and set different system boundaries which make them difficult to compare and evaluate for policy purposes. Even considering solutions to such issues, LCA has proven unable to meet expectations in solving societal debates, for instance regarding environmental protection policies for chlorine substances (Bras-Klapwijk, 1998; Tukker, 2000), and the current disagreement on the environmental impacts of biofuels.

It seems that one of the possible options for widening LCA applications in policymaking is expanding the method of LCA into other areas, and thus providing policymakers with larger frameworks for decisions. On this track, the ideal for sustainability assessment is to have a holistic view, building bridges between methodologies, much like the recent attempts to broaden the environmental pillar-focused LCA with social (Agyekum, Fortuin & van der Harst, 2017) and economic factors (Di Maria, Eyckmans & Van Acker, 2018). Other ideas for furthering the scope of LCA include combinations with other methodologies and types of information such as multi-criteria analysis (Koroneos, Nanaki, Dimitrios & Krokida, 2013) and geographic information systems (Hiloidhari et al., 2017).

As an alternative to expanding the scope of LCA, a different approach to LCA in governance and policymaking is to treat different methodologies as complementary and allow a multiplicity of perspectives before making the decision. As an example, Mancini, Benini and Sala (2015) compared three approaches: material flow analysis (MFA), LCA and "resource critical"-based impact accounting for a better-informed and science-based decision-making. The future for LCA in governance could thus take different and perhaps multiple paths.

14.6 Conclusions

The role of LCA as a tool for indicating which activities are more or less sustainable can be valuable in governing a transition towards a circular bioeconomy. It comes, however, with both opportunities and pitfalls. Viewing LCA as results, as a method and as a mindset has revealed different potential applications of LCA in relation to governance, and all three dimensions can be useful when applied to provide the right kind of support and to answer suitable questions. As LCA offers quantitative results, these can be used directly to benchmark products in order to choose less impacting alternatives. This dimension, however, blurs important methodological choices made, and more knowledge about the method applied can also provide broader knowledge on the issues, potential trade-offs and uncertainties at play. On a more general level, ideas based on life cycle thinking can guide governance towards certain patterns and priorities such as ideas of circular resource flows in a circular bioeconomy. Although LCA is based on life cycle thinking, following up broad ideas such as a future circular economy with quantitative assessments brings various methodological challenges to the method of LCA, which also reflect on results. In addition to this, LCA in relation to governance faces the challenge of providing robust and holistic grounds for decision-making while at the same time facing restrictions and uncertainties in terms of the data used, methodological choices made and challenges in broadening the scope to include several types of environmental impacts and both economic and social dimensions of sustainability. While it is clear that LCA for governance is not a simple project, it is also clear that the opportunities at hand can offer valuable guidance in transitions towards more sustainable modes of organising human activities, such as a circular bioeconomy.

Note

1 Although the number of environmental impact categories in general has grown, there has also been an increase in studies focusing on a single indicator, with climate change at the top of the list. Trends in impact categories' selection follow the general societal focus on adverse environmental issues.

References

Agyekum, E. O., Fortuin, K. P. J. & van der Harst, E. (2017). Environmental and social life cycle assessment of bamboo bicycle frames made in Ghana. *Journal of Cleaner Production, 143*, 1069–1080. doi:10.1016/j.jclepro.2016.12.012.

Bento, A. M., & Klotz, R. (2014). Climate policy decisions require policy-based life-cycle analysis. *Environmental Science and Technology, 48*, 5379–5387. doi:10.1021/es405164g.

Bidstrup, M. (2015). Life cycle thinking in impact assessment: Current practice and LCA gains. *Environmental Impact Assessment Review, 54*, 72–79. doi:10.1016/j.eiar.2015.05.003.

Brander, M., Tipper, R., Hutchison, C. & Davis, G. (2009). *Consequential and Attributional Approaches to LCA: A Guide to Policy Makers with Specific Reference to Greenhouse Gas LCA of Biofuels* (Vol. 44). Retrieved from https://ecometrica.com/white-papers/consequential-and-attributional-approaches-to-lca-a-guide-to-policy-makers-with-specific-reference-to-greenhouse-gas-lca-of-biofuels.

Bras-Klapwijk, R. M. (1998). Are life cycle assessments a threat to sound public policy making? *The International Journal of Life Cycle Assessment, 3*(6), 333–342. doi:10.1007/bf02979344.

Cherubini, F., Stromman, A. H. & Ulgiati, S. (2011). Influence of allocation methods on the environmental performance of biorefinery products: A case study. *Resources, Conservation and Recycling, 55*(11), 1070–1077. doi:10.1016/j.resconrec.2011.06.001.

Cohen-Rosenthal, E. (2004). Making sense out of industrial ecology: A framework for analysis and action. *Journal of Cleaner Production, 12*(8), 1111–1123. doi:10.1016/j.jclepro.2004.02.009.

Councilor of Vestfold. (2012). *Councilors Case and Decision Regarding the Establishment of a New Biogas Plant in Vestfold County, Norway. Utkast rådmannens saksforelegg og vedtak: Lokal gjenvinning av våtorganisk avfall i Vestfold og Grenland – en sak om investering i lokal avfallsinfrastruktur ("Grenland Vestfold Biogass AS").* Retrieved from www.vesar.no/filarkiv/File/biogass/Utkast_raadmannens.pdf.

Di Maria, A., Eyckmans, J. & Van Acker, K. (2018). Downcycling versus recycling of construction and demolition waste: Combining LCA and LCC to support sustainable policy making. *Waste Management.* doi:10.1016/j.wasman.2018.01.028.

EEA. (1997). *Life Cycle Assessment (LCA): A guide to approaches, experiences and information sources.* Retrieved from www.eea.europa.eu/publications/GH-07-97-595-EN-C/Issue-report-No-6.pdf/view.

Ekvall, T., & Tillman, A.-M. (1997). Open-loop recycling: Criteria for allocation procedures. *International Journal of Life Cycle Assessment, 2*(3), 155. doi:10.1007/bf02978810.

Erkman, S. (1997). Industrial ecology: An historical view. *Journal of Cleaner Production, 5*(1), 1–10. doi:10.1016/S0959-6526(97)00003-6.

EU. (2009). *Directive 2009/28/EC of the European Parliament and of the Council.* Retrieved from http://eur-lex.europa.eu/LexUriServ/LexUriServ.do?uri=Oj:L:2009:140:0016:0062:en:PDF.

European Commission – Joint Research Centre – Institute for Environment and Sustainability. (2010). *International Reference Life Cycle Data System (ILCD) Handbook – General guide for Life Cycle Assessment – Detailed Guidance.* Retrieved from Luxembourg.

European Commission. (2013). *Commission Recommendation of 9 April 2013 on the Use of Common Methods to Measure and Communicate the Life Cycle Environmental Performance of Products and Organisations*. Retrieved from http://ec.europa.eu/environment/eussd/smgp/pdf/Guidance_products.pdf.

European Commission. (2017). *Annexes to the Proposal for a Directive of the European Parliament and the Council on the Promotion on the Use of Energy from Renewable Sources (Recast)*. Retrieved from www.ebb-eu.org/EBBpressreleases/Annexes_REDII_Commission_Proposal_COM2016-767.pdf.

Fargione, J., Hill, J., Tilman, D., Polasky, S. & Hawthorne, P. (2008). Land clearing and the biofuel carbon debt. *Science, 319*, 1235–1238.

Finnveden, G., & Lindfors, L.-G. (1996). On the nordic guidelines for life cycle assessment. *The International Journal of Life Cycle Assessment, 1*(1), 45–48. doi:10.1007/bf02978635.

Finnveden, G., Hauschild, M. Z., Ekvall, T., Guinée, J., Heijungs, R., Hellweg, S., … Suh, S. (2009). Recent developments in Life Cycle Assessment. *Journal of Environmental Management, 91*(1), 1–21. doi:10.1016/j.jenvman.2009.06.018.

Frischknecht, R., Wyss, F., Knöpfel, S. B. & Stolz, P. (2015). Life cycle assessment in the building sector: Analytical tools, environmental information and labels. *The International Journal of Life Cycle Assessment, 20*(4), 421–425. doi:10.1007/s11367-015-0856-0.

Gentil, E. C., Damgaard, A., Hauschild, M., Finnveden, G., Eriksson, O., Thorneloe, S., … Christensen, T. H. (2010). Models for waste life cycle assessment: Review of technical assumptions. *Waste Management, 30*(12), 2636–2648. doi:10.1016/j.wasman.2010.06.004.

Hanserud, O. S., Cherubini, F., Øgaard, A. F., Müller, D. B. & Brattebø, H. (2018). Choice of mineral fertilizer substitution principle strongly influences LCA environmental benefits of nutrient cycling in the agri-food system. *Science of the Total Environment, 615*, 219–227. doi:10.1016/j.scitotenv.2017.09.215.

Hanserud, O. S., Lyng, K.-A., Vries, J. W. D., Øgaard, A. F. & Brattebø, H. (2017). Redistributing phosphorus in animal manure from a livestock-intensive region to an arable region: Exploration of environmental consequences. *Sustainability, 9*(4), 595. doi:10.3390/su9040595.

Heijungs, R., & Guinee, J. B. (2007). Allocation and "what-if" scenarios in life cycle assessment of waste management systems. *Waste Management, 27*(8), 997–1005. doi:10.1016/j.wasman.2007.02.013.

Herrmann, I. T., Hauschild, M. Z., Sohn, M. D. & McKone, T. E. (2014). Confronting uncertainty in life cycle assessment used for decision support. *Journal of Industrial Ecology, 18*, 366–379. doi:10.1111/jiec.12085.

Hiloidhari, M., Baruah, D. C., Singh, A., Kataki, S., Medhi, K., Kumari, S., … Thakur, I. S. (2017). Emerging role of Geographical Information System (GIS), Life Cycle Assessment (LCA) and spatial LCA (GIS-LCA) in sustainable bioenergy planning. *Bioresource Technology, 242*, 218–226. doi:10.1016/j.biortech.2017.03.079.

ISO. (2006). *Environmental Management – Life Cycle Assessment – Requirements and Guidelines* (ISO 14044:2006). Geneva: International Organization for Standardization.

Kloepffer, W. (2008). Life cycle sustainability assessment of products. *The International Journal of Life Cycle Assessment, 13*(2), 89–95. doi:10.1065/lca2008.02.376.

Korhonen, J. (2003). On the ethics of corporate social responsibility: Considering the paradigm of industrial metabolism. *Journal of Business Ethics, 48*(4), 301–315.

Koroneos, C., Nanaki, E., Dimitrios, R. & Krokida, M. (2013). *Life Cycle Assessment: A Strategic Tool for Sustainable Development Decisions*. Paper presented at the 3rd World Sustainability Forum, MDPI's conference platform, Sciforum.

Lazarevic, D. (2015). The legitimacy of life cycle assessment in the waste management sector. *The International Journal of Life Cycle Assessment.* doi:10.1007/s11367-015-0884-9.

Lazarevic, D., Buclet, N. & Brandt, N. (2012). The application of life cycle thinking in the context of European waste policy. *Journal of Cleaner Production, 29–30,* 199–207. doi:10.1016/j.jclepro.2012.01.030.

Lowi, T. J. (1972). Four systems of policy, politics, and choice. *Public Administration Review, 32*(4), 298–310. doi:10.2307/974990.

Lyng, K.-A., Stensgård, A. E., Hanssen, O. J. & Modahl, I. S. (2018). Relation between greenhouse gas emissions and economic profit for different configurations of biogas value chains: A case study on different levels of sector integration. *Journal of Cleaner Production, 182*(May), 737–745. doi:10.1016/j.jclepro.2018.02.126.

Lyng, K.-A., Modahl, I. S., Møller, H., Morken, J., Briseid, T. & Hanssen, O. J. (2015). The BioValueChain model: A Norwegian model for calculating environmental impacts of biogas value chains. *The International Journal of Life Cycle Assessment, 20*(4), 490–502. doi:10.1007/s11367-015-0851-5.

Mancini, L., Benini, L., & Sala, S. (2015). Resource footprint of Europe: Complementarity of material flow analysis and life cycle assessment for policy support. *Environmental Science & Policy, 54,* 367–376. doi:10.1016/j.envsci.2015.07.025.

McManus, M. C., & Taylor, C. M. (2015). The changing nature of life cycle assessment. *Biomass and Bioenergy, 82,* 13–26. doi:10.1016/j.biombioe.2015.04.024.

McManus, M. C., Taylor, C. M., Mohr, A., Whittaker, C., Scown, C. D., Borrion, A. L., ... Yin, Y. (2015). Challenge clusters facing LCA in environmental decision-making: What we can learn from biofuels. *International Journal of Life Cycle Assessment, 20*(10), 1399–1414. doi:10.1007/s11367-015-0930-7.

Modahl, I. S., Askham, C., Lyng, K.-A., Skjerve-Nielssen, C. & Nereng, G. (2013). Comparison of two versions of an EPD, using generic and specific data for the foreground system, and some methodological implications. *The International Journal of Life Cycle Assessment, 18*(1), 241–251. doi:10.1007/s11367-012-0449-0.

Modahl, I. S., Lyng, K.-A., Stensgård, A., Saxegård, S., Hanssen, O. J., Møller, H., ... Sørby, I. (2016). *Biogassproduksjon fra matavfall og møkk fra ku, gris og fjørfe. Status 2016 (fase IV) for miljønytte for den norske biogassmodellen BioValueChain. OR 34.16. Østfoldforskning AS.* Retrieved from www.ostfoldforskning.no/no/publikasjoner/Publication/?id=1987.

Norwegian Green Building Council. (2012). *BREEAM-NOR ver-1.1.* Retrieved from https://ngbc.no/wp-content/uploads/2015/09/BREEAM-NOR-Norw-ver-1.1_0.pdf.

Oldfield, T. L., White, E. & Holden, N. M. (2018). The implications of stakeholder perspective for LCA of wasted food and green waste. *Journal of Cleaner Production, 170*(Supplement C), 1554–1564. doi:10.1016/j.jclepro.2017.09.239.

Olofsson, J., & Börjesson, P. (2018). Residual biomass as resource: Life-cycle environmental impact of wastes in circular resource systems. *Journal of Cleaner Production, 196,* 997–1006. doi:10.1016/j.jclepro.2018.06.115.

Pelletier, N., Ardente, F., Brandão, M., De Camillis, C. & Pennington, D. (2015). Rationales for and limitations of preferred solutions for multi-functionality problems in LCA: Is increased consistency possible? *International Journal of Life Cycle Assessment, 20*(1), 74–86. doi:10.1007/s11367-014-0812-4.

Petit-Boix, A., Llorach-Massana, P., Sanjuan-Delmás, D., Sierra-Pérez, J., Vinyes, E., Gabarrell, X., ... Sanyé-Mengual, E. (2017). Application of life cycle thinking towards sustainable cities: A review. *Journal of Cleaner Production, 166*, 939–951. doi:10.1016/j.jclepro.2017.08.030.

Plevin, R. J., Delucchi, M. A. & Creutzig, F. (2014). Using attributional life cycle assessment to estimate climate-change mitigation benefits misleads policy makers. *Journal of Industrial Ecology, 18*, 73–83. doi:10.1111/jiec.12074.

Pradel, M., Aissani, L., Villot, J., Baudez, J. C. & Laforest, V. (2016). From waste to added value product: Towards a paradigm shift in life cycle assessment applied to wastewater sludge – A review. *Journal of Cleaner Production, 131*, 60–75. doi:10.1016/j.jclepro.2016.05.076.

Raadal, H. L., Stensgård, A., Lyng, K.-A. & Hanssen, O. J. (2016). *Vurdering av virkemidler for økt utsortering av våtorganisk avfall og plastemballasje. Ostfoldforskning AS, OR.01.16.* Retrieved from Kråkerøy, Norway.

Reed, D. L. (2012). *Life-Cycle Assessment in Government Policy in the United States* (Doctoral Dissertation, University of Tennessee).

Saavedra, Y. M. B., Iritani, D. R., Pavan, A. L. R. & Ometto, A. R. (2018). Theoretical contribution of industrial ecology to circular economy. *Journal of Cleaner Production, 170*, 1514–1522. doi:10.1016/j.jclepro.2017.09.260.

Sala, S., Reale, F., Cristóbal, J. & Pant, R. (2016). *Life Cycle Assessment for the Impact Assessment of Policies*. Luxembourg: Publications Office of the European Union.

Sandin, G., Røyne, F., Berlin, J., Peters, G. M. & Svanström, M. (2015). Allocation in LCAs of biorefinery products: Implications for results and decision-making. *Journal of Cleaner Production, 93*, 213–221. doi:10.1016/j.jclepro.2015.01.013.

Schrijvers, D. L., Loubet, P. & Sonnemann, G. (2016). Critical review of guidelines against a systematic framework with regard to consistency on allocation procedures for recycling in LCA. *International Journal of Life Cycle Assessment, 21*(7), 994–1008. doi:10.1007/s11367-016-1069-x.

Searchinger, T., Heimlich, R., Houghton, R. A., Dong, F., Elobeid, A., Fabiosa, J., ... Yu, T.-h. (2008). Use of U.S. croplands for biofuels increases greenhouse gases through emissions from land-use change. *Science, 319*, 1238–1240.

Suh, S., & Yang, Y. (2014). On the uncanny capabilities of consequential LCA. *The International Journal of Life Cycle Assessment, 19*, 1179–1184. doi:10.1007/s11367-014-0739-9.

Sund, K., Utgård, B. & Christensen, N. S. (2017). *Muligheter og Barrierer for Økt Bruk av Biogass til Transport i Norge*. Sund Energy, Oslo. Retrieved from http://presse.enova.no/documents/muligheter-og-barrierer-for-oekt-bruk-av-biogass-til-transport-i-norge-69550.

Svanes, E., Vold, M. & Hanssen, O. J. (2011). Effect of different allocation methods on LCA results of products from wild-caught fish and on the use of such results. *International Journal of Life Cycle Assessment, 16*(6), 512–521. doi:10.1007/s11367-011-0288-4.

Syversen, F., Lyng, K.-A., Amland, E. N., Bjørnerud, S., Callewaert, P. & Prestrud, K. (2018). *Utsortering og materialgjenvinning av biologisk avfall og plastavfall. Utredning av konsekvenser av forslag til forskrift for avfall fra husholdninger og liknende avfall fra næringslivet (2017/12503).* Retrieved from www.miljodirektoratet.no/no/Publikasjoner/2018/Oktober-2018/Utsortering-og-materialgjenvinning-av-biologisk-avfall-og-plastavfall.

Tonini, D., Hamelin, L. & Astrup, T. F. (2016). Environmental implications of the use of agro-industrial residues for biorefineries: Application of a deterministic model for indirect land-use changes. *Global Change Biology Bioenergy, 8*(4), 690–706. doi:10.1111/gcbb.12290.

Tukker, A. (2000). Philosophy of science, policy sciences and the basis of decision support with LCA based on the toxicity controversy in Sweden and the Netherlands. *The International Journal of Life Cycle Assessment, 5*(3), 177–186. doi:10.1007/bf02978621.

UNEP-SETAC. (2018). What is life cycle thinking? Retrieved from www.lifecycleinitiative.org/starting-life-cycle-thinking/what-is-life-cycle-thinking.

United Nations. (2015). *Resolution Adopted by the General Assembly on 25 September 2015: Transforming Our World: The 2030 Agenda for Sustainable Development* Retrieved from www.un.org/ga/search/view_doc.asp?symbol=A/RES/70/1&Lang=E (accessed 23.01.2018).

Wardenaar, T., van Ruijven, T., Beltran, A. M., Vad, K., Guinée, J. & Heijungs, R. (2012). Differences between LCA for analysis and LCA for policy: A case study on the consequences of allocation choices in bio-energy policies. *The International Journal of Life Cycle Assessment, 17*, 1059–1067. doi:10.1007/s11367-012-0431-x.

Weidema, B. (2014). Has ISO 14040/44 failed its role as a standard for life cycle assessment? *Journal of Industrial Ecology, 18*(3), 324–326. doi:10.1111/jiec.12139.

15 Conclusions

Antje Klitkou, Arne Martin Fevolden and Marco Capasso

15.1 Introduction

In the following sections we explain how the book has answered the raised expectations for the main thematic issues introduced in the first chapter, such as circularity across established sectors, regional embedding and geographical context of waste valorisation, resource ownership and interfirm governance, and policy and regulations of waste valorisation. We also highlight possible future research perspectives for these themes.

15.2 Circularity across established sectors and sustainability of valorisation

The circularity and sustainability of valorisation of residues have been addressed on many occasions in this book. The sectoral case studies in **Part II** discuss circularity and sustainability of valorisation from different angles. **Chapter 4** compares examples of industrial symbiosis at different stages of maturity: the emerging valorisation of residues from the pulp and paper industry and aquaculture for producing liquified biogas and the diversification of value creation of a former pulp and paper company into an integrated biorefinery, exploiting wooden resources for producing chemicals, materials and food and feed ingredients, crossing the sectoral boundaries from forestry to chemical industry, food industry and feed industry. While in the first example two companies collocate and integrate their streams of residues, the second example shows how the integrated biorefinery evolved over time and created many collocated and connected activities addressing the valorisation of residues and side-streams. The company even specifically addressed the sustainability of its products and performed an LCA. In that way the company could exploit the whole feedstock and could become more energy efficient and sustainable. In **Chapter 6** a comparison of different options for the valorisation of brewers' spent grain shows that there are other options to gain more economic value out of this side-stream than just giving it away for free to local farmers, but that the sustainability of switching to such a pathway has to be carefully assessed when the use as feed for local husbandry is an established practice and a lack of those feed resources would require the import of soya. The analysis of organic waste valorisation in land-based farming of salmon in **Chapter 8** shows that the salmon-producing companies are not driving innovation in this direction and they do little to assess the most sustainable use for the sludge from an LCA perspective. **Chapter 9** addresses valorisation of acid whey and explicitly reflects upon the sustainability of valorisation of side-streams. The authors stress that the valorisation of by-products has to also consider the context of traditional usage of those side-streams. Usage for human food or medicines may then be less sustainable than less technologically advanced solutions because the replacement of the established usage of side-stream by imports of feed from other continents might lead to more damage. The sustainability of valorisation pathways in terms of environmental, economic and social impacts has to be assessed through an LCA. Chapter 9 has also shown that new sustainability challenges can occur when the size of production increases.

The sectoral case studies have also shown that circularity and sustainable valorisation of residues are addressed differently in the private and the public realm. While in the public realm climate mitigation and sustainable development goals stand central, for companies the focus is on corporate social responsibility, and for that reason closed cycles and sustainable sourcing are addressed. For future research it would be important to address sustainable business models for the valorisation of residues and side-streams and how to compare the sustainability of different business models.

Part III showed us that quantitative studies can be an important source of insight into the bioeconomy, but that designing such studies is problematic and riddled with challenges. In **Chapter 10**, we learned that the CVs of researchers working on bioeconomy-focused research programmes under the Research Council of Norway can provide new insight into how knowledge production is organised in the bioeconomy and which knowledge bases are involved. A key finding in this chapter is that research on the bioeconomy is based on a wide range of disciplines, including agricultural and life sciences, engineering and the physical sciences, and social sciences, humanities and professional fields such as business administration and law. **Chapter 11** points out that the bioeconomy can also be studied using a wide range of different datasets such as R&D statistics, firm-level data and surveys. Nevertheless, all of these datasets have challenges related to defining the population of actors in the bioeconomy. Future research might study how the identified bioeconomy companies collaborate with firms outside their sectors and how careers of researchers are evolving through to collaboration with different types of companies.

Chapter 14 addresses LCAs as a tool for governance and, in this chapter, the difficulties in drawing the boundaries of an LCA are explained in more detail. They show that often LCAs are practised in a way that does not consider the circularity of the flow of resources properly but is based on an understanding which is dominated by a model of linear value chains.

Environmental impacts of recycling and reuse of resources can be studied either on an aggregated system level or at a more detailed level for the different products and processes of valorisation. When keeping the level of analysis at the aggregated level, many sustainability issues will become invisible and such analyses will not be useful for making the appropriate decisions. Therefore, future research avenues for addressing the circular and sustainable bioeconomy need to expand the scope of LCAs if they want to guide the governance of the bioeconomy.

15.3 Regional embedding and geographies of innovation

Geography is a recurrent theme in this book. **Chapter 4** explores how specific regional contexts affect the emergence of novel pathways for valorising forestry residues. In one of the Norwegian regions under analysis, in particular, forest-based value creation is reached by locating a biogas plant next to a pulp and paper plant, achieving a successful industrial symbiosis. The chapter also considers how regions can develop on the basis of their existing assets, including their endowment in terms of knowledge and institutions. Such a "path dependence" is theoretically framed in **Chapter 3**, together with the related concept of "lock-in". The territorial embeddedness of institutions also emerges as a relevant element in value creation in **Chapter 8**, which suggests a socio-technical transition to valorise waste in salmon aquaculture through land-based farming. The same chapter also shows that the current distance between traditional coastal farming districts and inland waste processors may constitute a barrier to waste valorisation. A similar barrier is considered in **Chapter 7**, which describes how the small and scattered slaughterhouses in Norway face more difficulties in collecting and handling animal by-products in comparison to the fewer and much larger Danish slaughterhouses. **Chapter 10** shows, in contrast, how geographic boundaries of the national knowledge base can be overcome by establishing international consortia, who apply for funding in Norway through a bioeconomy-focused research programme. The political interactions across different spatial scales are explored in **Chapter 13**, where a multi-level governance framework is applied to analyse policy efforts for food waste reduction. Finally, **Chapter 14**, on life cycle analysis and governance, considers how political implications of life cycle thinking depend on the spatial boundaries imposed on the industrial systems under analysis.

The spatial boundaries considered for environmental systems vary depending on the specific problem to be solved. In some cases, the concentration of activities in a small area can constitute a problem, as in the case of coastal aquaculture mentioned above. In other cases, the spatial boundaries for the analysis must be extended to enclose the whole world, as in the case of the consequences of greenhouse gas emissions on global warming. In the latter case, municipalities and counties can be asked to face global challenges, possibly with the intermediation of national state authorities. The connection between small-scale actions with large-scale challenges – "think global, act local", as the motto goes – can translate into the multi-level types of governance necessary to achieve the coordination of different geographic areas towards common goals. Actors able to take the appropriate actions, in order to promote waste valorisation in the face of global challenges, can indeed often be found by authorities through a small-scale search: in this way, the authorities can determine the potential for small-scale regions to develop specific valorisation pathways. But private companies can also explore the possibility of detecting and following leading actors located in their own region: they could connect to, establish partnerships with and provide intermediate goods for innovative leading firms in the region which have already implemented valorisation processes. Many different types of relations can occur between firms at a municipal or regional scale. Input–output relations can be established within a region, to give rise to industrial symbiosis, or even to generate non-priced interactions for waste valorisation: the case of breweries giving spent grain free to local farmers may be an example. The extent and success of the interactions between regional actors will also depend on the institutional background of the region: formal and informal institutions will define whether new interactions are possible, and whether old interactions can serve new functions towards environmental goals. Moreover, the skill-base to which firms can resort when restructuring themselves towards waste valorisation may also be searched within their region: new valorisation pathways can then take place if the skills present in the region are sufficiently flexible to accommodate such evolution. However, competences not available at regional, or even at national, level can sometimes be reached by crossing national borders: R&D consortia may, for instance, help to create bridges towards different types of knowledge, while imports can contribute to the acquisition of foreign knowledge as embedded in machinery for waste transformation. Exports, on the other hand, can increase a firm's market size, which in turn would translate into firm revenues, to be used for investments in side-stream valorisation.

Understanding how agreements on international trade affect the potential for waste valorisation, for the countries involved as well as for the world in general, constitutes an important line of future research. Tariffs, sanctions and other forms of trade restrictions have been shown, in this book, to have exerted a strong influence of the firms' choices towards valorisation. The compliance with EU regulations is another international element that has been mentioned and could be the topic of further study, together with the alignment of national and regional policies with supranational decisions. In order to tune environmental policies according to the regions and sectors involved, further research should be devoted to studying the institutional idiosyncrasies of regions and sectors, bringing institutional theory insights deeper into economic geography and industrial organisation subjects. At the same time, LCA, enriched with social elements and with insights from welfare economics, can stimulate researchers to define the correct spatial scale for answering questions on waste valorisation.

Research must explore the geographic dimension in order to define the opportunities for waste valorisation that are offered within specific regions or countries. Some geographic areas may indeed be more fertile for particular types of waste valorisation, in connection with particular economic sectors. Moreover, research can point to ways to overcome the limitations encountered in specific geographic areas, and show how economic actors, including firms and policymakers, can contribute to shaping new geographies of innovation which would favour waste valorisation.

15.4 Resource ownership and interfirm governance structures

In the introduction to this book, we suggested that the valorisation of waste streams often necessitates a high degree of coordination along and across value circles, but such coordination can be prevented by the intrinsic properties of waste. Indeed, both the amount and the production timing of waste depend on the needs of the value circle from which it originates, and are not usually planned on the basis of a potential waste valorisation. However, the book has shown several solutions that businesses can bring forward in order to avoid such a "waste puzzle". Organisational solutions are, for instance, highlighted in **Chapters 7** and **9**, about the valorisation of, respectively, animal and dairy by-products. Both chapters witness the successful establishment, by a large firm dominant in the sector, of a subsidiary firm which would focus specifically on the valorisation of by-products. **Chapter 11**, on actors and innovators in the circular bioeconomy, confirms that firms who actively seek to realise value from organic waste streams often describe organic waste activities as core activities, which constitute a distinctive mark of the same firms. Moreover, **Chapter 6** shows the dangers of considering waste activities as peripheral to the firm: breweries would, for instance, focus on a new bottling line rather than upgrading their brewing equipment for a more efficient use of spent grain. Waste valorisation could come back into focus when a main product is branded according to the firm's corporate social responsibility, as described in both **Chapters 3** and **6**. A firm's environmental reputation may indeed exert a strong push towards waste valorisation: environmental groups are currently playing an important role in the context of salmon farming (**Chapter 8**), a role which could later be taken over by national governments. Better coordination between public and private actors, as well as within the public sector, could also lead to improved waste valorisation, as suggested by **Chapter 5** about the municipality's management of urban waste in the city of Oslo. Within-firm coordination of different activities has instead been presented in **Chapter 4**, in the form of a Norwegian biorefinery which is able to optimise the cross-exploitation of side-streams from distinct firm activities.

If we put together results coming from the different sections of the book, two main suggestions can be brought to businesses. The first is based on incorporating waste valorisation explicitly in the firms' strategies. Without setting aside resources devoted to valorisation objectives, a firm will rarely improve the utilisation of current residues. By resources, we mean both human resources (for instance, in the form of R&D employees) and other physical resources (e.g. the equipment needed for waste transformation). Having one or more employees, within the firm, who explicitly take over the responsibility of waste valorisation, and get acquainted with themes that would not otherwise enter the firm's strategic discourse, can be an essential step. Waste valorisation may result from different technologies and can lead to different markets: expertise in this cannot be developed overnight, nor can the related subjects enter the firm's internal debates automatically. However, once expertise is developed within the firm, the subject is faced during regular meetings and investments take place as a consequence of long-term strategies, then waste valorisation can correspond to the promotion of the firm's main product: corporate social responsibility can become an element for product branding, and the firm itself can become an exemplary case in the eyes of environmental groups and national authorities, possibly helping to shape state regulations.

The second suggestion pertains to the cross-sectoral nature of waste valorisation. If a firm keeps a narrow view within its sectoral boundaries, it will rarely be exposed to the technical opportunities for waste transformation, or to the potential markets for future by-products. Keeping an innovative mindset within the firm, and letting incremental innovations shape the firm's development over time, would help to detect opportunities which sit outside the firm's comfort zone, and can lead to radical innovations in relation to new technologies and new markets. Once the opportunities have been explored, and cross-sectoral paths to waste valorisation have been established, the exploitation of those paths can be made stable by building organisational bridges between the sectors involved: for instance, entrepreneurs and managers from other sectors, who can have a specific interest in the waste valorisation processes of a firm, could be involved in the board or in the management of the same firm. Directorate interlinks or strategic partnerships could facilitate the connections needed for the valorisation of residues, and in particular contribute to ensure that resources can flow between sectors in a constant and efficient way. Such stability is often necessary to convince potential investors about the long-term profitability of waste valorisation.

To refine the two suggestions above, further research could proceed along two directions. First, it is important to understand whether agency within a firm can play a role in initiating processes of waste valorisation. While we can easily imagine firms reacting to changes in regulations, or in general reacting to traceable external impulses, it is more difficult to determine which forces can push a firm towards new valorisation pathways in the absence of a specific external impulse. Are there particular types of firms, and particular types of firm organisations, which constitute a fertile ground where an employee's idea about waste valorisation can arise and develop? Systematic qualitative research on this subject could shed light on how to bring and keep side-streams within the set of a firm's core strategic resources. Second, a better understanding is needed of how inter-firm governance leads to waste valorisation. How do boards of directors change following waste valorisation strategies? Can we identify recurrent patterns in the evolution of senior management and of ownership, as coupled with intersectoral flows of side-streams? Given the usually public nature of data describing ownership and top management, quantitative analyses could provide precise answers to the questions above.

Different solutions may be adopted in accordance with different types of firm management and of industry structure. The fact that the valorisation of waste from a firm often requires coordination with other activities that do not normally pertain to the firm's industry makes it necessary to investigate specific types of innovation, and in particular of organisational innovation. Moreover, research is needed to understand how the new value assigned to waste can be recognised by markets and be translated into firm profits.

15.5 Policy and regulation of waste valorisation

To improve the sustainability and economic viability of the bioeconomy, policymakers should try to both *increase* and *improve* the use of organic waste streams. We have seen in this book that waste streams in many sectors are put to poor use from both an economic and sustainability perspective. In **Part II**, we have, in contrast, seen that most organic waste streams were put to some sort of use: urban food waste was incinerated or turned into biogas, acid whey and farmers' spent grain were used as animal feed and waste from on-shore aquaculture operations was used as fertiliser. Nevertheless, most of these waste streams could have been used more effectively still. Food waste could have been prevented or the food could have been reused, the acid whey and farmers' spent grain could have been used in the production of food additives and pharmaceuticals and the sludge from aquaculture could have been used to produce omega-3 fatty acids (see **Chapters 5, 6, 8** and **9**). These alternative applications have the potential for greater economic returns and improved sustainability. Policymakers should therefore focus as much on *improving* as on *increasing* the utilisation of organic waste. Nevertheless, policymakers should also be aware that achieving greater economic returns on some residual streams can result in uses that are less sustainable.

Policymakers should consider three main pitfalls when they attempt to govern and regulate the bioeconomy. First, they should be aware that lock-ins can easily arise from investments in organisational capabilities and physical infrastructure and prevent further improvements in sustainability. The case study on urban organic waste, in **Chapter 5**, illustrated how investments in physical infrastructure – such as optical sorting plants, biogas facilities and incineration plants – required a steady flow of organic waste and provided little incentive for the municipality of Oslo to pursue more sustainable option, such as reuse or prevention of food waste. Second, they should be aware that regulations can have a strong effect on the innovative activity within a sector. In **Chapters 7** and **8**, we have seen that environmental and health and safety regulation influenced the innovative activity in the dairy and aquaculture sectors. The lesson is not that these types of regulations should be abolished. Rather, it is that these types of regulations should be implemented in a way that encourages rather than prevents innovation. Third, they should be aware that extensive outsourcing of services to private subcontractors can easily stifle innovation. In **Chapter 5**, we saw that the tender system the Oslo municipality used when it outsourced waste collection neither allowed for any mutual learning or exchange of knowledge between the private contractor and the municipality nor provided any incentives for the private contractor to improve the sustainability of the waste treatment system.

Policymakers should be aware that the bioeconomy consists of actors with varying and sometimes conflicting agendas. In **Chapter 2**, we learned that three different visions of the bioeconomy exist – a bio-technology, a bio-resource and a bio-ecology vision. We saw in **Chapter 12** that these three visions were represented by different groups of actors and that their different visions resulted in both conflicting and complementary interests and agendas. For policymakers it is important to note that if they try to introduce policies that run counter to one or more of these visions, they will meet considerable opposition. Conversely, if they can align these visions, they will receive support for their policies from the whole bioeconomy. Nevertheless, policymakers should be aware that regulating the bioeconomy will, in many cases, involve trade-offs between these visions and that they must balance the need for sustainability against demands for economic growth.

There are many interesting avenues for future research on policy and regulation of waste valorisation. There is a lack of systematic and comparative studies on how policymakers can increase and improve valorisation of waste, and studies that compare policies and regulations across different countries and across different waste resources should be very welcomed. Another interesting research avenue relates to how different actors try to influence waste valorisation policies. Do they attempt to form alliances with others of similar interests? Do they seek to affect policymakers through popular opinion or through concealed lobbying operations? Where and at what level are the most important policies developed – local, national or supranational?

15.6 Final remarks

This book has attempted to cover a broad range of issues related to the valorisation of organic residues and side-streams. It is situated within the wider scientific discourse about a circular and sustainable bioeconomy and tries to bridge different disciplinary approaches such as case studies, quantitative data-driven analyses, LCAs and policy analysis. We are convinced that this approach has helped to increase the relevance of the book and the scientific soundness of its studies. A recurring topic in this book has been that valorisation of organic residues is not just about making more money out of waste, but also about improving sustainability and social cohesion. The sustainability of valorisation pathways must be carefully considered, and such consideration has to guide the actions of both private firms and public organisations in order to avoid lock-ins into unsustainable and ineffective solutions. Innovation will be needed for both public and private actors, according to a broadened view on valorisation possibilities which crosses sectoral, disciplinary and geographic borders.

Index

Printed in the United States
by Baker & Taylor Publisher Services